Bioanalytical Approaches for Drugs,
Including Anti-asthmatics and Metabolites

METHODOLOGICAL SURVEYS IN BIOCHEMISTRY AND ANALYSIS

Series Editor: Eric Reid

Guildford Academic Associates
72 The Chase
Guildford GU2 5UL, United Kingdom

This series is divided into Subseries A: Analysis, and B: Biochemistry.
Enquiries concerning Volumes 1–18 should be sent to the above address.

Recent Titles

Volume 13(B):   Investigation of Membrane-located Receptors
Edited by Eric Reid, G.M.W. Cook, and D.J. Morré

Volume 14(A):   Drug Determination in Therapeutic and Forensic Contexts
Edited by Eric Reid and Ian D. Wilson

Volume 15(B):   Investigation and Exploitation of Antibody Combining Sites
Edited by Eric Reid, G.M.W. Cook, and D.J. Morré

Volume 16(A):   Bioactive Analytes, Including CNS Drugs, Peptides, and Enantiomers
Edited by Eric Reid, Bryan Scales, and Ian D. Wilson

Volume 17(B):   Cells, Membranes, and Disease, Including Renal
Edited by Eric Reid, G.M.W. Cook, and J.P. Luzio

Volume 18(A):   Bioanalysis of Drugs and Metabolites, Especially Anti-inflammatory and
Cardiovascular
Edited by Eric Reid, J.D. Robinson, and Ian D. Wilson

Volume 19(B):   Biochemical Approaches to Cellular Calcium
Edited by Eric Reid, G.M.W. Cook, and J.P. Luzio

Volume 20(A):   Analysis for Drugs and Metabolites, Including Anti-infective Agents
Edited by Eric Reid and Ian D. Wilson

Volume 21(B):   Cell Signalling: Experimental Strategies
Edited by Eric Reid, G.M.W. Cook, and J.P. Luzio

Volume 22(A):   Bioanalytical Approaches for Drugs, Including Anti-asthmatics and
Metabolites
Edited by Eric Reid and Ian D. Wilson

How to obtain future titles on publication

A standing order plan is available for this series. A standing order will bring delivery of each
new volume immediately upon publication. For further information, please write to:

The Royal Society of Chemistry
Distribution Centre
Blackhorse Road
Letchworth
Herts. SG6 1HN

Telephone: Letchworth (0462) 672555

Methodological Surveys in Biochemistry and Analysis, Volume 22

Bioanalytical Approaches for Drugs,
Including Anti-asthmatics and Metabolites

Edited by

Eric Reid
Guildford Academic Associates, Guildford, United Kingdom

Ian D. Wilson
ICI Pharmaceutical Division, Macclesfield, United Kingdom

ROYAL
SOCIETY OF
CHEMISTRY

CHEM
scplae

Based on proceedings of the Ninth International Bioanalytical Forum entitled
Bioanalysis of Drugs, including Anti-allergics and Anti-asthmatics, held September
3–6, 1991 in Guildford, UK.

Special Publication No. 110

ISBN 0-85186-236-5

A catalogue record for this book is available from the British Library

© The Royal Society of Chemistry 1992

Published by the Royal Society of Chemistry, Thomas Graham House,
Science Park, Cambridge CB4 4WF

Printed in Great Britain by Bookcraft (Bath) Ltd.

RS 189
B55
1992
Chem

Senior Editor's Preface

The **'Analysis' series** of which this is the ninth volume is, gratifyingly, attaining the status of an integrated reference collection*. (This volume, as for the fourth volume, features a Cumulative Analyte Index.) This is manifest by citations in the literature-at-large and back-references within the present volume (often inserted editorially) to still pertinent articles in earlier volumes. Where earlier material is being amplified, overlap has been minimized. Through the editing policies the book has escaped being a mere 'Conference Proceedings'. Yet it includes discussion remarks made at the Bioanalytical Forum that generated it, and editorial annotations, preserving a tradition that some reviewers have commended.

Editing principles.- Not all bioanalytical articles amongst the present-day flood (e.g. in *Journal of Chromatography*) have been strongly refereed and rendered 'reader-friendly'. That many fall short of this was brought home to the present Editor in harvesting key points for the 'Compendium' of assays which concludes Sect. #B. Besides general clarity of presentation one can reasonably expect, where applicable, the formula of the analyte, the reasoning underlying method choice, and the sensitivity attainable with biological samples. Editing in the present book series has aimed at completeness in such respects, coupled with terseness. In some articles there has been pruning of citations, especially references which were not to a 'hard' journal or which were to symposium presentations that never surfaced in a marketed publication. Altogether the editing policies left few publication texts unscathed; yet there were compliments from some authors, albeit protests from a few. Our busy authors do deserve thanks for having produced publication texts (promptly or belatedly), notwithstanding the editorial burden in some cases. This sometimes entailed remedial work on diagrams, or rectifying errors which the author had not detected in proofs. Hopefully few errors have survived.

Other acknowledgements.- Besides publishers who sanctioned use of already published Figs., as indicated in legends, two U.K. companies are warmly thanked for Forum support:- Glaxo and SmithKline Beecham. Various companies covered the costs for speakers from their staff. Topic planning for the Forum (which I.D. Wilson co-organized) was guided by Honorary Advisers including H. de Bree, U.A.Th. Brinkman, J.D. Robinson, R.J. Simmonds and, for anti-asthmatics, I.F. Skidmore.

*For procurement consult p. 170, and, for individual titles, the list facing the title page.

Analysts seeking guidance might firstly note that anti-asthmatics and kindred drugs (Sect. #C) comprise the 'therapeutic theme' that has been a feature of recent volumes, reflected in their titles and cross-referenced in the present book through the sub-headings in the 'Compendium' (pp. 149-171). The latter widens the coverage of the drug spectrum, including anti-asthmatics. Where a drug of interest does not feature in the book, ideas for assay or metabolite investigation may be gleaned from considering drugs that have 'chemical kinship', even if not in the therapeutic area concerned. Kinship in some chemical respects is the basis of the Analyte Index sub-divisions. Thus, 'IIa' drugs and many 'IIIa' drugs are amphoteric, implying difficulty with liquid-liquid extraction. The remedy may lie in solid-phase extraction ('SPE') - which features in a sketch of analytical approaches ('sub-preface', p. 43) and in a summary of analytical trends (p.172).

Terminology and abbreviations.- 'Bioanalysis' is a useful term that has become respectable, and likewise 'biofluids'. Whilst favouring abbreviations such as '3-D' where the meaning is clear, this Editor frowns on '2-D HPLC' (perhaps having a 'heart-cut' connotation). Whilst now seeing the advantage of 'L' over 'l' to connote litre, this Editor remains old-fashioned in denoting molar concentrations as (e.g.) '5 μM' rather than '5 μmol.L^{-1}'. In any case, for drug values a weight-per-ml basis is favoured.

Since all temperatures in the book are °C, the 'C' is consistently omitted. Column diameters are always internal, even where 'i.d.' is not specified. 'C-18' is felt preferable to 'ODS'. For HPLC packings in general, it has been endeavoured to help non-specialist readers who may be bewildered by bald use of a trade name or of a jargon term such as 'ICRP' (internally coated reversed phase) or 'PRP' (polymer reversed phase). Such abbreviations are listed at the start of each article. Other examples are 'Ab' (antibody), 'EC' (electrochemical; cf. GC use!) and 'TBA' (tetrabutylammonium).

Postscript: **proliferating primary journals.-** Librarians, already retrenching on established journals, can ill afford new ones (many brashly initiated by 'private enterprise'). Moreover, shelving elderly volumes is an escalating cost burden. Librarians in 1945 (Conf. Rept., *ASLIB*) warmly received a proposal (E. Reid *et al.*; fostered by J.D. Bernal) "to make the individual paper the basic unit"; the Society concerned would publish it and a separate Abstract, giving members the option of getting only selected papers. In 1991 a Librarian (F. Friend; Conf. Rept., *Eur. Sci. Editing*) "saw the trend towards electronic dissemination of the single article as unstoppable". Will this thistle ever be grasped?

Guildford Academic Associates,
72 The Chase, Guildford GU2 5UL, U.K.

ERIC REID

15 April 1992

Contents

The 'NOTES & COMMENTS' ('nc') items at the end of each Section include comments made at the Forum on which the book is based, along with some supplementary material.

List of Authors

Primary author

*Co-authors, with relevant name
to be consulted in left column*

J.D. Robinson - pp. 89-93
Hoechst Pharm'ls, Milton Keynes

P. Rossato - Grossi
H.M. Ruijten - de Bree

J.F. Sabot - (i) pp. 327-330;
(ii) pp. 331-332
Faculté de Pharmacie, Univ.
Claude Bernard, Lyon, France

J.A. Salmon - Land
R. Siemons - Maijer
K. Selinger - Hill (i)
J.P. Sharpe - Land

I.F. Skidmore - pp. 173-181
Glaxo Group Res., Ware

E.V.B. Shenoy - Jenkins
R.J. Simmonds - Burton;
    James; Rees; Wood

L.F. Statham - pp. 297-304
Glaxo Group Res., Ware

B.C. Sweatman - Nicholson

A. Strutton - pp. 321-324
Hoechst Pharm'ls, Milton Keynes

A. Taylor - Beerahee
B.E. Timmerman - de Bree
U.R. Tjaden - Niessen

H. Vik - pp. (i) 239-242; (ii) 243-246
Nycomed AS, Oslo, Norway

J.A. Troke - Gilbert
P.T. Tsang - James

R. Whelpton - (i) pp. 311-315;
(ii) pp. 317-320
Queen Mary Coll., London E1

M. Tugnait - Wilson (ii)
R.G. Turcan - Gilbert
N.A. Undre - Beerahee
J. van der Greef - Niessen
M.P.E. van Eck - de Bree

I.D. Wilson - (i) pp. 217-224;
(ii) pp. 291-296; & see Nicholson
ICI Pharm'ls, Alderley Park

J. Verstegen - Meijer
W.R. Vincent - de Bree
C.W. Vose - Gilbert

S.A. Wood - pp. 95-102 (& see Rees)
Upjohn Pharm'l Res., Crawley

A. Walhagen - Lindberg
A. Warrander - Wilson (i)
K.V. Watson - Gilbert

R. Wyss - pp. 79-88
Hoffmann-La Roche, Basel,
Switzerland

B.C. Weatherley - Leavens
D. Wilkinson - Gardner
L. Witherow - Wilson (i)

G.C. Young - pp. 233-234
Glaxo Group Res., Ware

A. Wittrock - Förster
P. Wright - Statham

Section #A

METABOLITE INVESTIGATION

#A-1

BIOFLUID NMR SPECTROSCOPY IN THE STUDY OF THE METABOLISM OF β-LACTAM ANTIBIOTICS

J.R. Everett

SmithKline Beecham Pharmaceuticals,
Chemotherapeutic Research Centre,
Brockham Park, Betchworth, Surrey RH3 7AJ, U.K.

The utility of biofluid NMR spectroscopy in studying the metabolism of high-dose drugs and other xenobiotics is now well established. The advantages and disadvantages of the method are now discussed and illustrated by studies on the metabolites of β-lactam antibiotics in unprocessed human and animal biofluids. The β-lactam antibiotics studied include amoxycillin and the potent novel penicillin BRL 36650.

NMR spectroscopy has undergone an explosive growth over the past decade in terms of both the range of its applications and its frequency of use in those applications. In biological applications, the technique has the unique ability to allow molecular structure and dynamics to be studied in systems as simple as a solution of a single biomolecule, e.g. a protein, or as complex as a whole human being. In drug metabolism the technique has a role of comparable diversity [1], covering*:
- structure elucidation of isolated drug metabolites;
- studies of drug metabolites in animal and human biofluids (the present focus);
- studies of drug metabolites in tissue extracts;
- *in vitro* drug metabolism studies in isolated cells or perfused organs;
- non-invasive *in vivo* drug metabolism studies in intact animals and humans.

Over the past 7 years we have been using NMR spectroscopy to obtain information on the metabolites of novel antibiotics in unprocessed human and animal biofluids [2-5]. For drugs

*Editor's note.- Readers should consult pertinent arts. and Index in earlier vols. in this *Methodological Surveys* ..... series [ed. E. Reid *et al.*; #16, #17 (nephrotoxin studies) & #18, Plenum; #20, Roy. Soc. Chem.], especially the detailed accounts in #16 (J.K. Nicholson; I.D. Wilson), complementing material in the present text.

which are administered at relatively high doses, biofluid NMR spectroscopy [6, 7] is a powerful and rapid means of obtaining information on metabolic disposition. The method is particularly useful in investigational metabolism studies in man and animals, and complements conventional analytical approaches.

NMR SPECTROSCOPY OF BIOFLUIDS

The wide range of unprocessed biofluids studied in the past decade includes blood plasma, serum, seminal plasma, CSF, milk, aqueous humour, synovial and amniotic fluids, sweat and tears [6, 7]. Each biofluid type has characteristic physicochemical and biochemical properties that must be taken into account in designing and executing NMR experiments [6].

'Water suppression'.- To obtain satisfactory ^{1}H NMR spectra, the large resonance from the water protons in biofluids must be suppressed in some way. In modern high-field spectrometers this is effectively achieved (suppression factors of >100) by low-power irradiation of the water resonance during the delay period between each FID* acquisition. The irradiation is gated off during the acquisition itself ('gated decoupling'). This method suffices in most experiment types, e.g. simple pulse-and-acquire, spin-echo and 2-D. Higher suppression factors (>1000) are obtainable by the WATR method [8–10]: a chemical relaxation agent, e.g. urea or guanidinium chloride, is added to the biofluid and the pH adjusted (if necessary) so that the water resonance is maximally broadened, through chemical exchange processes which result in water protons transferring to the agent and *vice versa*. With a spin-echo pulse sequence approach for acquisition, total water-suppression may be achievable [9, 10]. Having to add a chemical agent to the biofluid is a disadvantage, although with urea-rich urines pH adjustment alone may suffice.

Elimination of broad resonances from macromolecules.- The lipid and protein content of normal urine is relatively low, and the ^{1}H NMR spectrum is dominated by sharp resonances from low mol. wt. endogenous components. However, for other biofluids such as blood plasma [6] the protein and lipoprotein content is high and the ^{1}H NMR spectrum comprises sharp resonances for the mobile, low mol. wt. components superimposed on the very broad signals for the slow-tumbling macromolecules. So as to clearly observe the signals for the former, the latter must be suppressed, suitably by applying a Hahn spin-echo sequence, i.e. $90° - \tau - 180° - \tau - FID$. The broad macromolecule resonances are associated with very short spin-spin relaxation times (T_2) and their signals decay to zero in two τ-delays of the sequence if τ is sufficiently long. The pulse sequence also results in the phase modulation of multiplet signals

---

*Abbreviations.- FID, free induction decay; AMX, amoxycillin; DKP, diketopiperazine; PAc, penicillinoic acid; i.v., intravenous.

that remain in the spectrum, helping confirm signal assignment. Typically τ values of ~68 msec are used so that multiplets with $^3J_{H,H}$ of ~7.5 Hz are phase-modulated up (triplets) and down (doublets and quartets). Singlets are 'unaffected' by the pulse sequence and appear upright.

RATIONALE FOR USING BIOFLUID NMR IN DRUG-METABOLISM STUDIES

The main advantage of NMR spectroscopy in this context is that it non-selectively monitors the levels of *any* low mol. wt. components present in the biofluid above the detection limit of the experiment (~10-100 μM for ^1H NMR of urine at 400-500 MHz) - an extremely important attribute when looking for unknown metabolites (see below). Here NMR spectroscopy complements the commonly used chromatographic techniques such as HPLC, where the analysis is often much more sensitive but is limited to the detection of a particular class of related compounds (and which, unlike NMR spectroscopy, gives little structural information). The NMR spectrum thus provides an *analytical overview* of the biofluid. Other advantages include speed (5-10 min per sample), little sample preparation other than adding a small quantity of D_2O to 'lock' the magnetic field, and no method development.

Amongst disadvantages of NMR spectroscopy the most notable is its inherently low sensitivity. Since sensitivity rises with magnet field strength, operation at the highest possible field strength (400-600 MHz for ^1H NMR) is recommended. Similarly, spectral crowding is a problem with ^1H NMR, where the chemical shift range is only 10 ppm. Since signal dispersion (the *frequency* separation between one signal and another) increases linearly with field strength, further benefits accrue from operation at the highest possible field. For fluorinated drugs ^{19}F can be used (as well as ^1H NMR) to study the drug and its metabolites in the biofluid ([3, 4] & #D-4, this vol.). It has advantages over ^1H NMR: no water suppression is required; spectral crowding is not a problem (since ^{19}F chemical shifts span 1000 ppm *and* fluorinated bioconstituents are near-lacking); and the ^{19}F label acts comparably to a radiolabel, advantageously using *cold* material, in determining the metabolic fate of the entire dose.

A more insidious problem arises if a drug metabolite is significantly protein-bound in the biofluid, since spin-echo NMR will not show signals from bound metabolite even if its level exceeds the applicable detection limit. This occurs because the bound metabolite will adopt the slow motional character-istics of the protein (or other macromolecule) to which it is bound and hence its protons will have very reduced relaxation times. The short relaxation time signals of the bound species will thus be eliminated in the two τ delays of the spin-echo experiment. Finally, NMR spectrometers are expensive instruments and costs rise steeply with the field strength of the machine.

METHODS IN WORK NOW DESCRIBED

All ^{1}H NMR experiments were performed at 400.13 MHz on a Bruker WH400 NMR spectrometer at ambient temperature using standard Bruker software and a 5 mm ^{1}H probe. Hahn spin-echo spectra were acquired with a 60 msec echo time and with irradiation at the water frequency during the 2.4 sec delay between acquisitions. Typically, 144 FID's were accumulated into 16384 data points over a 4808 Hz spectral width. A 1 Hz line-broadening was applied to all FID's prior to Fourier transformation.

All urine samples were stored frozen at −40° or below prior to study. Thawed samples (450 µl) were mixed with D_2O (50 µl) and placed in a 5 mm NMR tube without filtration or centrifugation. Authentic samples of amoxycillin, BRL 36650 and degradation products were prepared in these laboratories. Confirmation of resonance assignment was achieved by adding small aliquots of authentic compounds as solids, directly into the biofluid sample, and re-running the NMR spectrum.

METABOLITES OF β–LACTAM ANTIBIOTICS STUDIED DIRECTLY IN BIOFLUIDS

The study of the metabolites of β-lactam antibiotics in intact biofluids by NMR-spectroscopic approaches has some serendipitous features. (1) β-Lactam antibiotics can be given in quite high doses, not uncommonly several grams per day in man; thus the insensitivity of NMR is less of an issue. (2) The penicillins (the main focus of the present work) possess a *gem*-dimethyl group which gives rise to a characteristic pair of sharp singlet ^{1}H NMR signals for the parent drug and most of its metabolites in the biofluid. (3) This group's signals are easily observed since they occur in a region of the ^{1}H NMR spectrum of control urine (the biofluid most often studied) that is relatively free from interference by the signals of endogenous urinary components. Initial studies in this area were concerned with the detection of the metabolites of ampicillin, an amino side-chain penicillin, in urine from rats given the sodium salt at high dosage (700 mg/kg i.v.). These initial studies at once showed the value of the technique and resulted in the detection of the diketopiperazine (DKP) and the two penicilloic acid (P Ac) metabolites of ampicillin [2]. No other analytical technique had previously disclosed the DKP metabolite.

Amoxycillin (AMX) was then studied (structure: Fig. 1; *p*–hydroxy in place of ampicillin's phenyl group). Fig. 2 shows the low–frequency region of the Hahn spin-echo ^{1}H NMR spectrum of control urine from a rat used in this study. Resonances are observed for endogenous constituents but, importantly, the region δ_H ~1.8-0.5 ppm is relatively free of these signals. This is fortunate since it is just here that the characteristic *gem*-dimethyl signals of the penicillins occur, as shown (Fig. 3A; expanded spectrum) for post-dose urine from the same rat. This spectrum is dominated by the almost coincident signals of the drug

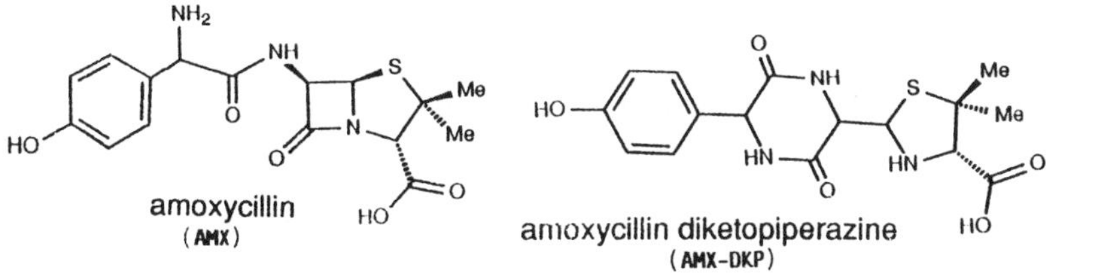

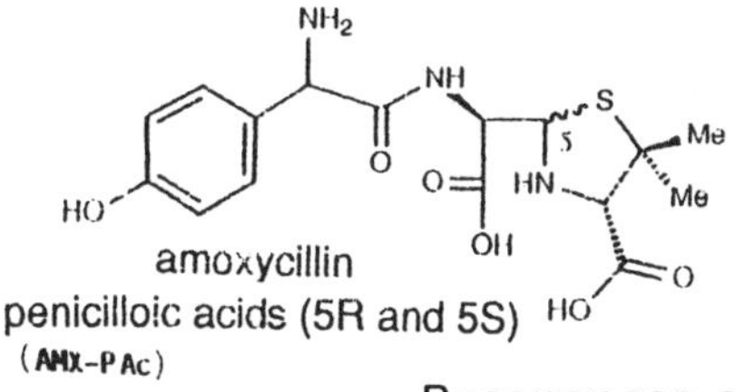

Fig. 1.
Structures of
AMX and major
metabolites.

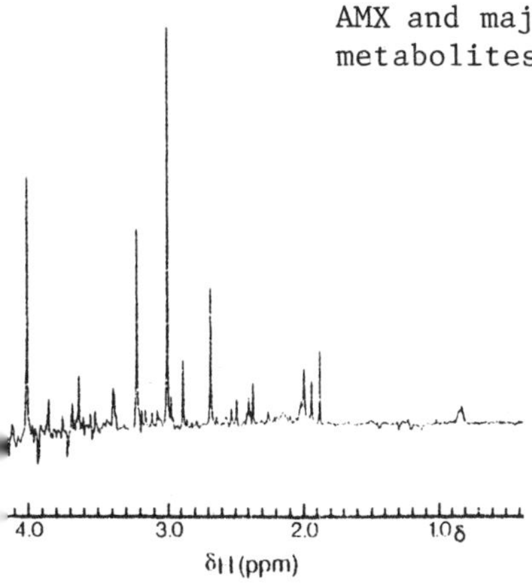

amoxycillin
(**AMX**)

amoxycillin diketopiperazine
(**AMX-DKP**)

amoxycillin
penicilloic acids (5R and 5S)
(**AMX-PAc**)

δH (ppm)

Fig. 2. The low-frequency
region of the 400 MHz Hahn
spin-echo ¹H NMR spectra of
control (pre-dose) rat urine.

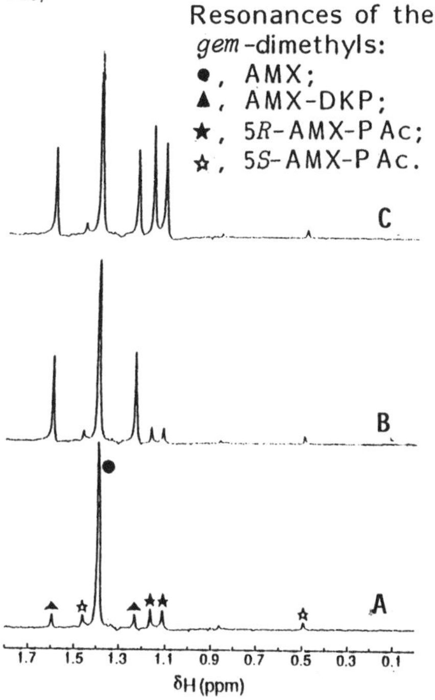

Fig. 3 *(right).* As for Fig. 2, but
3–6 h after dosing the rat i.v. with
AMX Na salt, 700 mg/kg. Unspiked (**A**),
and spiked with authentic AMX-DKP
(**B**; 547 mg) and 5R-AMX-PAc (**C**; 664 mg).

Resonances of the
gem-dimethyls:
●, AMX;
▲, AMX-DKP;
★, 5R-AMX-PAc;
☆, 5S-AMX-PAc.

δH (ppm)

and those of the 3 metabolites – which were identified very
rapidly on the basis of the ampicillin work [2] as 5R-AMX-PAc,
5S-AMX-PAc and AMX-DKP. Figs. 3B and 3C show, as marked, increased
intensities for two pairs of *gem*-dimethyl resonances after
spiking the same urine sample with authentic AMX-DKP and 5R-AMX-
PAc respectively, allowed the assignment of these twin signals
to AMX-DKP and 5R-AMX-PAc respectively. These metabolite identifications were also confirmed by the observation of exact resonance
overlap for the two aromatic signals from each of these metabolites
(not shown).   The *gem*-dimethyl resonances of the 5S-AMX-PAc
were similarly assigned.   All metabolite identifications were
further confirmed by HPLC (not shown; R_T basis).   AMX–DKP

had not previously been observed as a metabolite of AMX even
though its metabolism had been studied by non-NMR methods
for a number of years. Later workers confirmed the findings
[11]. AMX and these three metabolites were also detected
in human urine following administration of sodium AMX at typical
clinical doses, as well as in the urine of a volunteer after
co-administration of AMX (500 mg) and a novel β-lactamase inhibitor
being developed in these laboratories [J.R.E., unpublished; 1983].

BRL 36650.- NMR spectroscopy is very useful as an analytical tool in the early stages of metabolism studies in animals
and humans because of its non-selective detector characteristics,
and is extensively used as such in our laboratories. The
urinary excretion of BRL 36650 and its metabolites (structures:
Fig. 4) was monitored by 400 MHz NMR spectroscopy following
administration of 2-5 g i.v. to normal human volunteers in
the course of development studies on this novel penicillin.

As was noted for control rat urine, there are few strong
bioconstituent resonances in the low-frequency range (2.0-0.5 ppm)
of the ^{1}H NMR spectrum of normal human urine (Fig. 5). The
3-4 h post-dose spectrum from the same volunteer shows a
number of drug-related methyl signals in this region (Fig. 6).
The pattern of signals from the drug itself is more complicated
in this case since the resonance of the terminal methyl group
of the ethyl side-chain [Me(Et)] also occurs in this region,
as an upright triplet. Furthermore, due to slow rotation
about the NH-CH bond of the *N*-formyl substituent at position 6 of
the β-lactam ring, the drug exists in aqueous solution as
a mixture of *cis*- and *trans*-rotamers about this bond. Many
signals of the drug are thus doubled in the spectrum, although
those of the minor rotamer are only clearly visible for
H-3 and the low-frequency *gem*-dimethyl signal (Fig. 6). Other
drug-related signals are immediately apparent in Fig. 6,
especially the prominent pair of *gem*-dimethyl signals at
δ_H ~1.4 ppm (circles, ●). Earlier studies in these laboratories
[12] had shown that BRL 36650 undergoes an unusual degradation
in aqueous solution to form BRL 39256 and BRL 39280 (Fig. 4).
The circle-marked (●) signals in Fig. 6 were quickly recognized
as being due to BRL 39256, as was confirmed both by spiking
with authentic material and subsequently by a pre-column derivatization HPLC method using Ellman's reagent [unpublished work
with A.C. Marshall; 1983]. Further NMR experiments showed
that BRL 39280 was also excreted in the urines of dosed human
volunteers. The discovery of these two metabolites was important
for the development process on BRL 36650 as they accounted
for a significant proportion of the dose.

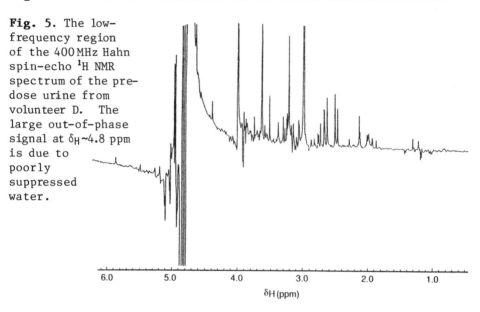

Fig. 4. Structures of BRL 36650 and two of its human metabolites.

Fig. 5. The low-frequency region of the 400 MHz Hahn spin-echo ¹H NMR spectrum of the pre-dose urine from volunteer D. The large out-of-phase signal at $\delta_H \sim 4.8$ ppm is due to poorly suppressed water.

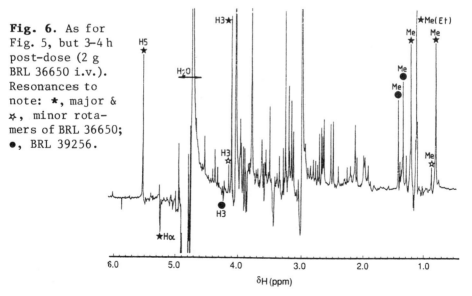

Fig. 6. As for Fig. 5, but 3–4 h post-dose (2 g BRL 36650 i.v.). Resonances to note: ★, major & ☆, minor rotamers of BRL 36650; ●, BRL 39256.

CONCLUSIONS

High-resolution NMR spectroscopy is a powerful method for detecting and identifying the metabolites of high-dose drugs in unprocessed biological fluids. The method complements chromatographic techniques and is especially useful in preliminary studies designed to obtain an analytical overview of the biofluid, since NMR is a non-selective detector. In the cases considered above there was a notable benefit: much NMR expertise existed on the compounds and their close relatives prior to starting the metabolism studies. Furthermore, authentic samples of many of the metabolites were either already available or were specially prepared. Even where this is not the case, NMR spectroscopy is still invaluable for detecting novel metabolites in unprocessed biofluids. However, the identification of the metabolites is likely to take longer and involve other spectroscopic techniques, and may necessitate isolation of the metabolites from the biofluid.

Acknowledgements.- The author is grateful to the many co-workers who have contributed to these studies, notably Malcolm Buckingham, Sue Connor, Keith Jennings, Jeremy Nicholson, John Tyler and Gary Woodnutt. Stimulating discussions with Ian Wilson are also gratefully acknowledged.

Human volunteers.- The experiments were conducted to a written protocol, after authorization by the local ethical committee.

References

1. Malet-Martino, M.C. & Martino, R. (1989) *Xenobiotica 19*, 583-607.
2. Everett, J.R., Jennings, K.R., Woodnutt, G. & Buckingham, M.J. (1984) *J. Chem. Soc. Chem. Comm.*, 894-895.
3. Everett, J.R., Jennings, K.R. & Woodnutt, G. (1985) *J. Pharm. Pharmacol. 37*, 869-873.
4. Everett, J.R., Tyler, J.W. & Woodnutt, G. (1989) *J. Pharm. Biomed. Anal. 7*, 397-403.
5. Basker, M.J., Finch, S.C. & Tyler, J.W. (1990) *J. Pharm. Biomed. Anal. 8*, 573-576.   [501.
6. Nicholson, J.K. & Wilson, I.D. (1989) *Progr. NMR Spectrosc. 21*, 449-
7. Bell, J.D., Brown, J.C.C. & Sadler, P.J. (1989) *NMR in Biomed. 2*, 246-256.
8. Rabenstein, D.L. & Fan, S. (1986) *Anal. Chem. 58*, 3178-3184.
9. Connor, S., Everett, J.R. & Nicholson, J.K. (1987) *Magn. Reson. Med. 4*, 461-470.
10. Connor, S., Nicholson, J.K. & Everett, J.R. (1987) *Anal. Chem. 59*, 2885-2891.
11. Haginaka, J. & Wakei, J. (1987) *J. Chromatog. 413*, 219-226.
12. Cutmore, E.A., Guest, A.W., Hatto, J.D.I., Moores, C.J., Smale, T.C., Stachulski, A.V. & Tyler, J.W. (1990) *J.C.S. Perkin Trans. I*, 847-853.

#A-2

MS-MS APPROACHES IN METABOLITE IDENTIFICATION, EXEMPLIFIED BY SALMETEROL

J. Oxford,   D. Higton and G.R. Manchee

Glaxo Group Research Ltd.,
Ware, Herts. SG12 0DP, U.K.

For identifying metabolites, typically present in low amounts amongst endogenous compounds in complex matrices, MS is the commonest approach. 'Soft' ionization techniques such as FAB* and TSP are informative concerning not only mol. wt. but also structure if MS-MS is employed. Types of scan are considered in relation to rapid identification of metabolites, as illustrated for salmeterol.

Rapid identification of drug metabolites is a challenging task which involves characterization of low-level (µg/ml to ng/ml) components in complex biological matrices. MS techniques have the requisite sensitivity and specificity, and have had considerable impact in this area during the last decade, quite apart from on-line use for chromatographic detection. Ideally MS techniques should rapidly provide firstly the metabolite's mol. wt. and secondly some structural information from the fragmentation pattern. Mol. wts. are obtainable by 'soft' ionization techniques - TSP ([1] & in present vol., articles #C-3, #D-1), FAB* [2] and, more recently, ion spray [3] - as surveyed in an earlier vol. [4]$^\oplus$. However, in general these methods of identification do not yield structural information unless conjoined with MS-MS - which encompasses a range of MS scanning techniques [5]$^\oplus$. Amongst the different types of MS-MS instrument [6], the type most commonly used for metabolic studies is the triple quadrupole [7].

FUNCTION OF THE FIRST (MS1) AND THIRD (MS2) QUADRUPOLES

The triple quadrupole instrument comprises three sequentially connected quadrupoles. MS1 transmits ions produced by the selected ionization process. The second quadrupole

---

*Abbreviations.- CID, collisionally induced dissociation; **FAB**, fast ion bombardment; **MS-MS**, tandem mass spectrometry; TSP, thermospray.
$^\oplus$Besides this art. [4] (A.P. Bruins), earlier vols. contain other pertinent arts. (indexed under 'MS'), notably an account of MS-MS (E. Houghton *et al.*) in Vol. 20, publ. Roy. Soc. Chem.- *Ed.*

acts as a collision cell, in which ions transmitted by MS1 undergo collision with molecules of an inert gas: through a 'CID' process, the ion (parent ion) undergoes dissociation yielding daughter ions which are mass-analyzed by MS2.

The three most commonly encountered scan modes in MS-MS are daughter ion, parent ion, and neutral loss. Daughter ion scans are of importance in obtaining structural information for a selected parent ion. Thus, protonated molecular ions of a specified m/z value produced by TSP ionization are transmitted by MS1. Following CID the daughter ions are mass-analyzed by MS2. Interpretation of the daughter ion spectrum will give structural information on the selected parent. Alternatively, information may be required on which parent ions give rise to specified daughter ions. In this instance, MS2 selectively transmits daughter ions of a specified mass whilst MS1 is scanned to determine which parent ions have given rise to the daughter ion. Finally, both MS1 and MS2 can be scanned with a fixed mass difference between them that is equal to the mass of a neutral loss, e.g. where a mixture of carboxylic acids loses CO_2 (44 mass units). Parent ion and neutral loss scans can be applied to screening complex biological matrices for metabolites containing a sub-structure of the drug from which they are derived. This approach has given rise to the term 'metabolite mapping' for the rapid identification of metabolites [8].

MS-MS STUDY OF SALMETEROL METABOLISM

We have used MS-MS techniques to aid in characterizing the major human metabolite of salmeterol, a long-acting β_2-adrenoreceptor agonist effective in the treatment of asthma [9]. It is administered as the xinafoate salt* by inhalation in very low dosage (50 μg twice daily). During its development, its metabolic disposition was studied in man following oral administration of 1 mg ^{14}C-salmeterol (as xinafoate). Salmeterol was extensively metabolized, being excreted predominantly in the faeces as a single metabolite (1 μg/g faeces).

The procedures required to isolate sufficient quantities of the metabolite from faeces were extensive (involving solid-phase extraction and semi-preparative HPLC). Hence preliminary MS analyses were carried out on a sample of a metabolite which had been produced by human hepatocytes *in vitro*. This metabolite had very similar chromatographic properties to the metabolite detected in human faeces.

Fig. 1 shows schematically the MS approach to identification of the metabolite. Initial analysis by TSP LC-MS

*It is irrelevant to metabolite study of salmeterol (formula at top of opposite p.) that, as dosed, it has the moiety xinafoate:-

ASSIGNMENT   COMMENT

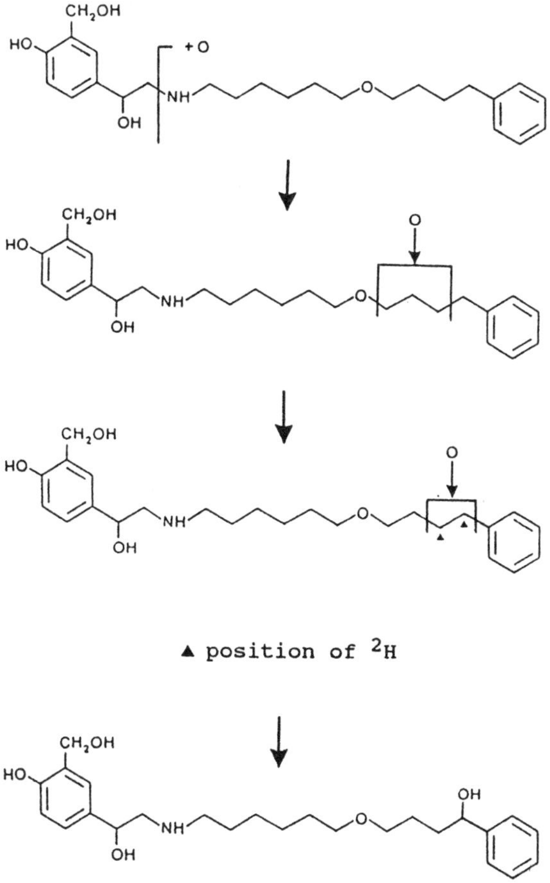

in vitro metabolite
TSP LC-MS
m/z 432 [M+H]⁺
[Parent-drug
formula: ignore [ ]⁺⁰

in vitro metabolite
NH₃ DCI
Daughter ions of
m/z 266

in vitro metabolite
(tetradeuterated
salmeterol)
Parent/daughter ions
Complex pattern at
m/z 91/92/93

▲ position of ²H

Faecal metabolite
GC-MS
Co-chromatography
with standard

Fig. 1. MS approach to identifying the position of oxidation
in the major human metabolite of salmeterol. For 'DCI' see below.

showed the presence of a protonated molecular ion at m/z 432,
thus indicating that the metabolite had a mol. wt. 16 mass
units greater than salmeterol.   Furthermore, the presence
of a fragment ion at m/z 266 showed that oxidation had occurred
on the lipophilic side-chain.

The metabolite was further investigated by ammonia desorption chemical ionization (NH₃ DCI)   using a Finnigan MAT
TSQ 70 quadrupole MS instrument.   Although the protonated
molecular ion was not detected, the fragment ion of m/z
266 was present (Fig. 2a). MS-MS experiments were then carried
out using argon in the collision cell with a collision energy
of -60 eV. The daughter ion mass spectrum derived from m/z 266
is shown in Fig. 2b, and Fig. 3 shows the proposed assignments.

Fig. 2. NH₃ DCI
mass spectra of the
metabolite formed
by incubation of
salmeterol with
human hepatocytes:
a, mass spectrum,
and **b**, daughter ion
mass spectrum of
m/z 266.

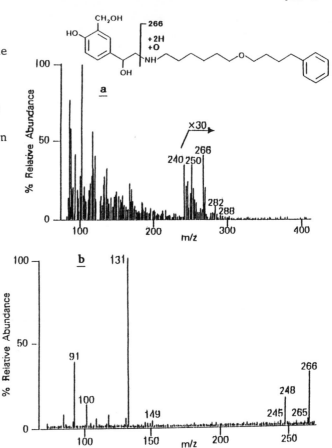

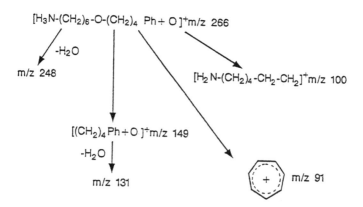

Fig. 3. Proposed assignments for the daughter ions shown in Fig. 2.

Analysis of the isolated faecal metabolite, of which ~2 µg was available, gave a similar daughter ion spectrum to the metabolite isolated from the hepatocyte incubation.

These data suggested that oxidation had occurred at one of the carbon atoms on the butyl side-chain. The presence of the daughter ion at m/z 91 tended to suggest, however, that oxidation had not occurred on the benzylic group of salmeterol.

MS-MS analysis of a hydroxylated metabolite prepared by incubating tetradeuterated salmeterol with human hepatocytes (cf. Fig. 1, 3rd formula) led to the proposal that oxidation had in fact occurred either α or β to the phenyl ring in salmeterol. Subsequent MS-MS analysis of the respective authentic compounds prepared synthetically showed that the two isomers were indistinguishable by this approach. In fact, the ion m/z 91 was present in the daughter ion spectra of m/z 266 derived from both isomers. However, the isomers were separated by GC-MS analysis of the trimethylsilyl, trifluoroacetyl derivatives (as prepared from salmeterol itself for an assay procedure: #ncC-4, this vol.). Thereby it was shown that, *in vivo*, salmeterol had been metabolized by oxidation of the carbon atom α to the phenyl ring.

Generalized conclusion.- In the foregoing example the major metabolite of salmeterol in man was isolated from a human hepatocyte preparation. This standard was important in that it enabled appropriate MS-MS techniques to be defined for obtaining structural information. In turn this information was used to define the standard compounds that had to be synthesized. Furthermore, MS-MS with GC-MS demonstrated the important role played by MS in identification of metabolites present at low levels in complex biological matrices.

References

1. Blakely, C.R. & Vestal, M.L. (1983) *Anal. Chem.* 55, 750-754.
2. Barber, M., Bordoli, R.S., Sedgewick, R.D. & Tyler, A.N. (1981) *Nature 293*, 270-275.
3. Bruins, A.P., Covey, T.R. & Henion, J.D. (1987) *Anal. Chem.* 59, 2642-2646.
4. Bruins, A.P. (1988) in *Bioanalysis for Drugs and Metabolites, Especially Anti-Inflammatory and Cardiovascular* [Vol. 18, this series] (Reid, E., Robinson, J.D. & Wilson, I.D., eds.), Plenum, New York, pp. 339-351.
5. McLafferty, F.W., ed. (1983) *Tandem Mass Spectrometry*, Wiley, New York.

6. Busch, K.L., Glish, G.L. & McLuckey, S.A. (1986) *Mass Spectrometry/Mass Spectrometry. Techniques and Applications of Tandem Mass Spectrometry*. VCH, New York, pp. 15-51.
7. Perchalski, R.J., Yost, R.A. & Wilder, B.J. (1982) *Anal. Chem. 54*, 1466-1471.
8. Straub, K.M., Rudewicz, P. & Garvie, C. (1987) *Xenobiotica 17*, 413-422.
9. Ullman, A. & Svedmyr, N. (1988) *Thorax 43*, 674-678.

#A-3

METABOLITE ISOLATION, PARTICULARLY PRE-HPLC
CONCENTRATIVE STRATEGIES

H.M. Ruijten, M.P.E. van Eck, B.E. Timmerman,
W.R. Vincent and H. de Bree

Solvay Duphar Research Laboratories, P.O. Box 900,
1380 DA Weesp, The Netherlands

The purpose and nature of metabolism studies is outlined.
Samples may be concentrated by freezing or extraction with 'cones'.

In drug development, especially safety assessment, metabolism studies are needed for several purposes:
- to select appropriate species for toxicology studies;
- to explain or predict interactions with other drugs;
- to detect reactive intermediates;
- to assess selectivity of the parent-drug assay.
For non-drug chemicals, e.g. crop protection agents, purposes include:
- to determine the effect of hydrolysis and sunlight;
- to examine the compound's metabolites;
- to determine the nature of any residue;
- to detect persistent metabolites.
To simplify metabolism research, the administered compound should be ^{14}C-labelled at a metabolically stable position.

Metabolite patterns, representing fingerprints of the drug/species/dose combination, are obtainable using a reversed-phase (RP) gradient from water to methanol, ensuring separation of all metabolites from extremely polar to non-polar. HPLC is highly selective, such that for each study a gradient can be chosen that elutes the parent compound at ~100 min and the metabolites beforehand.

The metabolite patterns are informative in several respects:
- similarities and differences between species and sexes;
- saturation of pathways at high doses;
- any change in kinetics with multiple rather than single dosage;
- unchanged drug; phase I and phase II (conjugation) metabolism, qualitative/quantitative;
- guidance on metabolite isolation.
Co-chromatography or isotope dilution can help in identifying expected metabolites. Definitive proof of identity is furnished by structural analysis of the isolated compounds, using UV, IR, MS, and notably NMR which, advantageously, does not need pure reference compounds.

In this article we consider tools used for sample preparation and pre-concentration prior to RP-HPLC with a combination of isocratic and gradient elution.   Pre-cones enable large volumes to be concentrated, up to 1 L with a large version [1-4], and serve for chromatography of solid samples by mixing them with silica; automated operation is described.   Very large volumes need an additional enrichment technique, suitably freeze-concentration [5].

SAMPLE CONCENTRATION BY FREEZING ('ZONE REFINING')

For large volumes (> 5 L) or strong hydrophilic compounds (difficult to concentrate by an RP approach), the bulk of the water is removed by freeze-enrichment - a technique based on the zone-refining principle as used in the metallurgical industry.   The vessel (Fig. 1) consists of two parts, the outer being a cooling jacket, not extending down to the lower part which has a stirring bar and a draining aperture. This lower part is surrounded by a heating medium. A temperature gradient is established from +4° at the centre to -10° at the wall.   Under these conditions, most of the water in the sample is frozen out in nearly pure form, starting from the inner wall and resulting in a high concentration of solutes.

There are three versions of the freeze vessels differing in dimensions (height x i.d., cm) and capacity: **I**: 46 x 13.5, 5500 ml; **II**: 36 x 9.5, 2300 ml; **III**: 20 x 7.5, 800 ml. Illustrating drug-solute recovery, this was 98% for 850 ml of liver homogenate reduced to 35% volume, 86% for 1700 ml of human urine reduced to 18%, and 70% for 1230 ml of hamster urine reduced to 23% (recoveries derived from radiolabel measurements).

The main advantage of the technique is that it operates unattended, under subtle conditions.   After the initiation, the freezing stops when an equilibrium has been reached; finally the enriched sample is drained off.   Freezing is complete in ~10 h with a $1-2\frac{1}{2}$ L sample and in ~30 h with a 5 L sample, furnishing a concentrate at ≤10°.   The low temperature prevents decomposition of the compound and its metabolites, and no losses of volatile metabolites occur. If the freeze concentration fails to produce clear ice and concentrated solutes, the sample is allowed to thaw, and the concentration procedure is re-started but with a higher cooling temperature.   The concentrates can be further processed on a pre-column for separation chromatography.

PRE-CONCENTRATION WITH CONE-SHAPED PRE-COLUMNS

As was described in Vol. 14 [2], conical pre-columns have been developed, in 3 size versions (25.5 ml for preparative work), to cope with volumes exceeding 1 ml.   The pre-column

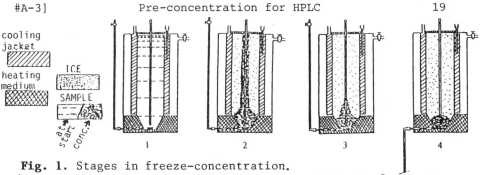

cooling jacket

ICE

heating medium

SAMPLE

at start conc.

1   2   3   4

Fig. 1. Stages in freeze-concentration.
(The central tube serves to admit air when draining.) See text.

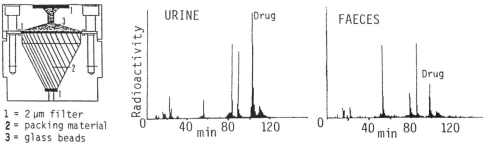

1 = 2 μm filter
2 = packing material
3 = glass beads

Fig. 2. Conical pre-column.

Fig. 3. Metabolite patterns: faeces (admixed with silica) *vs*. urine (no pre-cone). See text.

(Fig. 2) retains and distributes the material of interest uniformly on the packing material (C-8, C-18, silica or ion-exchanger), and effluent radioactivity serves to disclose any artifacts or selective losses. Smooth flow into the main column is ensured by an appropriate width (8 mm) at the foot of the pre-column and by filling its top portion with 100–200 mesh glass beads. A large pre-column may be able to cope with 2 L of urine.

Deproteinization is also achievable with RP material, the proteins being readily washed off. The subsequent metabolite pattern has been exemplified [2] for liver incubated with secoverine (1 ml or 650 ml of homogenate applied). RP-HPLC patterns have also been shown [2] for clovoxamine in rat urine (1 ml or 250 ml) and for pamoic acid in whole blood (1 ml). Fig. 3 shows patterns for a potential anti-hypertensive drug in mouse faeces, put into the pre-column in admixture with silica; the pattern tallied with that for urine (Fig. 3).

Automation, easily effected and allowing 24-h operation, is advantageous because otherwise only 2 or 3 patterns (each taking up to 5 h) can be obtained per working day. Fig. 4 shows the system, incorporating an exchanger (Fig. 5) for the cones (Fig. 2) which, with matching adaptors, are placed in a 40-place turntable. This is controlled by a Programmable

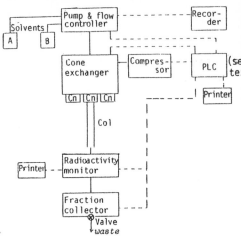

*(The column is situated not below the pre-cone, as Fig. 4 might suggest, but **above** it as in Fig. 5.)*

Fig. 4. Automated system for fraction collection (schematic).

Fig. 5, *right.* Diagram of cone exchanger. Cn, cone; Col, column; A, cone adaptor; T, turntable; Cy, cylinder with pneumatically operated piston (to push pre-cone and column together).

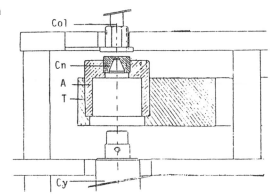

Logic Controller (PLC), which starts and controls all the equipment, keeps a continuous pressure (0.6 MPa) on the cone and column, identifies all types of cones and checks whether the cone has already been used for chromatography. The PLC activates the compressor, which presses the cone vertically onto the separation column (Fig. 5). For off- or on-line patterns, automation can also be used to collect fractions on a time or peak-detection basis.

References

1. Ruijten, H.M., van Amsterdam, P.H. & de Bree, H. (1984) *J. Chromatog. 314*, 183-191.
2. Ruijten, H.M., van Amsterdam, P.H. & de Bree, H. (1984) in *Drug Determination in Therapeutic and Forensic Contexts* [Vol. 14, this series] (Reid, E. & Wilson, I.D., eds.), Plenum, New York, pp. 65-70.
3. Ruijten, H.M., van Amsterdam, P.H. & de Bree, H. (1987) *Trends Anal. Chem. 6*, 134-138.
4. Ruijten, H.M., de Bree, H., Borst, A.J.M., de Lange, N., Scherpenisse, P.M., Vincent, W.R. & Post, L.C. (1984) *Drug Metab. Disp. 12*, 82-92.
5. Ruijten, H.M., *et al.* (1987) U.S. Patent #4697428.

#A-4

AN APPROACH TO A METABOLIC EXPERT SYSTEM: STRUCTURE-METABOLISM RELATIONSHIPS FOR BENZOIC ACIDS IN THE RAT

[1]F.Y. Ghauri, [1]C.A. Blackledge, [2]R.C. Glen, [2]B.C. Sweatman,
[2]J.C. Lindon, [2]C.R. Beddell, [3]I.D. Wilson and [1]J.K. Nicholson[⊕]

[1]Chemistry Department, Birkbeck College,
29 Gordon Square, London WC1H OPP, U.K.

[2]Physical Sciences Department, Wellcome Research
Laboratories, Beckenham, Kent BR3 3BS, U.K.

[3]Safety of Medicines Department, ICI Pharmaceuticals,
Mereside, Alderley Park, Macclesfield, Ches. SK10 4TG, U.K.

Computational chemistry and non-parametric statistics have been used to analyze the results for a series of congeneric benzoic acids in the rat. ^{1}H and ^{19}F NMR provided a rapid and convenient alternative to radiolabelled compounds for generating metabolic data. The approach was clearly able to differentiate, based on computed molecular properties, compounds forming glucuronide rather than glycine conjugates.

Without question, the development of methods for accurately predicting the metabolic fate of drugs and other xenobiotics would be of inestimable benefit in pharmaceutical research. Time and time again, otherwise promising compounds fail in development for reasons of metabolism. Although major metabolic routes are now well established, real guidance for the synthesis of replacement compounds is often not feasible, beyond 'expert' guesswork. However, the currently available computer-based 'expert' systems do not inspire confidence and, to date, there remains no substitute for the experienced scientist (who can at least list the potential metabolites even if he cannot confidently predict which ones will actually be found). The lack of success of the currently available commerical 'expert' systems for metabolite prediction probably lies in the basic philosophy approach which relies only on historical metabolic data. This approach is probably doomed to ultimate failure because it does not take into account the actual chemistry of the molecule. Thus all drug-receptor interactions, including those with metabolizing enzymes, are based on the recognition of the shape, charge and motions (including charge oscillations) of a molecule, and these must also be taken into account by the expert system.

[⊕]Any correspondence should be addressed to J.K.N.

We are currently attempting to determine QSMR's* for a series of congeneric substituted benzoic acids in the rat. These compounds have been chosen, at least partly, because in the rat benzoic acids are either excreted unchanged or are metabolized predominantly by Phase 2 metabolic pathways to either glycine or glucuronic acid conjugates. These compounds thus provide a relatively simple range of metabolic opportunities. Ideally, for QSMR, a large database is desirable. To obtain such a database by conventional means would require a large investment in the synthesis of radiolabelled compounds. As an alternative, we have obtained data from the literature and, in addition, have employed NMR-based methods, which we have previously shown to provide a rapid and effective means of studying drug metabolism and excretion ([1-4]; cf. art. by Everett, #A-1) to aid in building the requisite database. Besides drug metabolic information, ^{1}H NMR spectroscopy of biological fluids can show biochemical or toxicological effects of the compounds under investigation [5].

However, even with an ^{1}H NMR observation frequency of 600 MHz (the highest commercially available), there may be circumstances where signals from drug metabolites are partially or wholly obscured by resonances from endogenous compounds. As a result, we have devised a number of simple NMR-monitored chromatographic and enzymatic procedures to aid metabolite identification [6, 7]. We have also exploited the opportunities presented by the availability of a wide range of fluorinated compounds to use ^{19}F NMR where possible ([1, 3, 8] & arts. #ncB-3 & #C-5 in this vol.; cf. #A-1). ^{19}F NMR has 83% of ^{1}H NMR sensitivity, but there is negligible interference from endogenous fluorinated compounds. The organic ^{19}F chemical shift range is >200 ppm, and ^{19}F shifts may be sensitive to structural changes in molecules occurring up to 8 bonds from the ^{19}F atom [1, 8]. Advantageously for metabolic studies, the ^{19}F can act as a 'handle' for probing the 'metabolism-directing' effects of other functional groups substituted on an aromatic ring. Also, since there is seldom significant loss of fluorine during metabolism, the metabolism of diverse fluoroaromatic compounds can be explored with sequentially modified substituents.

As described below, we have then related this metabolic data to semi-empirical molecular orbital and structural physico-chemical parameters analyzed using computer pattern recognition methods.

---

*Abbreviations (others appear later in text).- D (as in multi-D), dimensional; FID, free induction decay (cf. GC usage!); FT, Fourier transformation; NLM non-linear mapping; PC(A), principal components analysis; QSMR, quantitative structure-metabolism relationship; SPE, solid-phase extraction; TSP, Na 3-trimethyl-silyl-[2,2,3,3-^{2}H]-1-propionate (reference compound).

EXPERIMENTAL

Compounds were all of analytical grade (Fluorochem and Aldrich) and were used without further purification. They were as follows, where the **R** substituent(s) in benzoic acid (Compound #1) are listed together with the urinary metabolites (* = fate determined by NMR), formed by **gly**cine conjugation or by **gluc**uronidation.-

#1:	H	**gly**	#8:	2-CH$_3$	**gly**
#2:	2-F*	**gly**	#9:	3-CH$_3$	**gly**
#3:	3-F*	**gly**	#10:	4-CH$_3$	**gly**
#4:	4-F*	**gly**	#11:	4-NH$_2$	**gly**
#5:	2-CF$_3$*	**gluc**	#12:	2-F, 4-CF$_3$*	**gluc**
#6:	3-CF$_3$*	**gly**	#13:	2-CF$_3$, 4-F*	**gluc**
#7:	4-CF$_3$*	**gluc**	#14:	3-CF$_3$, 4-F*	**gly**

Animals and treatments.- Each compound was given to 3 male Sprague-Dawley rats (200-250 g), housed individually in metabolism cages (12 h light, 12 h dark) and given food and water *ad libitum*. After acclimatization (2 days), a single i.p. dose was given (100 mg/kg in saline-phosphate). Urine was collected over ice for 24 h prior to dosing and at 8, 24 and 48 h postdosing. After volume and pH measurement and centrifugation (3000 rpm, 10 min, 4°), the samples were frozen and kept at -20° prior to analysis by NMR.

^{1}H NMR analysis of urine was performed on a Bruker WH400 spectrometer operating at a proton frequency of 400 MHz or a Jeol GSX500 spectrometer at 500 MHz, at ambient probe temperature (15 ±1°). For each sample, 64 FID's were collected into 16,384 computer points using a 45° pulse width, a spectral width of 5,000 Hz (centered on the water resonance frequency). This gave an acquisition time of 1.7 sec, and a further 3 sec delay was added between pulses to allow T$_1$ relaxation. Water suppression was achieved by a gated (off during acquisition) secondary irradiation field applied at the water resonance frequency. Exponential line-weighting functions were applied to the FID's prior to FT. Chemical shifts were referenced to TSP (0.0). Resonance assignments were confirmed by a combination of chemical shift, spin-spin coupling patterns, coupling constants, pH dependence of chemical shifts and ultimately by standard addition.

^{19}F NMR analysis of urine was performed with either a Bruker AM400 wide-bore or a Varian VX400 standard-bore spectrometer operating at a ^{19}F frequency of 376 MHz (probe at 15 ±1°). Samples were placed in 5 mm o.d. precision NMR tubes for measurement. For each sample, 256 FID's were collected into 32,768 computer points using a 45° pulse width, and a spectral width of 10,000 Hz. This gave an acquisition time of 0.5 sec, with a 6 sec delay intervening between pulses to allow full T_1 relaxation to occur. Exponential line weighting functions of 0.5 Hz were applied to the FID's prior to FT. Chemical shifts were referenced to the external standard reference 4-trifluoromethyl cresol (77 ppm). ^{19}F-(^1H-) NMR spectra were collected for monofluorinated aromatic compounds, but broad-band proton decoupling was not necessary for trifluoromethylated compounds. Resonance assignments were confirmed by a combination of standard additions, proton-fluorine coupling constants, enzymatic hydrolysis and cross-referenced proton NMR spectra [8].

NMR-monitored SPE 'chromatography' and enzymatic treatment of the samples.- In order to partially purify and identify xenobiotic metabolites from some samples, SPE followed by stepwise elution procedures was employed, with NMR monitoring. Acidified urine samples were applied to C-18 SPE columns (Bond Elut; Varian Assoc.) as previously described [6, 7]. Metabolites were eluted with a series of methanol/water mixtures (each 5 ml): 20:80, 40:60, 60:40 and 80:20 (by vol.), concluding with a 100% methanol wash. Methanol was removed with a dry N_2 stream, and residual water by freeze-drying. Samples were dissolved in 100% ^2H_2O prior to NMR analysis.

Calculation of physicochemical properties.- Molecular structures were built using the 'SYBYL' [9] molecular modelling software on a DEC-VAX 8550 system. The molecular geometries were optimized to 10^{-7} kcal/mole using 'MOPAC' (AM1 PRECISE [10]) on a Cray X-MP, and overlaid by a least-squares fit of the benzene ring and the principal functional groups (benzoic acid carbonyl carbon). These comprised atom-centered partial charges on ring carbons (ATCH) [10], energy of the highest occupied molecular orbital (EHOMO) [10], energy of the lowest unoccupied molecular orbital (ELUMO) [10], total dipole moment and its x, y and z components (DIP, DIPX, DIPY, DIPZ) [10], nucleophilic and electrophilic superdelocalizabilities (NSDL, ESDL) at each ring carbon [11, 12], and polarizability (POL) [10], all calculated by the semi-empirical method 'MOPAC'. In this version of the calculation of superdelocalizability, the algorithm as originally defined by Fukui & co-authors [11] is modified, the Hückel parameter (lambda) being replaced by the orbital energy **E**. **E** may tend to zero but in general is not zero. As such, this parameter itself apppears to be an indicative variable which correlates with biological activity [12].

In addition, a number of steric parameters were calculated using 'PROFILES' [13]. These comprised dead-space volume (DSV) [14], van der Waals volume (VDWV) [14], collision diameter (CD) [14], closest approach diameter (CAD) [14], surface area (SA) [15], moments of inertia in x, y and z directions (MIX, MIY, MIZ) [16], and principal ellipsoid axes in x, y and z directions (PEAX, PEAY, PEAZ) [16]. The octanol-water partition coefficient (CLOGP) and the molar refractivity (CMR) were also calculated from the 'MEDCHEM' program [17]. Molecular weight (MW) was also included as another descriptor.

Computer data analysis.- The calculated physicochemical data for the compound set were analyzed mainly by use of two unsupervised learning-pattern recognition methods, i.e. NLM [18, 19] and PCA [20] (p. 22 footnote) using 'ARTHUR' [21].

RESULTS AND DISCUSSION

The urinary metabolites of the detected benzoic acids under study were, as indicated above, determined using ^{1}H and ^{19}F NMR methods or were identified from the literature. A combination of directly applied NMR techniques was used to study urine samples or SPE-extracted metabolites as previously described for other compounds. Representative spectra for the respective NMR findings are shown in Fig. 1 for 4-fluorobenzoic acid, which is excreted in urine predominantly as a glycine conjugate in the rat, and in Fig. 2 for 4-trifluoromethyl benzoic acid, which is excreted predominantly as an ester glucuronide with multiple transacylation [2, 3] and mutarotation of the glucuronide adduct to give α and β anomers. The resolution of separate signals from the α and β anomers illustrates dramatically the sensitivity of the changes in ^{19}F chemical shifts of ^{19}F atoms which are remote from the site of metabolic transformation [3]. By using NMR in this way we were able to determine the metabolic fate of the test acids.

The results of these studies were then analyzed by NLM and PCA, which reduce the multi-D parameter space where each dimension (D) is a calculated physicochemical property. In simple terms, the NLM technique accepts the data values as coordinates in n-D space and then compresses these into an NLM which is a 2-D or 3-D approximation to the true multi-D inter-point distances. Two points which are close on the map should be more similar in terms of input variables than two distant points. PCA is a well-established multivariate technique for dimension reduction in which PC's are new variables created from linear combinations of the starting variables with appropriate weighting coefficients. The properties of these PC's are such that: (i) each PC is orthogonal with all other PC's; (ii) the first PC contains the largest part of the variance of the data set with subsequent PC's containing correspondingly smaller amounts of variance. PC's may be plotted one against another, to reveal the degree of classification contained in the PC's. Analysis of the factors contributing to the PC's in a plot which gives good classification of the data may reveal underlying trends which can be used to explain the classification mechanism. For example, in

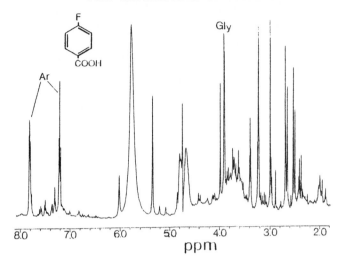

Fig. 1. A 400 mHz NMR spectrum of a urine sample from a male rat following a single i.p. dose of 4-fluorobenzoic acid. Signals for the glycine-conjugated metabolite are indicated (Ar, aromatic protons; Gly, methylene protons of the glycine).

this case it may be possible to gain insight into the physicochemical factors involved in the recognition, enzyme-binding and metabolism of the compounds.

As will be amplified elsewhere [same authorship], analysis of these data has enabled us to devise a PC map for all 14 compounds based on ATCH5 and ESDL5 alone which proved to be the most significant factors in determining which conjugate is excreted (Fig. 3). Evidently compounds prone to form glucuronides are well separated from glycine-conjugating compounds. Thus only a few physicochemical properties are required to classify benzoate recognition by the drug-metabolizing enzymes; hence knowledge of these properties for a particular compound should help in predicting which conjugation route predominates.

Benzoate metabolism may require recognition of the planar phenyl ring by a metabolizing enzyme system. As the structural information on the compound is largely only 2-D, the effects of molecular shape associated with recognition of chirality will be minimized. Electronic distributions around both the benzene nucleus and functional groups, and how these dictate the character of the enzyme-substrate transition states, will be critical in determining enzyme-ligand binding and hence the metabolic fate. Knowledge of the physicochemical parameters correlated with metabolic fate could give clues to the molecular recognition properties of the active site of the UDP-glucuronyl transferase isoenzyme responsible for benzoate glucuronidation.

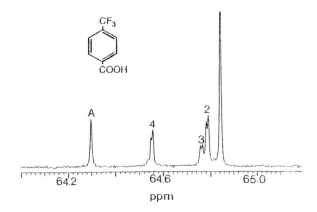

Fig. 2. A 376 MHz NMR spectrum of a urine sample from a male rat following a single i.p. dose of 4-trifluoromethylbenzoic acid. Signals for the unchanged parent (A) and the 1'-, 2'-, 3'- and 4'-*O*-acyl glucuronides (as α and β anomers; cf. split peaks) were detectable (1, 2, 3 & 4 respectively).

Fig. 3. A PC map produced for the 14 test compounds based on the 6 selected descriptors derived using 12 compounds as the training set: metabolism known to be dominated by glycine conjugation (●) or glucuronidation (■). Two compounds (▲) were tested to show the ability of the approach to predict metabolism: 2-trifluoromethyl-4-fluorobenzoic acid was mainly glucuronidated, whilst glycine conjugation predominated for 3-trifluoromethyl-4-fluorobenzoic acid.

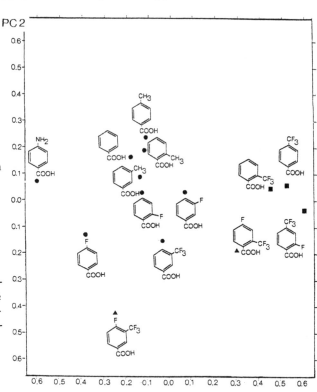

Concluding comments.- Although this work is at an early level of refinement with respect to detecting and calculating parameters, it shows that useful classification of metabolic-fate data is thereby achievable and that the PR approach to assessing structure-metabolism relationships may be highly revealing. Further development, now in hand, aims to explore the physicochemical property spaces for a wider range of benzoic acids with more diverse substituent groups so as to refine the prediction of metabolic fate based on pattern-recognition principles. Once this has been achieved and a robust set of descriptors found, then rule-induction methods [11] could be used to derive a chemically based 'expert system' for predicting metabolism. There could be extension to other compound classes, to different animal models and to comparisons between *in vivo* and *in vitro*. The present work further testifies to the usefulness of ^{19}F NMR, as an alternative to radiolabelling that co-provides balance and metabolite data.

References

1. Wade, K., Troke, J., MacDonald, C.M., Wilson, I.D. & Nicholson, J.K. (1988) in *Bioanalysis of Drugs and Metabolites* .....[Vol. 18, this series](Reid, E., *et al.*, eds.), Plenum, New York, pp. 383-388.
2. Ghauri, F.Y.K., Wilson, I.D. & Nicholson, J.K.(1990) in *Analysis for Drugs and Metabolites* ...... [Vol. 20, this series] (Reid, E. & Wilson, I.D., eds.), Roy. Soc. Chem., Cambridge, pp. 321-324.
3. Ghauri, F.Y.K., Blackledge, C.A., Wilson, I.D. & Nicholson, J.K. (1990) *J. Pharm. Biomed. Anal. 8*, 939-944.
4. Tulip, K., Timbrell, J., Wilson, I.D., Troke, J. & Nicholson, J.K. (1986) *Drug Metab. Disp. 14*, 746-749.
5. Nicholson, J.K. & Wilson, I.D. (1989) *Prog. NMR Spectros. 21*, 449-501.
6. Wilson, I.D. & Nicholson, J.K. (1987) *Anal. Chem. 59*, 2830-2832.
7. Wilson, I.D. & Nicholson, J.K. (1988) *J. Pharm. Biomed. Anal. 6*, 151-
8. Wade, K.E., Wilson, I.D., Troke, J.E. & Nicholson, J. K.   [165. (1990) *J. Pharm. Biomed. Anal. 8*, 401-410.
9. (1989) SYBYL, Molecular Modeling Software V 5.3, Tripos Associates, St. Louis, Missouri.
10. (1989) MOPAC Version 5, Quantum Chemical Program Exchange, Dept. of Chemistry, Univ. of Indiana, Bloomington, Indiana.
11. Fukui, K., Yonezawa, T. & Nagata, C. (1954) *Bull. Chem. Soc. Japan*
12. Gomez-Jeria, J.S. (1982) *J. Pharm. Sci. 71*, 1423-1424. [27, 423.
13. Glen, R.C. & Rose, V.S. (1987) *J. Mol. Graph. 5*, 79-86.
14. Edward, J.T. (1970) *J. Chem. Educ. 47*, 261-270.
15. (1981) SAREA. Pearlman, R.S. *QCPE Bull. 1*, 16.
16. Margenaux, H. & Murphy, G.M. (1956) *The Mathematics of Physics and Chemistry*, 2nd edn., Van Nostrand, New York, 604 pp.
17. (1984) CLOGP Version 3.54, Pomona Coll., Medicinal Chem. Project. Medchem Software, Daylight Chem. Info. Systems, Irvine, CA 92714.
18. Kowalski, B.R. & Bender, C.F. (1972) *J. Am. Chem. Soc. 94*, 5632-5639.
19. Sammon, J.W. (1969) *IEEE Trans. Comput. C-18*, 401-409.
20. Seal, H. (1968) *Multivariate Statistical Analysis for Biologists*, Methuen, London, pp. 101-102.   [Seattle, WA.
21. (1981) ARTHUR 81, Version 3. B.& B. Associates, PO Box 85505,

#ncA

NOTES and COMMENTS relating to
METABOLITE INVESTIGATION

Forum comments relating to the preceding main arts. and
to the 'Notes' that follow appear on pp. 39 & 40.

#ncA-1

A Note on

COLUMN-SWITCHING HPLC-MS FOR THE IDENTIFICATION OF LACIDIPINE METABOLITES IN RAT PLASMA

P. Grossi and P. Rossato

Drug Metabolism Department, Research Laboratories,
Glaxo S.p.A., via A. Fleming 2, Verona, Italy

Ld* is a novel calcium-antagonistic dihydropyridine derivative which has been developed as an antihypertensive drug [1] (formula shown later). Being potent in low dosage, its estimation in plasma (after acetonitrile treatment) called for subtle approaches which, with U.K. colleagues, we surveyed in a previous volume [2]. For a new drug, investigation of metabolic fate mainly entails evaluating metabolic profiles in animal species and in human volunteers. Oral administration of ^{14}C-Ld to rats and dogs has shown that it undergoes significant biotransformation into several metabolites differing widely in polarity (Grossi, unpublished) as shown in Fig. 1. To identify and profile the metabolites we have now used an approach which, for such purposes (cf. our Ld assay studies [2]), we have found to be rapid and efficient, namely the application to untreated biological samples of column-switching HPLC-thermospray MS.

SYSTEM DESCRIPTION

The chromatographic system comprised a binary solvent delivery system (#1090A, Hewlett Packard) with a UV detector

*Abbreviations.- Ld, lacidipine; s.i.m., selected ion monitoring.

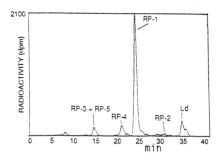

Fig. 1. HPLC ^{14}C peaks from rat plasma 6 h after 2.5 mg/kg ^{14}C-Ld orally (see text). Metabolite (RP) identity: see later (cf. Fig. 6).

Column:	Lichrosorb-ODS, 250 x 4.6 mm, 7 μm with a 30 x 4.6 mm guard-column, packed with 30-40 μm Perisorb RP 8.			
Flow rate:	1.5 ml/min.			
Temperature:	50 °C.			
Gradient mobile phase:				

t (min)	Methanol (%)	Acetonitrile (%)	Water (%)	0.1%Formic acid (%)
0	30	10	50	10
30	60	10	20	10
40	60	10	20	10
45	30	10	50	10
55	30	10	50	10

Flo-one radioactivity detector (Radiomatic Instruments):	
Cell size:	2.5 ml
Efficiency:	56.6%
Scintillation fluid:	Picofluor 40 (Packard)
Flow rate:	6 ml/min.

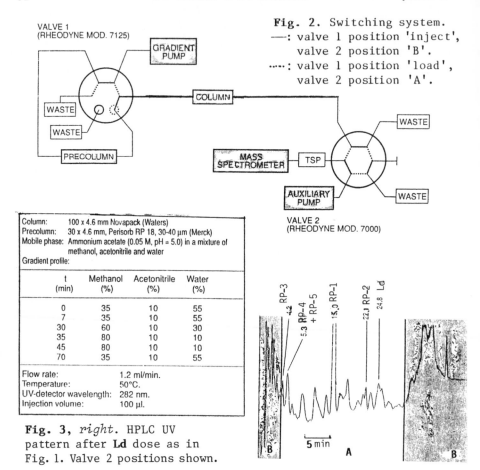

VALVE 1
(RHEODYNE MOD. 7125)

Fig. 2. Switching system.
———: valve 1 position 'inject',
valve 2 position 'B'.
·····: valve 1 position 'load',
valve 2 position 'A'.

VALVE 2
(RHEODYNE MOD. 7000)

Column: 100 x 4.6 mm Novapack (Waters)
Precolumn: 30 x 4.6 mm, Perisorb RP 18, 30-40 µm (Merck)
Mobile phase: Ammonium acetate (0.05 M, pH = 5.0) in a mixture of
              methanol, acetonitrile and water
Gradient profile:

t (min)	Methanol (%)	Acetonitrile (%)	Water (%)
0	35	10	55
7	35	10	55
30	60	10	30
35	80	10	10
45	80	10	10
70	35	10	55

Flow rate: 1.2 ml/min.
Temperature: 50°C.
UV-detector wavelength: 282 nm.
Injection volume: 100 µl.

Fig. 3, *right*. HPLC UV
pattern after **Ld** dose as in
Fig. 1. Valve 2 positions shown.

(#Uvidec-100, Jasco). Manual injection (valve #7125, Rheodyne) was by loop (100 µl). Alternatively, plasma was applied direct *via* a 30 × 4.6 mm pre-column packed with Perisorb RP 18 30-40 µm silica (Merck), pre-conditioned with methanol and water (each 3 ml) before each injection. Rat plasma (0.5-1.0 ml) was diluted before injection with 20% (v/v) aqueous acetonitrile (1 vol.; **Ld** protein binding thereby disrupted) and formic acid (conc., 100 µl). After each injection the column was washed with water (3 ml), then back-flushed by valve switching into an analytical column for gradient chromatography (Figs. 2 & 3).

Thermospray MS (single quadrupole instrument; Hewlett Packard #5988A with thermospray interface).- To reduce the amount of matrix-related material transferred to the thermospray probe, only the portion of the eluate that contained the main metabolite peaks (Fig. 3) was transferred, by switching (to position 'A', Fig. 2), a Rheodyne #7000 valve inserted between the UV detector and the thermospray interface. While

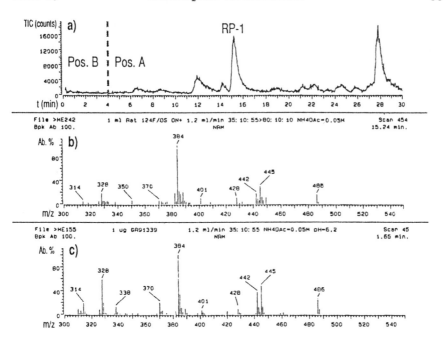

Fig. 4. MS examination of the HPLC eluate shown in Fig. 3. **a)** Total ion current (**A** & **B** are valve positions). **b)** Spectrum of the peak eluting at 15.2 min. **c)** Authentic RP-1 spectrum.

this valve was switched to divert the eluate to waste (position 'B', Fig. 2), a constant 1.2 ml/min flow of methanol/acetonitrile/ 0.05 M ammonium acetate was delivered to the mass spectrometer by an auxiliary pump (Waters, #M45). The interface probe temperatures were set initially at 102°/162–164° (stem/tip) and, after 7 min, lowered linearly to 98°/150–152° at 30 min. For positive-ion MS the source was kept at 276° (filament ON; other settings: analyzer manifold 2.5×10^{-6} torr, electron multiplier 2800 V, scan range 300–600 m.u., cycle time 1.87 sec for linear scan and 1.27 sec for s.i.m.).

RESULTS

Linear-scan MS spectra were obtained for rat plasma spiked with putative metabolites, then with plasma from dosed rats. Full thermospray spectra, agreeing with those for standards (Fig. 4), were obtained only for unchanged **Ld** and for its monodes-ethyl derivative, the main plasma metabolite (RP-1; m/z: $[M + H]^+$ 428, $[M + NH_4]^+$ 445, $[MH-CO_2]^+$ base ion 384). For the other metabolites detected in plasma (Figs. 5 & 6), more sensitive and specific target-compound analyses were done on eluate cuts (cf. Fig. 5) based on authentic-compound s.i.m.; ^{14}C and s.i.m. comparison with standards established identities, as shown in Fig. 6.

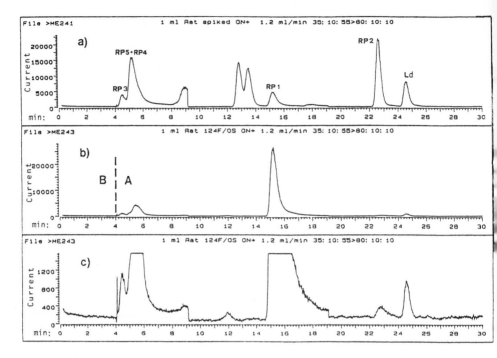

Fig. 5. MS–s.i.m. of HPLC eluates: **a)**, spiked plasma; **b)** & **c)**, from **Ld**–dosed rats. Eluate portions corresponding to standards were taken.

Compound Py = pyridine analogue	t_r (min)	M_w	characteristic ions (m/z)	monitored ions (m/z)
Lacidipine **Ld**	24.5	455	473,514,400,417,456	473,514
des–Et **RP-1**	15.2	427	384,445,486,442,428	384,445
Ld–Py **RP-2**	22.5	453	454,512,398	454,512
des–t-Bu **RP-3**	4.5	399	417,458,400,328	417,458
des–Et-Py **RP-4**	5.5	425	426,484,370	426,484
des– t-Bu-Py **RP-5**	5.0	397	398,456,354	398,456

Fig. 6. Ld metabolites. *Above:* s.i.m. conditions. *Right:* proposed pathways.

Concluding comments.– The above approach allowed specific detection of the main metabolites in plasma from rats dosed with ^{14}C–**Ld**. The structural identifications allowed pathways to be proposed (Fig. 6). The approach is being extended to bile and to other animal species and man.

References

1. Micheli, D., Collodel, A., Semeraro, C., Gaviraghi, G. & Carpi, C. (1990) *J. Cardiovasc. Pharmacol. 15 (Suppl. 4),* 666–675.
2. Evans, G.L., Ayrton, J., Grossi, P., Pellegatti, M., Maltas, J. & Harker, A.J. (1990) in *Analysis for Drugs and Metabolites including Anti–infective Agents* [Vol. 20, this series] (Reid, E. & Wilson, I.D., eds.), Roy. Soc. Chemistry, Cambridge, pp. 285–290.

#ncA-2

A Note on

^{19}F-NMR SPECTROSCOPIC IDENTIFICATION OF DOG URINARY METABOLITES OF IMIRESTAT, A SPIROHYDANTOIN ALDOSE REDUCTASE INHIBITOR

P.J. Gilbert, T.E. Hartley, J.A. Troke, R.G. Turcan, C.W. Vose and K.V. Watson

Department of Drug Metabolism,
Hoechst Pharmaceutical Research Laboratories,
Hoechst UK Ltd., Walton Manor,
Walton, Milton Keynes MK7 7AJ, U.K.

HOE 843 (imirestat) is an aldose reductase inhibitor currently under development at Hoechst. It has a fluorinated nucleus with a hydantoin ring attached. It was necessary to elucidate the human metabolite as quickly as possible. Likely routes of metabolism were hydroxylation, conjugation or hydantoin ring cleavage. Only a single reference compound was available, AL 03363, which has a single hydroxy group *meta* to one of the fluorine atoms.

HOE 843   AL 03363

In man the dose of radio-labelled drug was low (10 mg or less). In addition, the urinary elimination was prolonged, making man an unsuitable species for the isolation and identification of metabolites. Pilot studies, based on radio-HPLC, showed qualitative similarity between man and dog, with extensive metabolism to at least 13 metabolites. On this evidence, we decided to use dog as a model species, and gave ^{14}C-imirestat orally, 20 mg/kg, to an adult male dog.

PILOT STUDIES, WITHOUT METABOLITE ISOLATION

Urine collected during 0→48 and 48→72 h after this dosage contained respectively 17.7% and 12.5% of the ^{14}C dose (total for 0→96 h = 43%). Radio-HPLC of the 0→48 h urine revealed a complex mixture of metabolites and a small proportion of parent drug (1.6% of the dose). Direct ^{19}F-NMR spectroscopy of this urine showed fluoride ion, numerous metabolites which were predominantly glucuronide conjugates and, as a minor component, parent drug. This extensive conjugation was comparable to that in man.

After incubation with β-glucuronidase the 0→48 h urine gave a ^{19}F-NMR spectrum showing fewer signals. This finding is consistent with the main metabolic pathway being aromatic ring hydroxylation followed by glucuronidation. Examination of the deconjugated urine by proton-coupled ^{19}F-NMR and 2-D ^{19}F-^{19}F correlated spectroscopy indicated that major components included 3 monohydroxy derivatives of the parent drug, a diol with both phenolic functions in the same ring, and a phenolic metabolite containing only one fluorine atom. In the HPLC system we were using, much co-chromatography was evident after deconjugation by β-glucuronidase. Rather than incur much time in pursuing studies with unconjugated urine, we elected to isolate conjugates, and then use ^{19}F-NMR to detect and identify metabolites.

STUDIES BASED ON ISOLATION OF INTACT METABOLITES FROM URINE

Experimental approach, and outline of findings.- From 0-48 h dog urine, metabolites were extracted using either XAD-2 resin or large C-18 solid-phase cartridges. Semi-preparative HPLC gave ^{14}C peaks (also monitored by ^{19}F-NMR) comprising individual glucuronides as mixtures of epimers. Each fraction was deconjugated by β-glucuronidase treatment, and re-run by HPLC. The resulting aglycones were identified by multinuclear NMR and by GC-MS. The 3- and 4-hydroxy derivatives of imirestat were identified, and also the 2-hydroxy product formed during or following defluorination. The other major aglycone was postulated to be the 3-fluoro-2-hydroxy metabolite. This represents a novel pathway of 'NIH-shift' type for the metabolism of fluorobenzenes.

NMR spectra

The proton-decoupled ^{19}F-NMR of metabolites in the dog urine extract showed >50 signals. In the parent compound, the two fluorine atoms are equivalent, giving a single ^{19}F signal. After metabolism of one aromatic ring, the fluorines are rendered non-equivalent giving a pair of ^{19}F signals. After β-glucuronidase treatment the ^{19}F spectrum simplifies substantially, although almost every signal has shifted.

Hydroxylation of HOE 843 creates a chiral centre at the quaternary spiro carbon atom. Subsequent glucuronidation potentially produces pairs of diastereoisomers, which have different NMR spectra. Hence we can observe the stereochemical preference for hydroxylation.

We had originally hoped to assign structures by looking at proton-coupled ^{19}F spectra, without needing chromatography. Unfortunately it proved impossible to distinguish two of the three high-field signals in this way, although the third, highest, field signal showed only a single [19]F-^{1}H coupling, indicating a doubly substituted aromatic ring.

The asymmetric species showed a small ^{19}F homonuclear coupling. By the use of ^{19}F homonuclear COSY 2-D NMR we could pair the ^{19}F signals. Comparison with the coupled spectrum then told us that none of the major metabolites is substituted on both rings.

Assigned structures of particular interest

Several interesting structures were assigned to various HPLC-isolated species. Thus, the ^{19}F spectrum of HPLC 'peak 7' showed two very close peaks, collapsing to one after deconjugation. A proton-coupled ^{19}F spectrum showed apparently no aromatic hydroxylation, and at first we suspected modification of the hydantoin ring. However, a ^{13}C DEPT spectrum showed 6 aromatic signals, only 3 of which were ^{19}F-coupled.[θ] This led us to a structure where one fluorine atom has been replaced by OH. An authentic standard of this compound later became available, confirming our conclusion. In consequence of this replacement, fluoride ion is also a metabolite of HOE 843 and can be seen in ^{19}F spectra of neat urine.

The two compounds with different but similar high-field ^{19}F signals proved difficult to elucidate. Both showed substitution by a single OH, probably *ortho* to fluorine, so at first we identified them as two o-hydroxy derivatives of HOE 843. Eventually they were separated, deconjugated and cleaned up by further HPLC, using ^{19}F-NMR to monitor progress. The resulting samples were clean enough for ^{1}H-NMR. Both spectra showed proton signals for an unaltered aromatic ring, plus two doublets which were not mutually coupled and whose J values were the same as those obtained from the ^{19}F spectra.

These two compounds had identical mass spectra, but showed differences when subjected to GC analysis. We must therefore assign the two spectra to the following two structures:-

[θ]With ^{13}C NMR
the DEPT pulse
sequence shows
only protonated
carbons.

This shift of fluorine is very unusual, and we can find no precedent for it, although such a shift has been reported for chlorine and bromine. The defluorination product is also unusual in representing an important route, although not unprecedented.

These various metabolites, and others identified by analogous means, account for >90% of the urinary excretion.

CONCLUDING COMMENTS

We obtained clean samples from large volumes of urine very quickly using solid-phase extraction. The columns could be regenerated for repeated use.

Separation of conjugates by HPLC followed by deconjugation gave clean samples for spectral analysis.

Thereby we were able to identify several novel metabolites and observe an unusual F shift.

The work was completed within 10 weeks, and the useful interaction between metabolite elucidation and physical chemistry will be worth pursuing in future projects.

Acknowledgement

Angus Coe and Belinda Mann contributed to this work.

Comments on **#A-1**, J.R. Everett - β-LACTAM METABOLITES BY NMR
                #A-3, H. de Bree - METABOLITE ISOLATION/CONCENTRATION
                #A-4, I.D. Wilson - EXPERT SYSTEMS
                #ncA-1, P. Grossi - LACIDIPINE METABOLITES

I.F. Skidmore asked whether with very high dose levels
(say 1 g/kg) a metabolite that would be apparent at lower
doses might be obscured. **Everett's reply.**- Although this
is a risk, one should not tailor dose levels to NMR but
should attempt to get answers at the dose level the client
wants to use. In our laboratories (SKB) NMR is routinely
applied to biofluids to identify metabolites for all drugs
entering development (preclinical stage) and in the first
human (Phase I) studies, but not thereafter. In ICI (where,
perhaps significantly, drug dosages may be rather low for
NMR studies), NMR is used at the early preliminary stages
of development but usually not in human studies. **Question
from G.D. Land.**- Given that ^{19}F is a very useful moiety
for use in NMR investigation of metabolic products, can derivatization be applied to broaden the value of this technique?
Reply: indeed there is advantage in this approach, as tried
by us in a chiral study.

Editor's interpolation: excerpts from a Forum presentation
- I.D. Wilson (ICI Pharm.), with J.K. Nicholson (Birkbeck
Coll., London): AN NMR-BASED STRATEGIC APPROACH TO METABOLISM
STUDIES

In the context of ^{1}H- and ^{19}F-NMR in particular, consideration
was given to "factors such as the dose level required
for the performance of such studies, excretion balance
experiments, metabolite detection, profiling and identification.... [also considered:] the use of simple chemical and
enzymic treatments as an aid to the characterization of
metabolites, together with simple methods for rapid metabolite isolation".
A ^{19}F-NMR contribution from these authors' laboratories
appears later (art. #D-4).

de Bree, answering D. Schmidt: After freeze-concentration
urine samples show a rise in pH, but 8.0 is the highest
attained. **Reply to Ritter:** near-60 hamsters to get 9 ml of urine!

D. Bingham asked Wilson whether 'expert systems' could
be developed on an enzymic rather than chemical-structure
basis. "Yes" was the **reply**, e.g. for hepatocyte studies,
but this approach likewise has limitations. **Comment by Sue Rees.**-
Having been characterized for known reactions (including Phase II),
hepatocytes serve well where large numbers of related compounds
have to be screened (e.g. for metabolic fate), provided that

all data are carefully correlated and species comparisons made; judicious use of hepatocytes should not delay getting data for man. *SEE BELOW for 'hepatocyte continuation'.*

Question to Wilson [cf. art. #A-4] **by R.J. Simmonds.-** If all 'rule-based systems' are flawed and limited, what criteria does the 'human expert' use to predict metabolic fate? **Reply:-** mainly experience and intuition and, where these fail, literature precedents. We do not yet have 'rule-based expert systems' - because we don't really know the rules yet! - but only 'precedent-based systems'.

P. Grossi, answering G.S. Land.- Setting up the analytical system, by transfer to an existing HPLC system for ^{14}C assessment, entailed no particular problems.

Editor's interpolation, relating to a Forum talk (no publication text) which attracted discussion
- S.A. Wood & R.J. Simmonds [colleagues of Sue Rees]: STEREO-SELECTIVE METABOLISM IN HEPATOCYTES MONITORED BY HPLC

S.A. Wood, answering G.J. de Jong and G.C. Young who asked about the 'second-generation' AGP columns used for separating aminotetralin enantiomers.- Long-term stability was good, but a notable shortcoming was column-to-column variation: thus one lost its separatory efficiency and two others had lower and different efficiencies. **Reply to J.R. Everett,** concerning peak-splitting observed for one enantiomer with certain injection volumes:- the phenomenon is mysterious; it occurs, reproducibly, only with the Gilson auto-injector, conceivably due to analyte adsorption onto system surfaces after reaction with a plasma component (not pursued).

A metabolic point in a Forum talk (lacking in the text, #A-2):

W. Ritter asked J. Oxford about the metabolism of a tetrahydrocarbazole (an anti-emetic) to which she had referred.- Was the 2-hydroxy metabolite, predicted to be a major one, indeed found? **Reply (by I.F. Skidmore,** a colleague): no! - hydroxylation occurred only in the aromatic ring.

Citations contributed by Senior Editor
besides those featuring the term **invest**(igation) *in* **#ABC** *('Compendium')*

Electrochemical detection applied to delineating metabolic pathways, e.g. aromatic hydrocarbons, thiols, amines, heterocycles, quinones: Lunte, S. et al. (1990) J. Pharm. Sci. **79**, 557-567.
     Prediction by 'HPLC-Metabolexpert' of metabolite nature/R_T's for ecdysone: Kalasz, H. et al. (1990) Chromatographia **30**, 95-98. System limitations *(considered at Forum)*: its database? its 'inference engine' determining how rules are applied?

Section #B

ASSAY STRATEGIES FOR VARIOUS DRUGS

At end of Section, from p. 147:
   Section #ABC - *A Compendium* :

RECENT PAPERS ON DRUG ASSAY OR METABOLITE DELINEATION

#B

Section Preface

METHOD DEVELOPMENT AND OPTIMIZATION STRATEGIES

P.F. Carey

Glaxo Group Research, Ware, Herts. SG12 0DP, U.K.

EDITOR'S INTERPOLATION.- The author opened the Discussion on the title theme at the 1991 Bioanalytical Forum which gave rise to this book. There are implied allusions to contributions that appear in subsequent pages, and to articles in previous Forum-based volumes [see end of art.], discernible from Contents lists and from 'Assay' and 'Sample' Index entries; specific cross-referencing has not been attempted. The present survey will be especially helpful to bioanalytical novices.

Method Development and Optimization ('MDO') in bioanalysis encompasses the three general areas of sample preparation, analytical technique and detection (end-point). MDO cannot be discussed in isolation without first clearly defining the objectives of the proposed method of analysis. The extent to which the method is to be developed in terms of sensitivity, selectivity and throughput must be equated with the physico-chemical properties of the analyte and the end-purpose of the analysis.

As a simple illustration of varying analytical demands, we may compare the sensitivity and accuracy requirements of an assay designed to support two different types of study – those for general drug discovery and those for bioequivalence assessment in man. Where bioavailability is being screened with a view to selecting a drug candidate in the early stage of a drug-discovery project, an 'answer' within 10% may suffice for eliminating compounds, whereas the accuracy of the same assay used in a bioequivalence study in upwards of 20 volunteers needs to be far more stringently defined and controlled.

SAMPLE PREPARATION OPTIONS

In general, a sensible approach to sample preparation would be based on just enough work to achieve the objective. – In some cases it may be just as unacceptable to produce a method which goes far beyond these requirements as it is to produce one which is poorly designed and insufficiently reproducible. Approaches which may be used in preparing a biological sample for analysis include the following, listed with comments.-

#Protein precipitation.- Organic solvents (methanol, aceto-nitrile), acids (perchloric, trichloroacetic).
#Filtration.- CF-25 extraction cones.
#Dialysis (microdialysis).- 15 kDa cut-off.
#Salt-solvent pairs.- ISRP* columns; column switching.
#SPE*.- Cyanobonded phase (in a context mentioned below).
#Liquid-liquid extraction.

General objective: *to achieve the requisite sensitivity, accuracy and throughput; sample preparation has to take account of the analyte's physicochemical properties (need to derivatize?).*
Possible sample-preparation objectives.- *#Increasing selectivity for analyte isolation. #Removing unwanted macromolecules. #Trace enrichment. #Analyte stabilization.*

Protein precipitation methods may suit for compounds such as cephalosporins, and indeed the pre-clinical analysis for cefuroxime was done by just such a method (cited in [1]). Whilst membrane filtration is an option for removing macromolecules, although CF-25 cones (Amicon) may not suit for high throughput assay of single analytes; yet they may well have an application in preparing a limited number of samples for examining drug and metabolite profiles. We are particularly interested in applying microdialysis as a technique for on-line sampling of blood using microprobes, and within Glaxo we have a small group investigating the use of this technique for analytical and metabolic applications.

Drug analytes and metabolites may also be determined using salt-solvent pairs [as mentioned in a semi-obsolete sample-preparation review (1976)$^\theta$- *Ed.*]. The original method was designed to give a sample from urine containing a broad range of drugs and metabolites for identification by TLC in forensic applications [2]. The method used ethanol and solid potassium carbonate and was exothermic. A modification using ethyl acetate and ammonium carbonate was used to isolate drugs and metabolites from urine, plasma and breast milk for measurement by selective-ion detection in a variety of MS modes, giving sensitivities in the ng-to-pg range [3].

Analytes may also be determined, with virtually no sample preparation, by direct injection techniques, using ISRP columns or column-switching. Use may be made of a simple cartridge-in-loop device or of more complex systems involving up to 3 columns and 3 pumps for mobile phases.

Currently SPE is the most popular isolation technique, ranging widely in analyte and sorbent types. It has been

Abbreviations.- ISRP, internal-surface reversed phase; MS, mass spectrometry; SPE, solid-phase extraction.
$^\theta$Reid, E., *Analyst 101*, 1-18.

proposed that an 'expert system' for SPE based on cyanopropyl sorbents could cope with the extraction of acidic, neutral and basic drugs from aqueous samples or plasma [4]. Liquid-liquid extraction is still used for extraction of some analytes from plasma - particularly when the extraction is coupled with an electrochemical end-point [#ncB-3, this vol.- Borner].

OPTIONS FOR THE ANALYTICAL STEP

In developing the isolation strategy it is imperative that the analytical method be developed in parallel, if only to monitor and assess the efficiency of the extraction and isolation processes for non-radiolabelled analytes. Taking account of available instrumentation *vs.* analytical require-ments, one of the following may be chosen.-
#LC-MS ['LC' connotes HPLC] or GC-MS.
#Spectroscopy.
#Immunoassay: scintillation proximity (SPA); specific; non-specific.
#Polarography.
#Electrophoresis.
#Chromatography, not MS-coupled: GC, (HP)TLC, HPLC (NP or RP), ion-exchange; size-exclusion ('gel filtration').
Preliminary thought on available resources and the analytical objective may contra-indicate MS-based techniques as the first choice, but they may be the only choice when other methods have been exhausted. In principle, spectroscopic analysis could be applied directly to solvent-prepared extracts [5, & $\oplus$], but generally it is applied in the form of a detector coupled to a chromatographic column. Immunoassay is a powerful tool. SPA may be applied direct, or could be combined with a chromato-graphic isolation stage for specificity.

Polarography was a promising technique several years ago, particularly in the differential phase mode*. One wonders why interest in and use of this technique, and of TLC/HPTLC, is now confined to a core of enthusiasts. The answer may well be the popularity of HPLC and GC with their wider availa-bility, established automation and complete integration into data-handling systems. Today, however, for those laboratories wishing to maintain GC as an alternative option to HPLC, two questions arise.- Do they have the equipment readily available?   and the expertise for exploiting the technique to the full? HPLC is notable for its availability and wide use, especially with RP and silica (NP) columns. Possible detection modes include UV, fluorescence, electrochemical, MS and radiochemical. The choice will depend on availability and analytical requirement.

*Editor's note.- See the Index entry 'Polarography' in early vols. (enquire about procurement! for second, see [5]); the first (1976) has a review art. | $\oplus$1976 review: see $\oplus$ opposite.

OVERALL STRATEGY

Method development is multi-stage.-
1. Assemble all physicochemical data.
2. Optimize extraction (recovery, interference). *HPLC chosen?* -
3. Establish primary retention mechanism
   (re-establish with a new column).
4. Use solvent optimization software for
   analyte resolution/metabolite separation.
5. Record *all* results; establish a data base.

The three depicted parts have to be integrated, maybe by painful fine-tuning and 're-
cycling' until the required analytical aim is
achieved; the whole process may even have
to be repeated from start to finish. Using
a column with a good plate no., usually a retention ratio between 2 and 5 is achieved
intuitively by mobile-phase adjustment; then
co-elution problems may entail changing α.

THE PROBLEM
Optimization; resolution

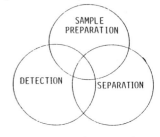

$$R_s = \tfrac{1}{4}\,\frac{\alpha - 1}{\alpha}\cdot\sqrt{N}\cdot\frac{1}{(1 + k')}$$

resolu-   selecti-   col. retention   vity   effi- tion
                      ciency ratio
[α = relative
retention,   effective
k'_B/k'_A]   plates

The Bioanalytical Forum format is unique in that there
is discussion of methods that were poor, or were made to
work only after several attempts and false starts, or did
not work consistently. Such information is helpful in assay
development or adaptation: it serves as encouragement and
a reference point to anyone who has ever 'failed', for whatever
reason, to develop an analytical method. Whilst we now
have reliable, specific instruments and processors which were
unavailable 5-10 years ago, a glance through past Forum-based
books suggests that formerly there was greater application
of chemical principles to achieve similar or even better
goals. Perhaps present-day instrument availability is masking
some decline in thorough exploitation of the analyte's chemistry
for optimizing and establishing analytical methods. Automation
should lighten the analyst's burden, giving more time to design
methods with quality built in; but often that time is fully
used in processing the automation-generated mass of data,
or in coping with new GLP requirements and internal 'QA'
audits, or in writing-up and reporting. Moreover, notwithstanding 'expert systems' it may be better to rely on experience
and expertise than to follow slavishly an ill-understood software
package in pursuit of an analytical solution: unexpected facets
may emerge from the development process.

References

1. Ward, J.B. & Harper, P.B. (1987) *Drugs*
   34 (Suppl.2) [cephalosporins], 253-258.
2. Bastos, L.M., Kananen, G.E., Young, P.M., Nonfort, J.R.
   & Sunshine, I. (1970) *Clin. Chem. 16*, 931-940.
3. Horning, M.G., Gregory, P., Nowlin, J., Stafford, K.,
   Lertratanangkan, K., Stillwell, W.G. & Hill, R.M.
   (1974) *Clin. Chem. 20*, 282-287.
4. Moors, M. & Massart, D.L.
   (1990) *TRAC 9*, 164-169.
5. Martin, L.E., Carey, P.F.
   & Bland, R.E. (1978) in
   Blood Drugs ..... [Vol. 7, this
   series] (Reid, E., ed.),
   Horwood, Chichester, 227-242.

#B-1

ASPECTS OF METHOD DEVELOPMENT[⊕]

P.H. Degen and G.Flesch
Research and Development,
Ciba-Geigy Ltd., CH-4002 Basel, Switzerland

Drug development entails performing key studies with the aid of bioanalysis: kinetics in animals, toxicokinetics, biotransformation (animals, man), kinetics in human volunteers, kinetics in patients, bioavailability, dosage form evaluation, dosage regimen. The development of reliable, specific and sensitive analytical procedures is therefore paramount. The analyte may be partly or completely transformed (degradation, biotransformation) and is often in low concentration. The originally developed analytical method may have to be altered or adapted to separate enantiomers. The measurements performed must be clinically relevant; otherwise the plasma concentration curves are of little value.

For reasons of cost, ease of handling and transfer, and availability, GC and HPLC are mostly used, as in 70-80% of all bioanalytical methods in our laboratories. There have been major instrumental and column-technology advances in the last 20 years. Examples of method development or adaptation, confined to GC and HPLC (not with MS), are discussed for drugs in different therapeutic areas (anti-rheumatic, anti-depressant, anti-amoebic, anti-epileptic, anti-tumour).

The enormous increase in publications on analytical methods has diverse causes including better instrumentation, escalating demands, low-dose action that calls for high sensitivity, and enantiomeric differences. Bioanalysis has been reviewed in relation to pharmacokinetics [1, 2]. Consideration is given here to illustrative GC and HPLC approaches, some for difficult analytes, taking 'LOQ'[*] into account.

[⊕]Pertinent arts. in earlier vols. (ed. E. Reid or E. Reid *et al.*) include:- R.G. Cooper ('Development of analytical methods: general philosophy'), pp. 1-5 in V. 7 (Horwood, 1978); A. Bye, pp. 97-105 in V. 18 (Plenum, 1988); C. Town, pp. 109-116 in V. 20 (R. Soc. Chem., 1990). V. 20 also features method validation and quality in the 'GLP' context. Some introductory material in the present text has been curtailed.- *Ed.*

[*]*Abbreviations*.- ECD/NPD, electron-capture/nitrogen-phosphorus detector; LOQ, limit of quantitation; SPE, solid-phase extraction.

In a good analytical method (guidelines: [3]) the LOQ
should lie at ~10% of the peak plasma concentration and
still, by definition, have C.V. <10% in multiple determinations.
Reproducibility of results and simplicity are aimed at. One
option is straightforward chromatography after the analyte
has been carefully enriched, purified and specifically derivatized.
The other option, more technological than chemical, is elaborate
chromatography after simple, non-specific enrichment. Options
for metoprolol (e.g.) include packed-column GC after derivatization, and direct injection and complicated column-switching.
An individual analyst may have a particular preference. All methods
must meet present-day 'GLP' criteria.

CRITICAL POINTS (with allusion to analytes in EXAMPLES, below)

Stability.- The stability of a new compound in aqueous
solutions, important for setting up a bioanalytical method,
is generally known by that time. However, as part of the
method validation, the stability of the compound and possibly
its metabolites in biological matrices such as blood, plasma
or urine needs to be evaluated before pharmacokinetic studies
are planned; moreover, spiked plasma and actual samples may
differ in stability (hydralazine [4, 5], dihydralazine [6])*.

Extraction.- Some compounds are not amenable to liquid-
liquid extraction (hydralazine [4, 5], dihydralazine [6],
phanquone [7], pamidronate [8], CGP 37 849 - unpublished; see
later). Extraction properties may also vary from one biological
fluid to another (diuretics [9]). SPE with cartridges,
effective for highly polar compounds, is increasingly used
and facilitates automation (levoprotiline, below). SPE has
both advantages and disadvantages (review: [10]); performance
is often hampered by poor reproducibility and recovery.

One approach is to derivatize directly in plasma and
extract and chromatograph the derivative [4-7, 11]. A very
powerful and widely used technique for polar compounds is
extractive alkylation, the combination of ion-pair extraction
and alkylation [12-14].

Conjugated metabolites, e.g. glucuronides and sulphates,
are normally determined in their free form after enzymatic
or chemical hydrolysis, calling for very harsh conditions
if, as with levoprotiline [15], stability is very great. In
such cases, the direct determination of the conjugate is
the better solution, although it does necessitate access to
authentic conjugate.

*Note by Ed.- In the General Index of each earlier 'A' vol.
(including V. 10, out-of-print), 'Lability' entries are pertinent.
Other entries evident from Index 'browsing', e.g. Solvent,
bear on other points, e.g. extraction. Glucuronide hydrolysis:
V. 12, p. 268.

DERIVATIZATION (anticipating EXAMPLES, below)

Derivatization for GC, widely used to improve volatility, sensitivity, separation or stability, commonly entails alkylation or acylation. To benefit ECD detection, halogen-containing reagents are advantageous [16-18] but not obligatory [19].

Derivatization for HPLC is of lesser importance, since many compounds can be chromatographed directly (diclofenac [20]). It is needed when the compounds have no chromophore (UV or visible) or fluorophore. Generally the derivatives formed are less polar than the parent compounds and so are easier to separate. In general it is easier to derivatize pre-column than post-column. Formation of diastereoisomers of racemic drugs is a cardinal and often-used pre-column derivatization step. The reagent used to derivatize the drug enantiomers is a single enantiomer, and an achiral stationary phase is used (CGP 37 849 as analyte example).

CHROMATOGRAPHY AND DETECTION

For both GC and HPLC the extent of purification of the sample influences the quality of the chromatography. Generally the sample preparation is more complicated for GC than for HPLC. The lifetime of columns, especially capillary, is increased if the samples are well purified. In HPLC the direct injection of plasma calls for complex column-switching devices; yet this may be the method of choice if numerous samples of the same type are to be analyzed [21].

Few GC analytes have good chromatographic properties and detectability without derivatization, often done specifically for ECD or NPD. In HPLC, however, many compounds are UV-detectable directly, albeit often with limited sensitivity. Better sensitivity is achievable fluorometrically, usually needing specific derivatization (often easy; pamidronate [8], CGP 37 849), or electrochemically if the compound is electro-reducible or -oxidizable.

EXAMPLES

Hydralazine, dihydralazine (cardiovascular):- *précis by Ed., alluding to Degen's hydralazine review in an earlier vol.*[22]; *legend to* Fig. 1, *overleaf, helps clarify.*- The background to the extensive assay literature [22] is that these drugs, in pure solution and especially in plasma, manifest stability problems as amplified in [22], notably in respect of side-chain $-NH-NH_2$. Its spontaneous loss from the parent molecule in spiked plasma calls for expeditious assay; some loss occurs even in the thawing of samples stored at $-20°$ (for formula of one product formed from hydralazine, *viz.* phthalazine, see **III** in [22], Fig. 1).

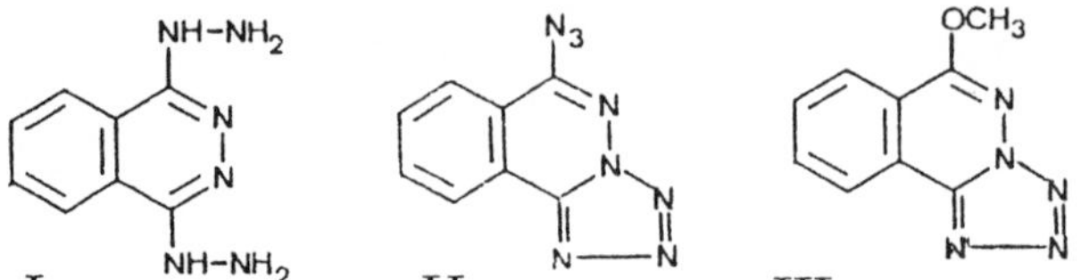

Fig. 1. Formulae relevant to assay of
dihydralazine, *I*, and similarly of hydralazine
to which the () entries that follow refer; it
has only one -NH-NH$_2$ side-chain (formula: **I** in [22], Fig. 1).
Nitrous acid treatment of *I*, or of its pyruvic acid hydrazone
(**II** in [22], Fig. 1), for GC-ECD assay [4], yields the azido-tetra-
zolophthalazine, *II* (**IV** in [22], Fig.1; thermolabile), convertible
by methanolic sodium methoxide to **III** (stable).
2,4-Pentanedione gives, from *I* but not its hydrazone, a 3,5-
dimethylpyrazolylphthalazine, *IV* (**VI** in [22], Fig. 1), allowing
GC-NPD assay of the free drug specifically [5].
Explanatory note: in *II*, 'N$_3$' illustrates the tautomeric variant
of the 'bridge' form, in the context of nitrous acid treatment.

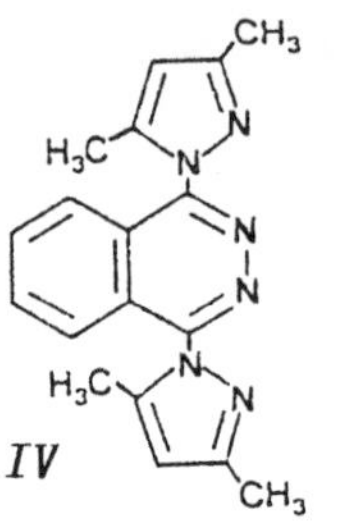

However, plasma incubated after spiking or taken after drug
administration shows substantial loss of free drug due to
'metabolic' phenomena, notably a reaction with endogenous
pyruvic acid yielding an acid-labile hydrazone (see legend to
Fig. 1). This is co-estimated with free drug in 'apparent-drug'
GC methods where the plasma is treated with nitrous acid
at the outset, thereby furnishing a solvent-extractable ECD-
responsive derivative (*II*, & see legend). As illustrated
in [22] for plasma from dosed volunteers, application of
an 'apparent hydralazine' procedure showed no analyte loss
after plasma storage for 1 month at -20°. This was also evident
in a 24-h study (peak at 3 h) after a 100 mg oral dose of
hydralazine hydrochloride. [*Excluded by Ed.:* a Fig. showing this,
and another showing, for spiked plasmas containing 0.1-1µM hydralazine, ~90% recovery of 'apparent drug' after 3 days at -20°.]

Besides 'apparent-drug' methods (e.g. [4]), there are
various methods that exclude the hydrazone and measure only
free drug (tabulation for hydralazine: [22]). Our approach
[5], with final GC-NPD, is to treat the plasma with 2,4-pentane-
dione to give an extractable pyrazole (*IV*, Fig. 1).

Phanquon (anti-amoebic; 4,7-phenanthroline-5,6-dione), which
resists conventional extraction, was derivatized directly in
plasma with a high excess of methoxyamine hydrochloride (Fig. 2);
the product is stable and extractable. Two extraction steps
give sufficient clean-up for GC-ECD detection down to 70 nM [7].

4-Hydroxyandrostenedione (aromatase inhibitor/anti-tumour;
4-OHA) exemplifies the usefulness of extractive alkylation

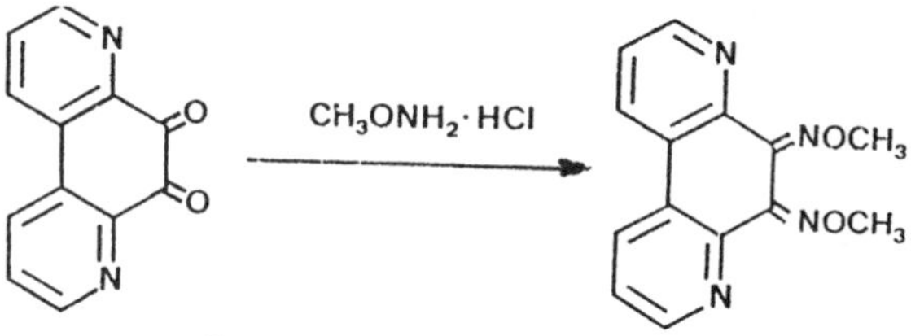

Fig. 2. Conversion of phanquone to a stable dimethoxime by treatment of the plasma with methoxy amine [7].

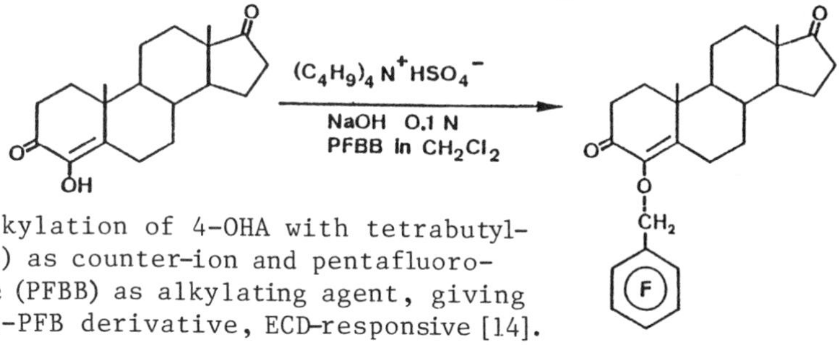

CH₃ONH₂·HCl

Fig. 3.
Extractive alkylation of 4-OHA with tetrabutyl-ammonium (TBA) as counter-ion and pentafluoro-benzyl bromide (PFBB) as alkylating agent, giving a stable mono-PFB derivative, ECD-responsive [14].

$(C_4H_9)_4 N^+ HSO_4^-$

NaOH 0.1 N
PFBB in CH_2Cl_2

(Fig. 3). Thereby it can be estimated in plasma down to ~20 nM, with no endogenous steroid interferences. Greater sensitivity, although not needed in our pre-clinical and early clinical studies, is achievable by use of GC-MS [23] or RIA [24, 25].

Halometason (topical corticosteroid) already contains two F and one Cl atoms. Being a preparation intended for percutaneous use, low plasma levels were expected, demanding high sensitivity. Since the C-17 side-chain precluded reproducible derivatization, it was oxidatively removed and a dimethoxime was formed, stable and giving very high ECD sensitivity (Fig. 4) [26].

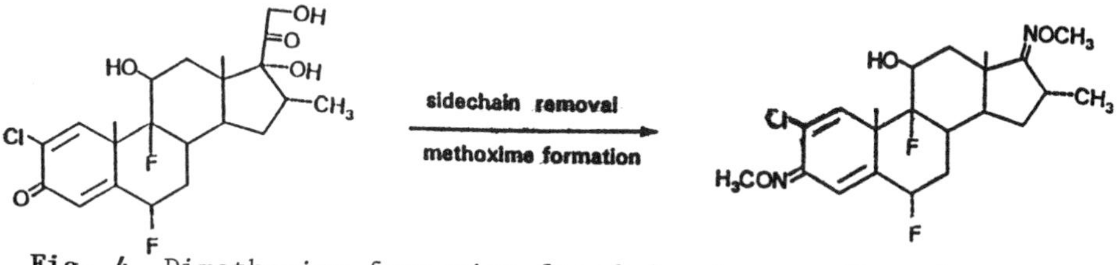

sidechain removal

methoxime formation

Fig. 4. Dimethoxime formation from halometasone after cleavage.

Metoprolol (β-receptor blocker) is readily extractable and also easily determined either by GC after derivatization or by HPLC. Numerous analytical methods have been published, with two totally different approaches as now touched on. The older methods are usually based on solvent extraction followed by haloacylation [16], or by HPLC with a column-switching system allowing direct injection of plasma samples. Such systems suit well for large sample numbers, but have to be thoroughly standardized and validated to ensure problem-free running [21]. Enantioselectivity is now achievable, by diastereo-isomer formation or using chiral columns [27, 28].

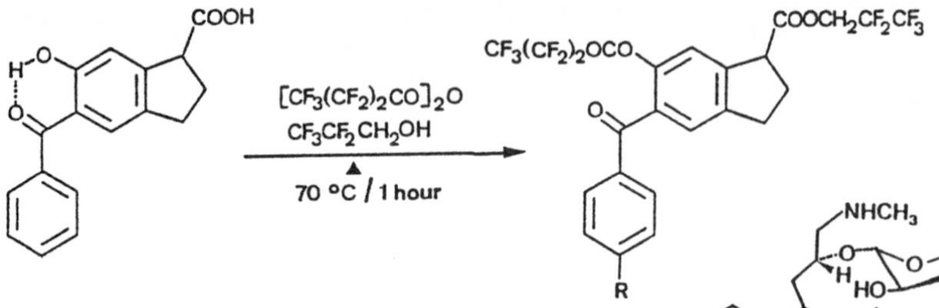

Fig. 5. Formation of an acylated and esterified derivative of oxindanac by a 'one-pot' reaction with heptafluorobutyric anhydride (HFBA) and pentafluoropropanol.

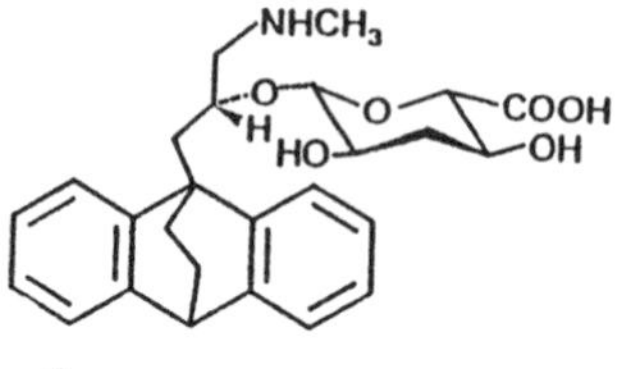

Fig. 6.
Levoprotiline glucuronide.

Oxindanac (analgesic), which has a carboxyl as well as a hydroxyl group, could be derivatized as in Fig. 5 with a reagent mixture. Following a single extraction, 2,2,3,3,3-pentafluoro-1-propanol (PFP) served for carboxyl-group esterification, catalyzed by HFBA as used to acylate the hydroxyl group [17]. The resulting derivative was stable and highly ECD-responsive. The derivative need not be purified, and altogether the method is very simple, and can detect <30 nM levels, <1% of the peak plasma level after the lowest therapeutic dose [17].

Levoprotiline (antidepressant), renally excreted predominantly as a glucuronide (Fig. 6), is itself easily determined by GC-MS [15] or by GC. The classical approach to measuring conjugates is hydrolysis and assay for free drug. However, levoprotiline is very stable and only hydrolyzable using harsh conditions (12 M HCl, 150°, 1.5 h). Analysis of the intact conjugate is therefore the better approach, hinging on availability of authentic conjugate. Extraction was accomplished by SPE, followed by HPLC. The entire procedure was fully automated using a Gilson ASPEC sample preparation unit, as in the following protocol (the % values are v/v).-
1. SPE cartridge conditioning: Bond Elut silica gel, 500 mg; + 2 ml methanol, + 2 ml water.
2. Sample application: +0.2 ml urine previously diluted with water (1:4) and spiked with internal standard (0.1 ml).
3. Washing: + 2 ×2 ml 5% aq. methanol, + 0.9 ml water/acetonitrile/methanol, 70:16.7:13.3.
4. Elution: + 0.75 ml water/acetonitrile/methanol, 10:50:40.
5. Dilution of eluate: + 0.85 ml water.
6. Mixing, by bubbling air; 2 ml.
7. Injection: 0.3 ml diluted eluate.
8. Chromatography: Lichrospher C-8, 75 x 4 mm; mobile phase pH 6 ammon. acetate buffer/methanol/acetonitrile, 56:22:22.
9. Detection: UV at 227 nm.

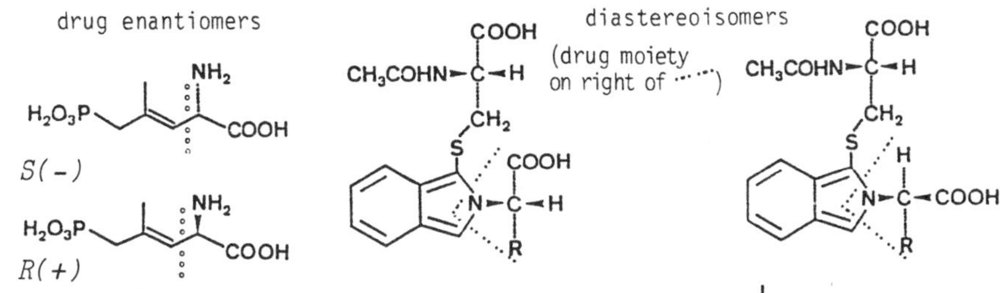

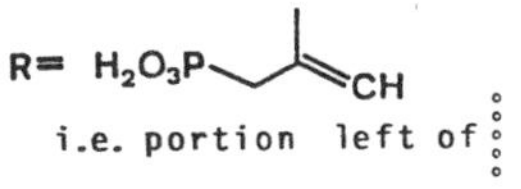

drug enantiomers   diastereoisomers

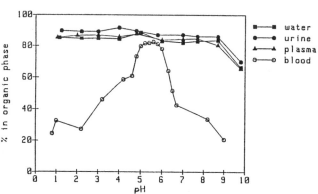

S(-)

R(+)

Fig. 7. CGP 37 849: the two
enantiomers, and the diastereoisomers formed by reaction with
OPA and **N**-acetyl-L-cysteine.

R= H₂O₃P...
i.e. portion left of ...

   CGP 37 849 (NMDA antagonist) is a racemate, (E)-2-amino-4-
methyl-5-phosphono-3-pentenoic acid. CGP 40 116 is the R-enantio-
mer (eutomer) and CGP 40 117 the S-enantiomer (distomer) (Fig. 7).
CGP 37 849 has disconcerting physicochemical properties, being
extremely polar with high water-solubility, and it lacks a
sensitive chromophore. The compound was isolated from plasma
by SPE and derivatized with o-phthaldialdehyde and N-acetyl-L-
cysteine (Flesch & Degen, Symposium presentation), as described
previously for amino acids [29]. The fluorescent diastereoisomers
being easily separable by HPLC, this method will allow a
rational approach for the generation of stereoselective data
from the enantiomers.

   Diuretics.- The three most commonly used diuretics,
chlorthalidone, hydrochlorothiazide and furosemide, are all
easily determined using extractive alkylation [9].   However,
unexpected extraction behaviour has been encountered with chlorthalidone.   It is easily extractable from water, plasma or
urine over a wide pH range.   However, due to binding to
erythrocytes it is extractable from blood only between pH 5
and pH 6 (Fig. 8). This exemplifies the importance of evaluating
extractability from all biological fluids used and over a
wide pH range.

Fig. 8. Extraction
of chlorthalidone
using methyl-iso-
butyl ketone: pH
dependence. See
text for comment.

■—■ water
●—● urine
▲—▲ plasma
⊙—⊙ blood

% in organic phase

pH

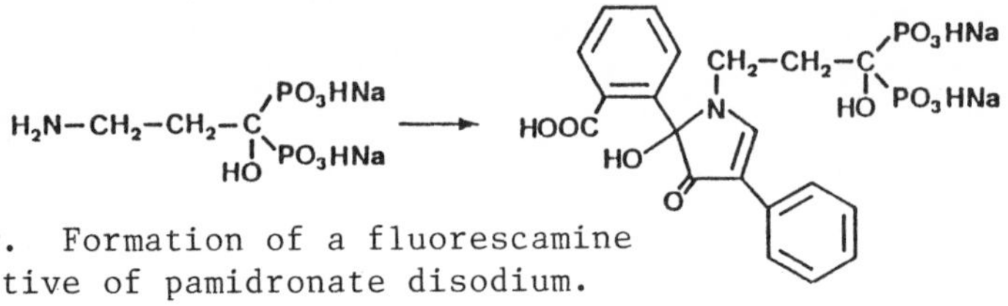

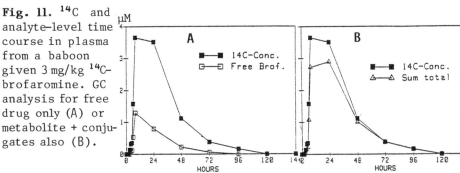

Fig. 9. Formation of a fluorescamine derivative of pamidronate disodium.

Pamidronate disodium (bone resorption inhibitor).- Because pamidronate is not readily detectable by UV it has to be derivatized before or after chromatography. An assay method for urine and plasma has been reported [8]. The compound's strong chelating capacity was exploited to isolate it from biological fluids, through co-precipitation with calcium phosphate, derivatization with fluorescamine (Fig. 9) and HPLC with fluorescence detection. Detection of phosphorus is an obvious alternative [30].

METHOD ADAPTATION: TWO EXAMPLES

Brofaromine hydrochloride (a putative antidepressant) is under development. For pharmacokinetic study in baboons, the ^{14}C-drug was given orally, and plasma samples scintillation-counted. Unchanged drug was determined by a GC-ECD method [18] (Fig. 10). Much of the plasma radioactivity was unaccounted for (profiles: Fig. 11A). With adaptation of the method to include the desmethyl metabolite, besides the conjugates of both compounds, >80% of the radioactivity was now accounted for (Fig. 11B) [report by F. Waldmeier & P. Degen].

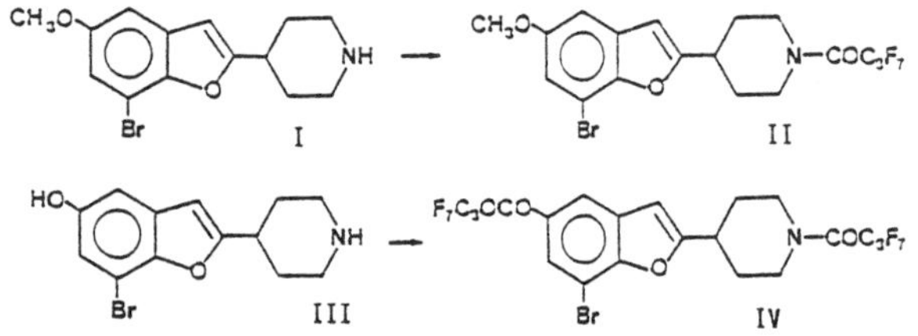

Fig. 10. Conversion of brofaromine (I) and O-desmethylbrofaromine (III) to HFB derivatives (II, IV).

Fig. 11. ^{14}C and analyte-level time course in plasma from a baboon given 3 mg/kg ^{14}C-brofaromine. GC analysis for free drug only (A) or metabolite + conjugates also (B).

Diclofenac sodium (NSAID):- *précis by Ed., alluding to Degen's review in an earlier vol.* [31].- When percutaneous drug absorption came to be studied, published assay procedures [31] (recent example: [20]) for the drug and its conjugate were precluded since there is rapid formation of 5 metabolites (mono- or di-hydroxy, and hydroxy-methoxy [31]). The need was met by a capillary-GC method which further served to assess the bioavailability of a slow-release tablet merely by assaying urine for total excretion including the metabolites (W. Riess & colleagues). The GC pattern for these is shown for plasma along with an outline of the extractive-alkylation approach ([31]; see [13] for details).

CONCLUDING COMMENTS

Optimal method development can be performed only if the exact requirements are known:- stereospecificity (racemic drugs); sensitivity (LOQ and detection) required; estimated time for developing the analytical method; types of biological matrix; laboratory capacities. In the pharmacokinetics field, method development is an ongoing process. It requires close cooperation with other disciplines (chemical, metabolic, pre-clinical, clinical) involved in the drug development process. It must be recognized, however, that no described method is a final one. The use of an appropriate method is essential, because the information derived from biological fluid concentrations is crucial for the further development of the drug.

References

1. Peng, G.W. & Chiou, W.L. (1990) *J. Chromatog. 531*, 3-50.
2. Hirtz, J. (1986) *Biopharm. Drug Dispos. 7*, 315-326.
3. American Chemical Society (1980) *Guidelines for data acquisition and data quality evaluation in environmental chemistry. Anal. Chem. 52*, 2242-2249.
4. Jack, D.B., Brechbühler, S., Degen, P.H., Zbinden, P. & Riess, W. (1975) *J. Chromatog. 115*, 87-92.
5. Degen, P.H. (1979) *J. Chromatog. 176*, 375-380.
6. Degen, P.H., Brechbühler, S., Schneider, W. & Zbinden, P. (1982) *J. Chromatog. 233*, 375-380.
7. Degen, P.H., Brechbühler, S., Schäublin, J. & Riess, W. (1976) *J. Chromatog. 118*, 363-370.
8. Flesch, G., Tominga, N. & Degen, P.H. (1991) *J. Chromatog. 568*, 261-266.
9. Degen, P.H. & Schweizer, A. (1977) *J. Chromatog. 142*, 549-577.
10. Furton, K.G. & Rein, J. (1990) *Anal. Chim. Acta 236*, 99-114.
11. Vessman, J., Karlsson, K.E. & Gyllenhaal, O.F. (1986) *J. Pharm. Biomed. Anal. 4*, 825-834.
12. Degen, P.H., Schneider, W., Vuillard, P., Geiger, U.P. & Riess, W. (1976) *J. Chromatog. 117*, 407-413.

13. Schneider, W. & Degen, P.H. (1986) *J. Chromatog. 383*, 412–418.

14. Degen, P.H. & Schneider, W. (1991) *J. Chromatog. 565*, 67–73.

15. Ackermann, R., Kaiser, G., Dieterle, W. & Dubois, J.P. (1987) *Proc. 3rd Eur. Congr. Biopharmaceutics & Pharmacokinetics*, Vol. II (Aiche, J.M. & Hirtz, J., eds.), Imp. de l'Univ., Clermont-Ferrand,

16. Degen, P.H. & Riess, W. (1976) *J. Chromatog. 121*, 72–75. [pp. 278–285.

17. Degen, P.H. & Schneider, W. (1983) *J. Chromatog. 277*, 361–367.

18. Schneider, W., Keller, B. & Degen, P.H. (1989) *J. Chromatog. 488*, 275–282.

19. Vessman, J. (1980) *J. Chromatog. 184*, 313–324.

20. Sioufi, A., Richard, J., Mangoni, P. & Godbillon, J. (1991) *J. Chromatog. 565*, 401–407. [Cf. (1985) *338*, 151–159.]

21. Lecaillon, J., Souppart, C., Dubois, J.P. & Delacroix, A. (1988) in *Bioanalysis of Drugs and Metabolites* [Vol. 18, this series], (Reid, E., Robinson, J.D. & Wilson, I.D., eds.), Plenum, New York, 225–233.

22. Degen, P.H. (1988) *as for* 21., 193–200.

23. Guarna, A., Moneti, G., Prucher, D., Salerno, R. & Serio, M. (1989) *J. Steroid Biochem. 32*, 699–702.

24. Dowsett, M., Goss, P.E., Powles, P.J., Hutchison, G., Brodie, A.M.H., Jeffcoate, J.L. & Coombes, R.C. (1997) *Cancer Res. 47*, 1957–1961.

25. Khubieh, J., Aherne, G.W. & Chakraborty, J. (1987) *Cancer Chemother. Pharmacol. 19*, 175–176.

26. [Murphy, J.E., ed.] (1983) *J. Internat. Med. Res. 11*, *Suppl. 1*, 1–57 ('Sicorten® and Sicorten Plus®').

27. Persson, B.A., Balmer, K., Lagerström, P.O. & Scagill, G. (1990) *J. Chromatog. 500*, 629–636.

28. Ahnoff, M., Chen, S., Green, A. & Grundevik, I. (1990) *J. Chromatog. 506*, 593–599.

29. Buck, R.H. & Krummen, K. (1984) *J. Chromatog. 315*, 279–285.

30. Daley-Yates, P.T., Gifford, L.A. & Hoggarth, C.R. (1989) *J. Chromatog. 490*, 329–338.

31. Degen, P.H. (1988) *as for* 21., 107–114.

#B-2

A STRATEGIC APPROACH TO THE ANALYSIS OF
BASIC DRUGS BY HPLC

Brian Law

ICI Pharmaceuticals plc, Drug Kinetics Group,
Alderley Park, Macclesfield, Cheshire SK10 4TG, U.K.

RP-HPLC* is frequently unsuited to the simultaneous analysis of a parent drug and its metabolites, especially with basic analytes. For strong bases (pKa > 8) especially, it has proved advantageous to employ a silica stationary phase in the ion-exchange mode with aqueous/methanolic eluents. This system works well even for quaternary ammonium compounds and, moreover, allows prediction of retention for both drugs and metabolites from a number of easy calculable physico-chemical properties. These also help in choosing, for basic drugs in biological fluids, an extraction procedure - liquid-liquid or solid-phase (SPE) - that gives clean chromatograms and good selectivity with the chosen HPLC system. For weak bases, 'SCX' in place of silica allows good ion-exchange separation.

The present approach, still preliminary, could lead to an expert system for .the analysis of basic drugs. It should be possible to predict, merely from the structure, the extraction and chromatographic characteristics for most basic compounds.

In diverse areas of drug analysis such as forensic toxicology and emergency drug screening there is a frequent need to provide data rapidly. This is particularly true in the early phases of drug discovery and development where the analyst is often required to provide metabolic and pharmacokinetic data on a range of novel compounds. To have maximum impact on the drug development programme the analyst needs to minimize or totally eliminate the method-development aspect of the work. This article focuses mainly on this area.

One obvious answer to these problems is to use an expert system to give guidance as to the choice of chromatographic and other conditions. Hitherto, however, most expert systems have been mere collections of data. If the compound of

---

**Abbreviations.*- MS, mass spectrometry; RP, reversed-phase. Some terms used (& see text): k', capacity factor; $\log K_\beta$, hydrogen bond acceptor value; V, molecular volume; $\log P$, octanol/water partition coefficient.

interest has not been studied and does not appear in the
data base, then the system is of little value. Also, with
such systems the user is frequently required to provide the
expert system with information such as the molecule's UV
properties and the presence or not of electroactive groups
for electrochemical detection. This would appear to be a
serious under-utilization of the computer system, since, given
the appropriate software, most of the relevant molecular parameters can be readily recognised or calculated.

A more rational approach would be to provide the expert
system with the compound's structure and let it calculate
the properties of interest such as k', GC retention or solvent
extractability. Many properties of interest to the analytical
chemist can be described as a linear function of basic molecular
parameters, viz.

$$Property = f(X_1 + X_2 + X_3 + \ldots\ldots X_n)$$

where X_1, X_2 etc. are quantities such as log P, pKa, molecular
volume (V), dipole moment, highest occupied molecular orbital
or log K_β.   Parameters such as log P can be obtained from
various literature sources [1], calculated using manual methods
[2], or calculated from computer programs [3]. Dissociation
constants (pKa) can also be obtained from the literature
[1, 4, 5], calculated using manual methods [6] or determined
by analogy with known compounds. The quantum chemical parameters
such as atomic charges, molecular dipoles, etc., require sophisticated software which until recently has been run on mainframe
computers. A glance at any current chemistry journal will
reveal a number of suitable programs which will now run
on PC's. The ability to calculate the necessary parameters
is thus within the reach of most analytical chemists.

CHOICE OF CHROMATOGRAPHIC SYSTEM

The methods chosen for this type of work need to be
versatile, in that they should be able to chromatograph a
wide range of drug compounds including metabolites, which,
compared with the parent drug, may differ widely in lipophilicity, with implications for selecting the separation mode.
Most drug analyses are carried out using RP-HPLC. Reflecting
this, >80% of the column and column packing sales of one
U.K. supplier in 1990 consisted of C-8 and C-18 materials.
Although highly versatile, RP-HPLC is often less than optimal
for the analysis of basic drugs. Peak tailing and poor efficiency
are often in evidence even when silanol blocking agents or
ion-pair reagents are used. The adoption of such strategies
also results in the use of complex eluents which can require
careful optimizing. Besides, these eluents contain non-volatile
constituents which are incompatible with coupled HPLC-MS

analysis employing the widely used thermospray interface. The dependence of RP-HPLC separation on analyte lipophilicity handicaps simultaneous isocratic separation of a drug and its polar metabolites.

For the analysis of strong bases the use of silica in the cation-exchange mode with RP-type eluents has proved particularly useful. The preferred system [7] gives high efficiencies (2<h<4 where h = reduced plate height), excellent symmetry (As <1.2) and has broad applicability. This is clearly demonstrated in the separation (Fig. 1) of a wide range of both structural and pharmacological drug types including central stimulants, anti-histamines, tranquillizers, anti-emetics and benzethonium, a quaternary ammonium bactericidal. Many of these compounds, particularly the quaternary amine, can be difficult to assay with good chromatographic performance by RP-HPLC.

RETENTION PREDICTION

Whilst retention in this silica system was found to depend mainly on solute pKa, for a set of arylalkyl amines only 62% of the variation was explained by pKa [8]. However, by using additional parameters and modelling the data using multiple linear regression much improved fit can be obtained (Fig. 2). In this instance the use of three parameters – pKa, V and FPM1_2 (the polarizability of the bond of the first substituent on the amine) – greatly improves the fit of the data, explaining 82% of the variability. Although this is still not good enough to allow accurate prediction of retention times, further work is in progress to improve the correlation and hence retention predictability.

Note by Ed. - For 'lore' on silanol groups, see Index entry 'Solid...' in Vol. 18, this series (ed. E. Reid *et al.*, Plenum, 1987).

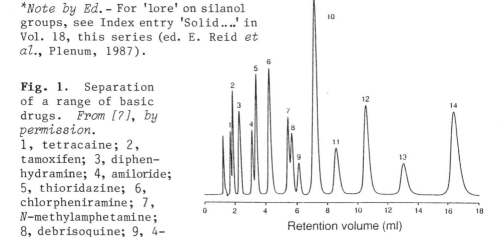

Fig. 1. Separation of a range of basic drugs. *From [7], by permission.*
1, tetracaine; 2, tamoxifen; 3, diphenhydramine; 4, amiloride; 5, thioridazine; 6, chlorpheniramine; 7, N-methylamphetamine; 8, debrisoquine; 9, 4-hydroxydebrisoquine; 10, quinacrine; 11, strychnine; 12, betahistine; 13, benzethonium; 14, pyrantel. Column (100 × 4.6 mm): Spherisorb SSW; eluent: methanol/ammonium acetate buffer pH 9.1 (9:1 by vol.).

Fig. 2. Correlation between observed and predicted log k' for a series of basic compounds, using multiple linear regression. pKa = log of the dissociation constant; V = molecular volume; FPM1_2 = polarizability of the bond between the N and the first substituent. HPLC conditions as for Fig. 1.

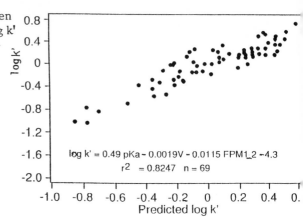

$$\log k' = 0.49\ pKa - 0.0019V - 0.0115\ FPM1\_2 - 4.3$$
$$r^2 = 0.8247 \quad n = 69$$

Using the simple arylalkyl amines as model compounds the retention properties of typical drug metabolites have also been studied. Irrespective of the structure of the core of the molecule, a particular metabolic transformation always produces the same characteristic change in retention [7, 8]. Using the functional group contribution approach [9], retention increments (Tau) associated with various metabolic transformations have been generated. Tau is defined as:

$$Tau = \log(k'_m / k'_p)$$

where m denotes metabolite and p parent compound. Typical metabolic transformations of the compounds shown *(right)* have been studied; Tau values, ±S.D. (& n):-

amitriptyline

N-methyl-amphetamine

- 2'-hydroxylation   +0.26 ±0.025   (3);
- 4-hydroxylation   +0.035 ±0.017 (10);
- *N*-oxidation   +0.12 ±0.0273 (3);
- *N*-demethylation   +0.47 ±0.08   (4);
- *N*-didemethylation   +0.22 ±0.015   (3).

From a knowledge of the retention of the parent compound and of the Tau value for a particular transformation it is thus a simple matter to calculate k' for the metabolite.

Most of the compounds discussed so far have been relatively simple arylalkyl amines that cannot really be reckoned truly representative of most drugs which generally contain a range of polar substituents. To accurately predict the retention of these types of compounds, other factors need to be considered. For example, Fig. 3 shows the separation of a range of β-blockers. Four of these have the same amine-bearing side-chain and almost identical pKa's of ~9.5, yet differ in retention characteristics because of the presence of additional polar functional

groups in the more highly retained compounds atenolol and
practolol both of which have polar amides attached to the
ring. Amide groups have strong hydrogen bond acceptor properties
($\log K_\beta$ ~2.8 on a scale of ~0 to 4), and [to be published]
the cause of the increased retention is believed to be a hydrogen
bonding interaction with unionized silanols.

Further support for this view has come from experiments
where a series of analogues with various ring substituents
have been chromatographed. The substituents were chosen to
cover a wide range of $\log K_\beta$ values [10]. The change in
retention associated with each substituent was determined
by reference to a parent unsubstituted compound. Some provisional
data for these compounds appear in Fig. 4 where Tau has
been correlated with $\log K_\beta$ and substituent volumes, using
multiple linear regression. Although there is a considerable

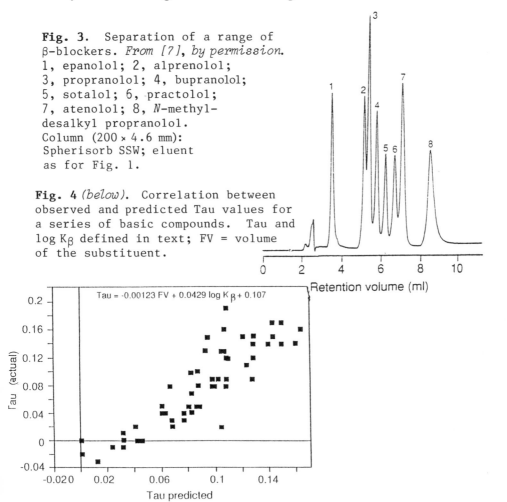

Fig. 3. Separation of a range of
β-blockers. *From [7], by permission.*
1, epanolol; 2, alprenolol;
3, propranolol; 4, bupranolol;
5, sotalol; 6, practolol;
7, atenolol; 8, *N*-methyl-
desalkyl propranolol.
Column (200 × 4.6 mm):
Spherisorb SSW; eluent
as for Fig. 1.

Fig. 4 *(below).* Correlation between
observed and predicted Tau values for
a series of basic compounds. Tau and
$\log K_\beta$ defined in text; FV = volume
of the substituent.

Tau = -0.00123 FV + 0.0429 log K_β + 0.107

scatter in the data, there is a clear indication that smaller substituents with strong hydrogen bond acceptor ability result in greater retention. It is hoped that further analysis of these data using additional molecular parameters will significantly improve the fit of the data.

The above work, some of which is provisional in nature, has demonstrated that there is a clear correlation between the retention properties of bases on silica and various easily calculable molecular parameters. Further work now in progress should lead to refinement of the correlations, thereby allowing accurate prediction of solute retention.

ANALYSIS OF WEAK BASES

The only major drawback of the silica-based system is its limited application to compounds with pKa >8.0 and minimum steric crowding around the basic nitrogen. Since compounds with low pKa are also troublesome when assayed on RP columns, attention was again turned to cation-exchange columns as a way of overcoming these problems. In designing the system, emphasis was again placed on the ability to chromatograph a wide range of drugs and their metabolites with a single eluent composed of volatile components, thus facilitating the use of MS coupled to HPLC for metabolite identification. The system tried comprised an SCX column and an acidic eluent (details in Fig. 5 legend).

This system is still under development but looks quite promising. It can handle a wide range of compound types including strong as well as weak bases. Initial results suggest that chromatographic efficiency and peak symmetry are very good. The selectivity also differs markedly from that seen on either an ODS (C-18) or a silica column. Fig. 5 exemplifies a comparison (with 'RP' mobile phases; see legend) of SCX and ODS, for an amide pro-drug (ICI D7114) that is hydrolyzed *in vivo* to an active acid metabolite which, as for the parent drug, can also be ring-hydroxylated to give the polar hydroxy metabolite. The acidity of the ODS eluent (pH 3) suppresses the ionization of the acidic metabolite such that it runs at a similar retention to the amide. The hydroxy metabolites, however, are only just separated and, as they run very early, would probably be occluded by endogenous substances in a biological sample. Furthermore, even with use of the silanol blocking agent hexylamine there is evidence of peak tailing. In contrast, on the cation system all four compounds chromatograph with good efficiency, are reasonably well separated and are also well retained, such that there would be minimal interference from endogenous compounds. As

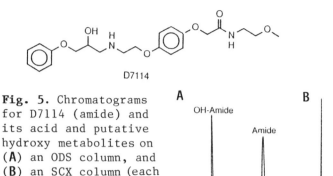

D7114

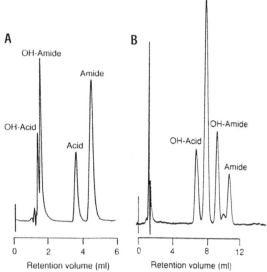

Fig. 5. Chromatograms for D7114 (amide) and its acid and putative hydroxy metabolites on **(A)** an ODS column, and **(B)** an SCX column (each 100 × 4.6 mm). **A:** Hypersil 5 ODS; acetonitrile/water/ phosphoric acid/hexyl-amine (20:80:0.1:0.14 by vol.). **B:** Spherisorb 5 SCX; methanol/water (80:20) containing 0.02 M ammonium acetate, taken to pH ~3 with formic and trifluoro-acetic acids.

all the eluent components are volatile, the system can be readily used with MS, or for preparative work where metabolites are being purified for identification by NMR.

CONCLUSIONS

For the analysis of basic compounds it has proved particu-larly useful to employ cation-exchange chromatography, allowing many of the problems associated with RP-HPLC to be overcome. The systems described (with details given in Fig. legends) are highly versatile, give excellent chromatographic perfor-mance, and are compatible with coupled MS. Retention on the silica-based system is also predictable both for parent drug and metabolite(s). Although the retention models are still under development, the computer hardware and software is well advanced and able to take advantage of any developmental advances. Work in progress is aimed at developing an 'expert system' which, given only the structure of the compound, can predict the best chromatographic conditions to use as well as the retention of the compound and its likely metabolites.

References

1. Moffat, A.C., ed. (1986) *Clarke's Isolation and Identification of Drugs*, The Pharmaceutical Press, London, 1223 pp.
2. Rekker, R.F. (1977) *The Hydrophobic Fragmental Constant*, Elsevier, Amsterdam, 389 pp.
3. Computer Program — *USA:THOR Masterfile 351; CLOGP version 3.51*, Pomona College Medchem Project, Claremont, CA 91711.
4. Perrin, D.D. (1965) *Dissociation Constants of Organic Bases in Aqueous Solution*, Butterworths, London, 473 pp.
5. Perrin, D.D. (1972) *as for* 4., Supplement, 235 pp.
6. Perrin, D.D., Dempsey, B. & Serjeant, R.P. (1980) *pKa Prediction for Organic Acids and Bases*, Chapman & Hall, London, 146 pp.
7. Law, B. (1990) *Trends Anal. Chem. 9*, 31–36.
8. Law, B. (1987) *J. Chromatog. 407*, 1–18.
9. Chen, B-K. & Horvath, C. (1979) *J. Chromatog. 171*, 15–28.
10. Abraham, M.H., Duce, P.P., Prior, D.V., Barratt, D.G., Morris, J.J. & Taylor, P.J. (1989) *J. Chem. Soc. Perkin Trans. 2*, 1355–1375.

#B-3

BIOANALYSIS OF A SERIES OF AMINOTETRALINS:
A 'CORE' METHOD APPROACH

S.A. Rees, S.A. Wood and R.J. Simmonds

Upjohn Laboratories - Europe,
Fleming Way, Crawley, West Sussex RH10 2NJ, U.K.

For a range of CNS-active aminotetralin analogues, sensitive assays were required for plasma, whole blood and brain in small-animal (e.g. rat) pharmacokinetic studies. Initially, various HPLC columns and mobile phases were investigated for some analogues and the extraction of a representative aminotetralin from biomatrices with a spectrum of SPE* cartridges was examined. A 'core' method, involving extraction with a diol or CBA phase, RP-HPLC on base-deactivated phases and UV or fluorescence detection was then validated for a particular aminotetralin. This method, with minor modifications, has been used to analyze a wide variety of substituted analogues, with good results. Experience of the 'core' assay has enabled identification of a suitable i.s. for each analyte and speedy adaptation. Reduced validation suffices for each assay without compromising GLP guidelines.

The aminotetralins currently under investigation are structural analogues of 8-hydroxy-2-(di-n-propylamino)tetralin (formula depicted; 8-OH DPAT), a centrally acting 5-hydroxytryptamine (5-HT) agonist [1].

Early studies [2] demonstrated low bioavailability due to rapid metabolism. *In vitro* and *in vivo* pharmacokinetic models, to screen a diverse range of aminotetralin analogues for their 'metabolic stability', were established to expedite drug-candidate selection. In order to support these activities, many sensitive (100 ng/ml or less) bioanalytical assays were required, suitable for biofluid samples as small as 50 µl.

RATIONALE

Rather than develop novel bioanalytical methods for each new analogue, it was proposed that a representative aminotetralin

---

*Abbreviations.- A_s, analyte peak asymmetry; CBA, carboxylic acid; DEA, diethylamine; GLP, Good Laboratory Practice; i.s., internal standard; LOQ, limit of quantification; RP, reversed-phase; SPE, solid-phase extraction; TFA, trifluoroacetic acid.

be selected for HPLC assay development. The 2-OH compound
was chosen. A range of detection techniques, chromatographic
conditions and methods of sample preparation were evaluated,
leading to a possible 'core' assay involving efficient chromatography, good selectivity for biological extracts and either
UV or fluorescence detection. This assay was fully validated
to fulfil GLP guidelines. The developed assay was robust
in routine use, capable of analyzing serial biomatrix samples
(plasma, whole blood and tissue) from small animals (e.g.
rat) and had adequate sensitivity (5 ng/ml, 0.2 ng on-column).

The 'core' assay was adapted to assay new analogues.
Data obtained during the validation and application of the
'core' assay formed a database, which was applied to subsequent
assays, removing the need for extensive validation of each
new assay.

'CORE' ASSAY DEVELOPMENT

Detection.- Typical UV spectra for aminotetralins showed
absorption maxima at 230 and 270 nm, the latter proving more
sensitive for biological extracts. Fluorescence detection
(ex 230, em 298 nm) gave greater selectivity and sensitivity.

Chromatography

The aminotetralins are basic compounds (pKa >8) and tend
to exhibit peak tailing on RP columns if amine modifiers,
e.g. *N,N*-dimethyloctylamine, are omitted from the mobile phase
- as was the case here so as to allow easier exploration
of electrochemical detection if greater sensitivity became
desirable. Initially a wide range of bonded silica phases
was screened,[θ] with mobile phases containing TFA and, in one
system, DEA also (Table 1).

Zorbax Rx-C8, a base-deactivated phase (where residual
unreacted silanol groups have been chemically modified), was
singled out for further investigation. The highly acidic
mobile phase (TFA; pH 2-3) gave excellent chromatographic parameters: efficiency typically >13,000 plates per column, A_s
1.17 with an adequate k' (column capacity) value of 3.1 (Fig. 1A).

Zorbax phenyl was more retentive for several aminotetralin
analogues, though similar in other chromatographic respects,
e.g. efficiency and the elution order of analytes with different
functional substituents (Fig. 1B). However, the relatively
high percentage of TFA necessary in the mobile phase would
preclude use of the phenyl phase for any analogues with UV
maxima <220 nm (the UV cut-off for TFA).

θ Columns (250 x 4.6 mm) were from Hichrom, Reading, Berks.

Table 1. Summary of initial column screen (% values are v/v).

Column*	Mobile phase	k'	A$_s$	Plates/col.(N)
Zorbax C8	35% CH$_3$CN, 0.1% TFA (aq.)	3.0		p o o r
Zorbax CN	30% CH$_3$CN, 0.2% TFA	2.2	1.07	9,500
Zorbax TMS	30% CH$_3$CN, 0.1% TFA	3.8		p o o r
	+ 0.1% DEA	2.8	1.30	13,500
Zorbax Phenyl	30% CH$_3$CN, 0.4% TFA	3.8	1.25	14,000
Zorbax Rx-C8	20% CH$_3$CN, 0.05% TFA	3.1	1.17	13,500

*from Hichrom; Dupont stationary phases.

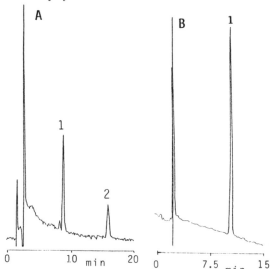

Fig. 1. HPLC of an aqueous aminotetralin standard (peak **1**), accompanied in **A** by internal standard (peak **2**).
A: Zorbax Rx-C8 column; mobile phase 20% acetonitrile/0.05% TFA (in water; v/v); fluorescence detection (ex 230, em 298 nm).
B: Zorbax phenyl column; mobile phase 30% acetonitrile/0.4% TFA; detection at 230 nm.
Oven temp. 40°; flow rate 1.5 ml/min.

Concern about the robustness and supply of the Zorbax Rx-C8 led to the evaluation (Table 2) of other phases of base-deactivated or high-% carbon type, and, respectively, YMC Basic and Kromasil C8 (Eka Nobel) seemed possible substitutes for Zorbax Rx-C8. However, efficiency values are lower and the robustness of these phases has not been fully explored.

Sample preparation

SPE was the method of choice for sample preparation, being rapid, potentially more selective than liquid/liquid methods and easier to automate. A wide variety of bonded sorbents was investigated, with both aqueous standards and spiked control biofluids. Varian Bond Elut (Jones Chromatography, Hengoed, U.K.), C-2, phenyl, diol (2 OH) and CBA phases all showed good distinction of the analyte from endogenous interferences. These phases advantageously differed from the C-8 HPLC phase in the presumed mechanism of adsorption.

For the representative aminotetralin, SPE using a weak non-polar partition mechanism appeared the most promising,

Table 2. Comparison of some base-deactivated and high % carbon phases for the chromatography of a representative. All mobile phases contained   acetonitrile/TFA as for Table 1.

Column	k'	A_s	Plates/column (N)
Zorbax Rx-C8	2.5	1.1	11,000
YMC Basic	3.1	1.0	10,000
Hichrom HI-RPB	2.4	1.0	10,000
Kromasil C8	3.3	1.0	10,000
Supelco Suplex pkb-100	3.8	1.0	9,000

giving good recovery and selectivity for all biomatrices. The inclusion of the eluting solution in the cartridge-priming sequence improved the LOQ by removing contaminants, produced in cartridge manufacture, that chromatographed with the analyte. The aqueous methanol wash also improved selectivity.   The sequence of extraction steps, using 100 mg/1 ml   diol (2-OH) cartridges, was as follows:-

(1) Prime cartridge with: 2 × 1 ml acetonitrile; 2 × 1 ml 60% acetonitrile/0.1% aq. TFA (v/v); 2 × 1 ml water or buffer.
(2) Load *dilute* (20:1) sample.
(3) Wash cartridge with: 2 × 1 ml water or buffer; 2 × 1 ml 30% aq. methanol (v/v).
(4) Elute with 0.6 ml 60% acetonitrile/0.1% aq. TFA.

The dilution step, with water or 0.05 M pH 7.1 phosphate buffer, was performed whatever the sample - plasma, whole blood or brain homogenates.   It lysed erythrocytes and, by weakening drug-protein complexes, facilitated adsorption of the analyte onto the solid phase.   Because the dilution is so great, the actual biomatrix itself has little effect upon adsorption; this allowed rapid adaptation of the method for other biomatrices and for other animal species.

When the method came to be applied to other aminotetralins, the chosen sorbent was CBA, a weak cation-exchanger of wide applicability because of its mechanism of analyte adsorption and elution.   An i.s., a close structural analogue, was included to minimize intra- and inter-assay variation.

'CORE' ASSAY VALIDATION

Analyte stability.- The aminotetralin was stable in newly prepared rat plasma for at least 3 h at room temperature. Extracts were stable for 6-7 days at -20°, as shown by repeat analysis.   The analyte was also stable in whole rat blood and plasma after repeated freezing/thawing.

Extraction efficiency from plasma and whole blood was 89% in all cases. Recovery from brain tissue homogenate was difficult to assess without radiolabelled compound, but appeared to be quantitative.

Linearity.- A graph of back-calculated concentrations *vs.* spiked plasma concentrations was linear over two orders of magnitude (10-995 ng/ml). The correlation coefficient (r) was >0.99 (n = 15) in all cases. At the **LOQ,** 10 ng/ml, the inter-assay C.V. was <6.9%. Because of the preliminary nature of some of these early experiments, concentrations as low as 6 ng/ml were reported.

APPLICATION OF 'CORE' ASSAY FOR NEW AMINOTETRALIN ANALOGUES

For each new analogue, the elements of the 'core' assay were modified and optimized. Early analogues were substituted with electron-donating or weak electron-withdrawing moieties, and detection was by fluorescence (ex 200-210 nm). Though sensitivity was little better than by UV detection, selectivity was much superior. Later analogues, substituted with more strongly electron-withdrawing groups, were detected at 230 nm although UV absorption was poor. This wavelength was a compromise between the UV cut-off of the mobile phase and the requirement for selectivity *vs.* UV-absorbing endogenous components.

The chromatographic conditions were modified for each new analogue. This entailed modifying the concentrations of the organic modifier and acid so that for optimized assays k' was between 2.5 and 4.5. Generally efficiency was >8,000 plates per column and A_s was ~1. For SPE there was similar optimization, the acetonitrile, TFA and methanol concentrations for washing and elution being adjusted for each new analogue.

Wherever possible, an i.s., a close structural analogue of the analyte, was selected from those already tried, using the database, aiming at a separation factor (α) of at least 1.3 from the test compound.

If the extraction scheme, chromatographic parameters and detection methods were satisfactory, the assay was 'validated'. This comprised a careful check of recovery (at least 80% expected). Parameters such as linearity, LOQ and precision were estimated retrospectively from calibration and sample data obtained during application of the assay. These were satisfactory in all cases.

CONCLUSIONS

The 'core' method was well characterized during extensive method development and validation [3]. This enabled rapid development of further methods for new aminotetralin analogues without compromising GLP guidelines. Knowledge of the 'core' method enabled judgements to be made about the validity of the data produced which would not have been possible with a unique assay. In this instance, the application of the 'core' method approach improved the candidate selection process by providing and applying 'new' assays rapidly.

References

1. Arvidsson, L-E., Hachsell, U., Nilsson, J.L.G., Hjorth, S., Carlsson, A., Lindberg, P., Sanchez, D. & Wickstom, H. (1981) *J. Med. Chem. 24*, 921-923.
2. Duncan, J.N., Parton, T. & Enos, T. (1990) *Biochem. Soc. Trans. 18*, 1200-1201.
3. Wood, S.A., Rees, S.A. & Simmonds, R.J. (1991) *J. Liq. Chromatog. 20*, 3761-3782.

#B-4

USEFULNESS OF CHROMATOGRAPHIC OPTIMIZATION SOFTWARE IN BIOANALYTICAL METHOD DEVELOPMENT

C.A. James, R.J. Simmonds and P.T. Tsang

Upjohn Laboratories – Europe, Upjohn Ltd.,
Fleming Way, Crawley, West Sussex RH10 2NJ, U.K.

In developing an HPLC bioanalytical system, even restricting the choice to RP-HPLC* there is a vast range of columns and mobile phases to choose from. Two commercially available computer programs have been investigated to optimize the separation of certain steroid analytes, so as to improve their assay in a biomatrix such as plasma. Consideration is given to the usefulness of this software and to the problems and limitations encountered in bioanalytical applications.

Devising suitable chromatographic conditions is important in the development of any HPLC bioanalytical method. Even if the choice is restricted to RP-HPLC, there is still a vast range of choice, in respects such as columns and packing types, organic modifiers, ion-pair reagents and buffers. Relatively simple theoretical or empirical relationships can describe the effect of certain parameters (e.g. organic or ion-pair concentration, or pH) on retention [1, 2] and, given data from a limited number of trial experiments, can be used to model the behaviour of analytes over a range of chromatographic conditions. The effect of changes in conditions on R_S for the various analytes can then be calculated and optimum conditions suggested.

Computer-aided optimization techniques have particularly been applied to optimization of ternary or quaternary mixtures of solvents used in RP-HPLC. These techniques can also be applied to NP-HPLC, and to other mobile phase conditions differing in respects such as pH or ion-pair concentration.

Abbreviations.*- NP-, RP-: normal-, reversed-phase (HPLC). ACN *(as in Figs.)*, acetonitrile; DMOA, dimethyloctylamine; TFA, trifluoroacetic acid; THF, tetrahydrofuran. CNRP, Calibrated Normalized Resolution Product; R_S, peak resolution. MPL & other analytes: see Fig. 1 (i.s., internal standard) [PM and SM feature also in art. #B-5 (S.A. Wood & co-authors) - *Ed.*]. *All **% values for compositions are v/v.*

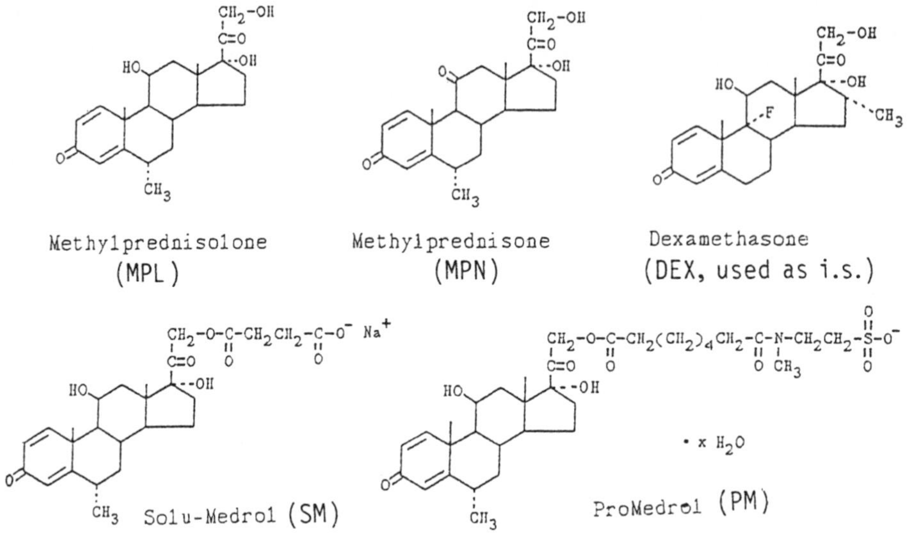

Fig. 1.   Structures of steroids and methylprednisolone prodrugs.

Optimization techniques can be divided into *sequential* methods e.g. 'Simplex',[†] and *simultaneous* methods usually involving factorial design of a series of experiments followed by modelling of the behaviour of the analytes to predict optimum conditions. There are also hybrid or iterative, approaches, which combine a sequential approach and simple modelling of analyte retention.[φ]

There are commercial computer-aided optimization systems available which exploit both approaches, or indeed aim to automate the entire optimization procedure (e.g. Hewlett Packard ICOS software, Waters WISE system, Phillips Diamond system, DryLab).

To explore the use of computer optimization software for development of a bioanalytical method, two programs have now been used   - Waters 'OPSIM' (Millipore UK Ltd., Middx.) and 'HiPac TQ' (Phase Separations Ltd., Deeside).   Both have the advantage of running on IBM-compatible PC's   and   are available at relatively low cost, although retention data had   to be entered manually by us.

THE PROBLEM, AND PROGRAMS INVESTIGATED

An improved method was needed to measure MPL, its metabolite MPN, and two MPL prodrugs (Fig. 1) in human plasma and in incubates from *in vitro* experiments.   Although a separation was achieved (Fig. 2), an impurity present in samples of SM interfered with measurement of MPL, and the retention of prodrugs was excessive, leading to very long analysis

†See refs. [3] and [4].   φSee ref. [5].

Fig. 2. Chromatogram of a
mixture of the 5 compounds
indicated (abbreviated as in
Fig. 1, *opposite*). I is an
impurity noted in supplies
of Solu-Medrol.
Mobile phase: 20% THF,
0.3% TFA, 0.4% DMOA.

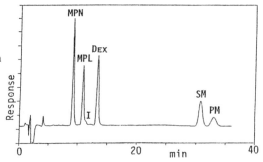

times.   Accordingly, experiments 'recommended' by the following
two   programs   were   conducted   to   improve   the   separation   of
the steroid drug mixture.

'**OPSIM'**.- This   program   offers   binary   optimization   of   %
organic component (RP-HPLC), pH, and ion-pair concentration.
Binary solvent optimization requires data from 2 or 3 experiments;
4 experiments are needed for pH optimization, and 2 for ion-pair
optimization.   The resolution (R_S) between the closest eluting
pair   of   peaks   is   calculated   over   a   range   of   conditions   and
displayed graphically, and 'optimum' conditions identified.

'**HiPac TQ'** offers optimization of 2 or 3 factors simultaneously,
and   requires   data   from   up   to   12   experiments   for   quaternary
solvent optimization.   Simultaneous optimization of two factors,
e.g.   pH   and   organic   modifier   concentration,   or   of   ion-pair
concentration   and   pH,   requires   7   experiments.   The   program
uses   a   factorial   design   [6]   for   experiments   to   be   conducted,
and   requires   data   from   a   trial   gradient   run,   or   else   input
from   the   analyst   to   set   appropriate   limits   (e.g.   lowest   and
highest pH).   'HiPac TQ' offers four types of optimization.-
   (1) Global R_S Minimum.- The   conditions   which   maximize   the
resolution   (R_S)   between   the   closest   eluting   pair   of   peaks
are calculated.   This procedure is also used by 'OPSIM'.
   (2) CNRP [2], which aims to provide a more even distribution
of peaks throughout the chromatogram.
   (3) Peak R_S Min.-   This maximizes the separation of a single
selected peak from all others in the chromatogram.
   (4) Threshold R_S Min.-   This aims to provide a combination
of   resolution   and   short   analysis   time,   whilst   preserving   a
minimum R_S value between the closest eluting pair of peaks.

METHODS

HPLC was performed with a Waters injector (WISP #710B),
pumps (#510; flow-rate 1.5 ml/min) and gradient controller,
a Shimadzu column oven (set at 45°; #CTO-6A) and a Spectra
Physics #773 UV detector operating at 254 nm.   The column
was Spherisorb ODS II, Excel 5 µm, 250 × 4.6 mm (from HiChrom Ltd.).
The original mobile phase was 20-22.5% THF/0.3% TFA/0.4% DMOA.

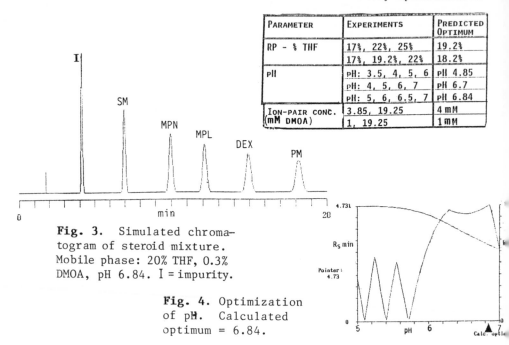

Table 1. 'OPSIM' - Binary optimization.

Parameter	Experiments	Predicted Optimum
RP - % THF	17%, 22%, 25%	19.2%
	17%, 19.2%, 22%	18.2%
pH	pH: 3.5, 4, 5, 6	pH 4.85
	pH: 4, 5, 6, 7	pH 6.7
	pH: 5, 6, 6.5, 7	pH 6.84
Ion-pair conc. (mM DMOA)	3.85, 19.25	4 mM
	1, 19.25	1 mM

Fig. 3. Simulated chroma-
togram of steroid mixture.
Mobile phase: 20% THF, 0.3%
DMOA, pH 6.84. I = impurity.

Fig. 4. Optimization
of pH. Calculated
optimum = 6.84.

Mobile phase compositions in the optimization experiments
are indicated for each. When pH adjustment was necessary,
0.05 M Na_2HPO_4 and 0.05 M NaH_2PO_4 were mixed in appropriate
proportions. Addition of DMOA always preceded pH adjustment,
and organic solvent was added finally.

Plasma extraction was performed on 0.25 ml samples (human
plasma; EDTA/NaF anticoagulant) which were diluted to 1 ml with
water and, after adding i.s. (DEX), loaded onto primed C-2
solid-phase cartridges (Varian, 1 ml/100 mg size)[θ]. Then, after
washing (2 ml water, 2 ml 20% methanol), analytes were eluted
with 0.6 ml 60% acetonitrile/0.1% TFA. Eluates were partially
evaporated *in situ*, and 50 μl chromatographed.

RESULTS

With the 'OPSIM' program, which offers optimization of
single parameters, varying % THF or DMOA concentration gave
little benefit (Table 1). However, pH optimization predicted
an improved separation (Fig. 3), the retention of both SM
and impurity (I) being significantly affected by pH. Fig. 4
shows the 'resolution map' for pH optimization.

At this time the 'HiPac TQ' program became available.
Since 'global' optimum conditions are better predicted by

[θ]See end of refs. list.

Table 2. 'HiPac TQ' optimization: **A**, ion-pair and pH;
B, ternary solvent.

Optimization Method	Predicted Optimum: A	Predicted Optimum: B
Global R_S min.	2.14 mM, pH 6.6	1.12% ACN, 19.2% THF
Peak R_S min.	1.0 mM, pH 6.4	0.56% ACN, 19.68% THF
Threshold R_S min.	2.14 mM, pH 7.0	11.2% ACN, 14.4% THF
CNRP	4.04 mM, pH7.0	1.68% ACN, 18.7% THF

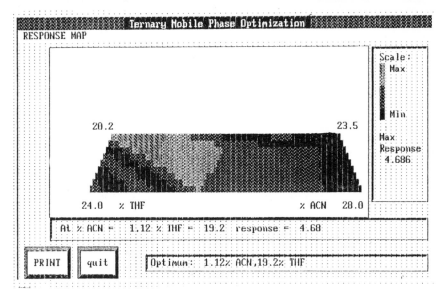

Fig. 5. 'HiPac TQ': ternary mobile phase optimization.

simultaneous procedures than by optimizing each parameter
separately [2], further experiments were conducted using this
program. Optimization of pH and ion-pair concentration was
performed first (Table 2A), followed by ternary solvent optimization (Table 2B). Results are shown for all the optimization
options available with this program. Fig. 5 illustrates the
'response map' obtained for ternary mobile phase optimization,
by the 'Global R_S Min' approach.

The predicted conditions (Table 2, A & B) were then tried
in practice. Figs. 6-9 show chromatograms obtained using
the predicted 'optimized' mobile phase, both for 'Global R_S Min'
optimizations (i.e. greatest resolution of peaks; Figs. 6 & 7),
and for 'Threshold R_S Min' (i.e. minimum resolution of R_S = 1.5
in shortest time; Figs. 8 & 9).

Table 3.
'HiPac TQ':
comparisons of
predicted and
actual chroma-
tography after
optimization
of (**A**) ion-
pair and pH,
(**B**) ternary
mobile phase.

COMPOUND	PREDICTED		ACTUAL		%ACTUAL/ PREDICTED	
	RT (MIN)	K'	RT (MIN)	K'	RT	K'
A IMPURITY	4.88	1.71	4.76	1.64	97.5	95.7
SOLU-MEDROL	9.66	4.37	9.33	4.18	96.6	95.7
METHYLPREDNISONE	12.44	5.91	12.11	5.73	97.4	97.0
METHYLPREDNISOLONE	15.48	7.60	15.70	7.72	101.4	101.6
DEXAMETHASONE	19.28	9.71	19.53	9.85	101.3	101.4
PROMEDROL	26.91	14.0	25.81	13.3	95.9	95.6
B IMPURITY	4.81	1.67	4.80	1.67	99.79	100.00
SOLU-MEDROL	8.52	3.73	8.75	3.86	102.70	103.49
METHYLPREDNISONE	12.23	5.79	11.56	5.42	94.52	93.61
METHYLPREDNISOLONE	15.17	7.43	14.49	7.05	95.52	94.89
DEXAMETHASONE	18.72	9.40	17.62	8.79	94.12	93.51
PROMEDROL	23.67	12.2	25.22	13.01	106.55	106.64

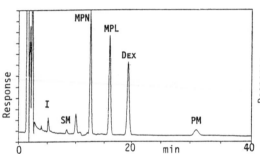

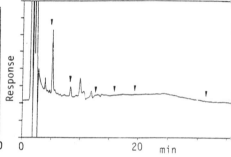

Fig. 6. 'HiPac TQ': chromatogram
for steroid-spiked plasma; mobile phase
optimized by 'Global R$_S$ Min' approach.

Fig. 7. Blank plasma. *For
Fig. 6 also:* 1.12% ACN, 19.2%
THF, pH 6.6, 2.14 mM DMOA.

Fig. 8. 'HiPac TQ':
as for Fig. 6, but
'Threshold R$_S$ Min'
approach, and 11.2%
ACN, 14.4% THF (still
pH 6.6, 2.14 mM DMOA).

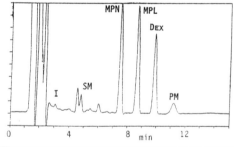

Fig. 9. Blank plasma;
otherwise as for
Fig. 8.

*In the blanks, Figs. 7 & 9,
the analyte positions are
marked; for* I *and* SM *there
would evidently be inter-
ferences.*

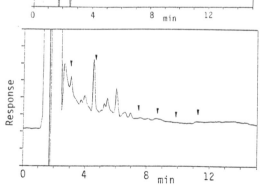

CONCLUSIONS

Optimization software proved useful in improving the separation (and speed of separation) of a mixture of steroids and steroid prodrugs. Attempts were made to include unidentified peaks of endogenous origin in the optimization experiments by injection of blank plasma extracts (Figs. 7 & 9); but, in practice, tracking and clear identification of these peaks under different conditions proved impossible.

The best approach appeared to be to optimize the separation of the analyte mixture, and then test the 'optimized' system with biofluid extracts. In the example considered, the predicted optimum conditions led to a satisfactory separation for analysis of MPL, MPN, PM and DEX in plasma. Further modification of the mobile phase would be required to analyze SM and its impurity in plasma, although 'HiPac TQ' can be used to predict other conditions that still provide a separation of all the analytes, which can then be tested with biofluid.

The ability to predict analyte retention with various mobile phases can also be used to ensure that minor changes in mobile phase composition do not significantly compromise the separation. This can normally be assessed from resolution maps (e.g. Figs. 4 & 5).

The accuracy of the predicted chromatograms as compared with actual chromatograms was good, the differences in actual k' rarely exceeding the predicted value by >5-6%. Retention order was always accurately predicted.

For bioanalytical methods requiring the separation of several analytes, optimization software appears useful in obtaining an adequate and rapid separation of the analytes, although problems of peak tracking make it more difficult to apply to biofluid extracts. Unless it proves possible to track down minor peaks of endogenous origin, software such as the two systems now investigated is less useful for many typical bioanalytical methods where only one or two peaks of interest occur in the chromatogram (i.e. drug and i.s.).

Improved methods for peak tracking [2, 4, 7, 8] could be useful in this respect, although, given the complexity of biofluid extracts and the high sensitivities often required (precluding use of diode array detectors), peak tracking is likely to remain a problem.

Acknowledgement

Brian King (Phase Separations Ltd., Deeside, U.K.) is thanked for supply of 'HiPac TQ' software.

References

1. Berridge, J.C. (1985) *Techniques for the Automated Optimization of HPLC Separations*, Wiley, Chichester.
2. Schoemakers, P.J. (1986) *Optimization of Chromatographic Selectivity* [Vol. 35, *J. Chromatog. Library*], Elsevier, Amsterdam.
3. Berridge, J.C. (1989) *J. Chromatog. 485*, 3-14.
4. Glajeh, J.L. & Kirkland, J.J. (1989) *J. Chromatog. 485*, 51-63.
5. de Galan, L. & Billiet, H.A.H. (1986) *Adv. Chromatog. 25*, 63-104.
6. Hu, Y. & Massart, D.L. (1989) *J. Chromatog. 485*, 311-323.
7. Molnar, I., Boyser, R. & Jekow, P. (1989) *J. Chromatog. 485*, 569-579.
8. Lankmayr, E.P., Wegscheider, W., Daniel-Ivad, J., Kolossrány, I., Csonka, G. & Otto, M. (1989) *J. Chromatog. 485*, 557-567.

Extraction method: included in a presentation on MPL analysis (James, C.A., Burton, N.K. & Simmonds, R.J.) to *15th Int. Symp. Column Liquid Chromatography*, Basel, June 1991.

#B-5

DETERMINATION OF NATURAL AND SYNTHETIC RETINOIDS IN BIOLOGICAL SAMPLES BY HPLC WITH ON-LINE SOLID-PHASE EXTRACTION

R. Wyss$^{\oplus}$ and F. Bucheli

Pharma Division, Preclinical Research,
F. Hoffmann-La Roche Ltd, CH-4002 Basel, Switzerland

A fully automated system is described for determining 1st, 2nd and 3rd generation retinoids using on-line SPE* with automated column switching and RP-HPLC with UV detection. This technique is especially apt for retinoids because of their sensitivity to photoisomerization and oxidation. However, special injection conditions had to be used to overcome strong protein binding:- plasma samples were either injected directly after acetonitrile addition, or deproteinized with ethanol. Methods are presented for determining isotretinoin, tretinoin and their 4-oxo metabolites, and etretinate and its metabolites acitretin and 13-cis-acitretin in plasma. Similar conditions were used for skin and other tissue homogenates. The sensitivity of the plasma methods was improved by routinely injecting up to 1 ml of plasma, with automated pre-column replacement when a pre-column became clogged. LOQ's of 0.3-0.5 ng/ml with simple UV detection were obtained for acitretin, isotretinoin, sumarotene and their metabolites. The 3-valve column-switching system allowed forward- and back-flush purging of the pre-column, purging of capillaries to prevent memory effects, and on-line addition of acetonitrile or water to improve the recoveries.

The retinoids include both naturally occurring substances with vitamin A activity and synthetic analogues with or without biological activity. The retinoids are currently an area of spectacular progress and great interest [1]. In medicine, they have been used mainly in dermatology and oncology. They can be classified into three groups (structures: Fig. 1). The first-generation retinoids tretinoin (all-*trans*-retinoic acid) and isotretinoin (13-*cis*-retinoic acid) are metabolites of retinol and are used to treat acne and photo-damaged skin.   Second-generation retinoids such as etretinate or its

---

$^{\oplus}$contact for any correspondence; to PRPK, 68/121A at company address.
*Abbreviations.- i.s., internal standard; LOQ, limit of quantification; SPE, solid-phase extraction; RP, reversed-phase. M1A, M2 etc. are liquid phases (Fig. 2 legend).

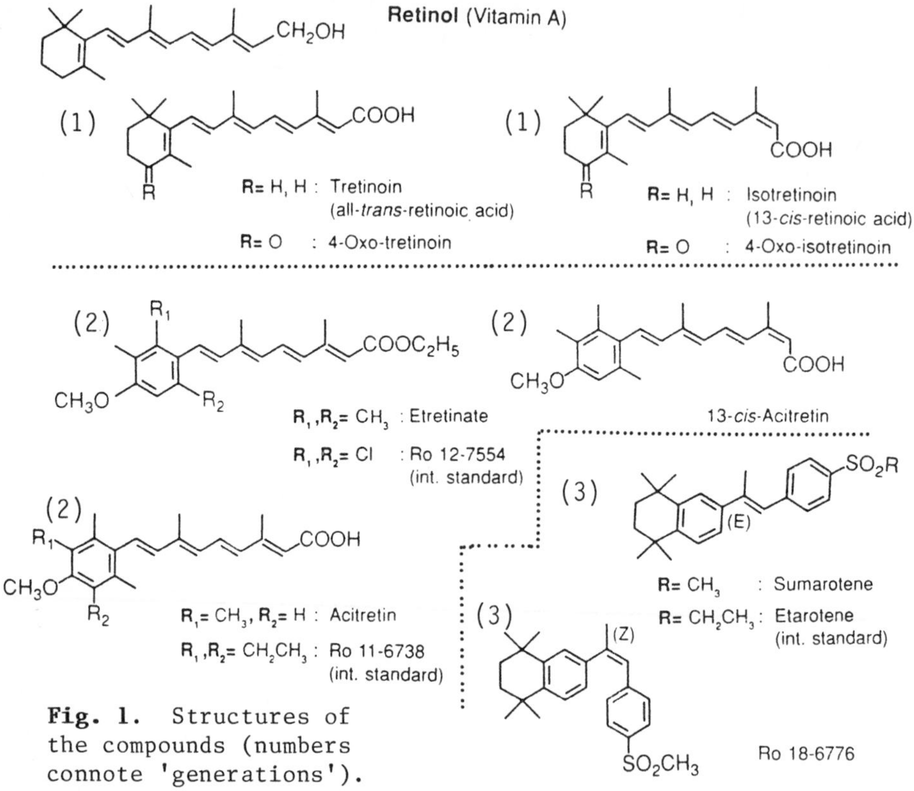

Fig. 1. Structures of the compounds (numbers connote 'generations').

metabolite acitretin are effective against psoriasis and other keratinizing disorders, whereas the third-generation retinoids sumarotene and etarotene, in clinical development, may counter-act skin ageing.

As recently reviewed [2], HPLC-UV is the method of choice for determining retinoids in biological samples, being rapid and sensitive and allowing separation of geometrical isomers and metabolites within a wide range of polarity. Accordingly, HPLC with on-line SPE using automated column switching is especially useful for determining retinoids, because the analytes are totally protected from light and sample work-up is simple. Cumbersome extraction in a darkened room or under yellow light, as in many published methods, can thereby be avoided. Moreover, high sensitivity can be obtained by direct injection of large plasma volumes. Here we describe our experiences with HPLC column-switching as applied to the determination of retinoids in biological samples.

COLUMN-SWITCHING SYSTEM

Fig. 2 shows schematically the column-switching system used in the highly sensitive method for acitretin [3]. During the development of methods for the various retinoids (Fig. 1)

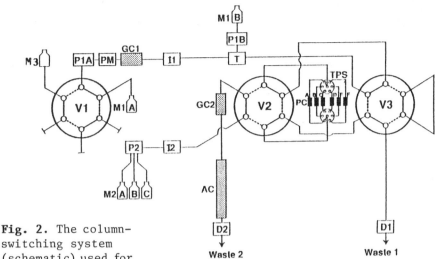

Fig. 2. The column-
switching system
(schematic) used for
the acitretin assay. *From [3], by permission.*
Components besides those in text:- **V1-V3**, switching valves; **P1A**,
HPLC pump (1.4 ml/min); **PM**, pressure monitor,limit 80 bar; **P1B**,
HPLC pump (0.7 ml/min); **T**, T-piece; **TPS**, tandem **PC** selector; **P2**,
gradient HPLC pump (1 ml/min); **I2**, manual injector. *Columns (mm):*
GC1, 14 × 4.6, Bondapak C18 Corasil 37-50 µm; **GC2**, 30 × 4, Spherisorb
ODS 1, 5 µm; **PC**, 14 × 4.6, as for **GC1**; **AC**, 3 columns each 125 × 4, as
for **GC2**. *Mobile phases* (vol.basis; AN = acetonitrile, AA = ammon-
ium acetate): **M1A**, 1% AA/AN (100:2); **M1B**, 1% AA/AN (6:4); **M3**, AN/
water (8:2); **M2: A**, 0.1% AA/AN/acetic acid (40:60:3); **B**, 0.8% AA/
AN/acetic acid (5:95:1); **C**, water/AN/acetic acid (20:980:1).

various systems were used, from a single valve to the elaborate
system (Fig. 2) employed in the final methods discussed below.
In accordance with W. Roth *et al.* [4], plasma samples were
injected by autosampler (**I1** in Fig. 2) onto a C-18 pre-column
(**PC**).   Plasma proteins and polar compounds were washed out
by mobile phase **M1A**, and the analytes were pre-concentrated
on **PC** and finally eluted in the back-flush mode by switching
V2.   The additional valve **V3** allowed forward- and back-flush
purging of the **PC** [5].   Thereby proteins and solid particles,
which would have been partly adsorbed onto the sieves on
the top of the **PC**, were transferred to waste instead of
onto the analytical column (**AC**), preventing its pressure becoming
increased.   The purging of the **PC** could be monitored by
the UV detector **D1** (set at 230 nm; 360 nm for **D2**, post-**AC**).

After switching **V2**, the retained components were transferred
from the **PC** to the **AC** and separated by gradient elution.
Meanwhile the capillaries between **V1** and **D1** were purged with
M3, so obviating memory effects due to adsorption of the
lipophilic retinoids onto the steel capillaries. In a previous

column-switching system this capillary purging was performed
by the gradient mobile phase (**M2**), which was connected to
V1.   In contrast, the configuration in Fig. 2 allowed simultaneous gradient elution from the **AC** and purging of the
capillaries, which reduced the run time by 8 min.   Two guard
columns (**GC1, GC2**) were used, the first preventing pre-concentration of interferences from mobile phase **M1A** during equilibration
of the **PC**.   The other components of the column-switching
system (see Fig. 2 legend) are discussed below.

DETERMINATION OF ISOTRETINOIN/TRETINOIN AND 4-OXO METABOLITES

Initially a method for isotretinoin and metabolites, tretinoin, 4-oxo-isotretinoin and 4-oxo-tretinoin [6, 7], was developed.
However, when plasma samples were directly injected, using
water or a buffer as mobile phase (**M1A**),   the recoveries
of isotretinoin and tretinoin were only ~40-50%.   Accordingly,
10%* acetonitrile in 1% ammonium acetate was used as mobile
phase, and   9 mM NaOH/acetonitrile (8:2;* 3 vol.) was added
to plasma (2 vol.) before the injection (0.5 ml).   The NaOH
addition as well as cooling the autosampler to 10° improved
the stability of the samples during overnight injections.
Thereby >90% recoveries and 2 ng/ml LOQ's were obtained, with
a linearity range 2-2000 ng/ml and inter-assay precision
2.3-3.2% (range 20-1000 ng/ml).

Fig. 3 shows isomer-pair separations in plasma (tretinoin,
isotretinoin, the 4-oxo metabolites and, as i.s., acitretin spiked
in) and endogenous peaks, e.g. retinol (R_T ~28 min).   This
separation was achieved by two coupled columns (each 125 mm
long) of Spherisorb ODS 1 (5 µm), a non-end-capped material,
in combination with an acetonitrile-containing mobile phase.
Since endogenous levels of the 4 analytes (0.5-4 ng/ml) were
found in blank plasma (Fig. 3), individual calibration curves
for all 4 analytes were obtained from spiked-in standards.
The pre-column was replaced after 40-50 injections.

DETERMINATION OF ETRETINATE, ACITRETIN AND 13-*CIS*-ACITRETIN

With the above injection conditions, etretinate recovery
was only 15%, calling for new injection conditions.   Evidently
pre-concentration by on-line SPE of plasma may be problematical
with very lipophilic and highly protein-bound drugs, besides
polar drugs.   Retinoids (1st and 2nd generation) were therefore
systematically investigated [5] for the influence on recovery
of injection-solution composition, the pre-column and the purge
mobile phase.   The final method [7-9] comprised protein precipitation with ethanol (1 ml/0.5 ml plasma) and injection of
the supernatant (0.5 ml).   The retinoids were pre-concentrated
on the pre-column, and the polar plasma components were washed
out using 1% ammonium acetate and then 1% acetic acid/acetonitrile

*based on volume (throughout the text).

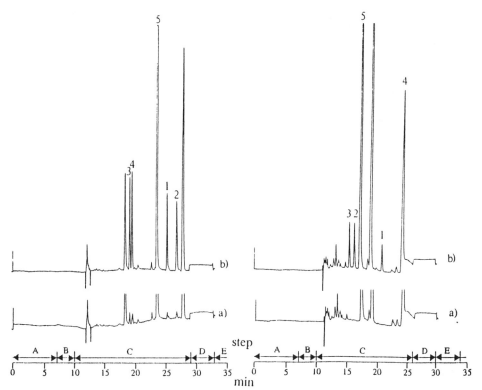

Fig. 3. Isotretinoin and metabolites. Peaks for the spiked-in compounds: **1**, isotretinoin; **2**, tretinoin; **3**, 4-oxo-isotretinoin; **4**, 4-oxo-tretinoin - each 20 ng/ml spike; **5**, acitretin (i.s.) - 150 ng/ml, into blank also, which showed 1.9–2.8 ng/ml endogenous **1–4**.

Fig. 4. Etretinate and metabolites. **1**, etretinate; **2**, acitretin; **3**, 13-*cis*-acitretin - each 20 ng/ml spike; **4**, Ro 12-7554 (i.s. for **1**) and **5**, isotretinoin (i.s. for **2** & **3**) - each 200 ng/ml, into blank also.

(8:2). Thereafter the isotretinoin method was followed, and comparable results obtained (patterns as in Fig. 4) for linear range, precision and accuracy; the LOQ was 2 ng/ml, and recoveries were 63–80%. Rather than a single i.s., isotretinoin was used for acitretin (an acid) and Ro 12-7554 for etretinate (an ester), because of the difference in chemical properties.

DETERMINATION OF SUMAROTENE AND ITS Z-ISOMER

The column-switching technique was also successfully applied to the 3rd-generation retinoids, or so-called arotinoids - sumarotene (Ro 14-9706) and its Z-isomer, in rat, dog and human plasma [10]. Although the extinction coefficient and detection wavelength (303 nm) are less favourable for arotinoids,

an even higher sensitivity was obtainable by direct injection
of large plasma volumes. This technique as first introduced
for acitretin, isotretinoin and sumarotene entailed injection
of 1 ml samples of plasma, containing ~17% acetonitrile [11].
This amount of acetonitrile avoided protein precipitation and
sufficed to give a good recovery.   From the literature,
it is known that ~15 ml of undiluted plasma can usually be
injected onto one pre-column before it becomes clogged. This
would have allowed only 15-20 injections of 1 ml. Therefore,
a tandem pre-column selector (**TPS** in Fig. 2) and a pressure
monitor (**PM**) were incorporated into the column-switching system,
allowing automated switching to a new pre-column, either after
a defined number of injections or when a certain pressure
(70-100 bar) was reached [11].

     For the routine determination of sumarotene, only limited
volumes of plasma were available from toxicokinetic studies.
The method was therefore validated for an injection volume
of 0.5 ml [10]. **M1A** comprised 1% ammonium acetate/acetonitrile
(85:15).   Because the analytes and the i.s. are neutral
compounds and possess no carboxylic group, the presence of
ammonium acetate in **M2** (legend to Fig. 2) had no observable
effect.   **M2** therefore comprised three acetonitrile/water
mixtures: **A**, 70:30; **B**, 95:5; **C**, 99:1.   Fig. 5 shows that
sumarotene was adequately separated from its Z-isomer and interferences, through use of three coupled columns (Fig. 2 legend).
The LOQ was 1 ng/ml for sumarotene and 2 ng/ml for the Z-isomer
with 0.5 ml injection volumes.   It could be improved to
0.5 ng/ml by injecting 1 ml plasma (containing 17% acetonitrile).
Nearly quantitative recoveries and excellent inter-assay
precision (2.3-7.2% over the whole 1-1000 ng/ml calibration
range) were obtained with 0.5 ml injection volumes.   One
pre-column could tolerate >40 injections, allowing overnight
injections.   The method could also be adapted to etarotene
(Ro 15-1570) which was used as i.s.

HIGHLY SENSITIVE DETERMINATION OF ACITRETIN AND 13-*CIS*-ACITRETIN

     There was, and still is, a need for more sensitive
methods for some retinoids. One example is acitretin, which
may be used instead of etretinate in the oral treatment
of psoriasis. To improve the LOQ to <1 ng/ml, a large volume
of plasma, only minimally diluted with acetonitrile, was
injected (see above) [11]. Whereas no isomerization during
storage in the autosampler was observed with arotinoids (see
the method for sumarotene), *cis-trans* isomerization occurred
in some of the patient samples containing acitretin and 13-*cis*-
acitretin to which acetonitrile had been added to ~17%. This
phenomenon was not encountered during method validation with
blood-bank plasma from different volunteers. A detailed investigation revealed that ~10-20% of the blood-bank plasmas and

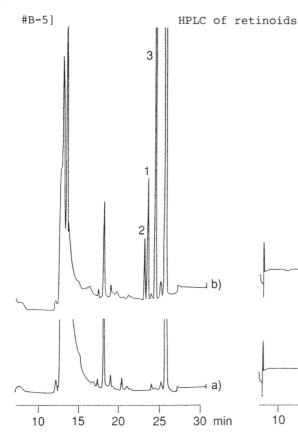

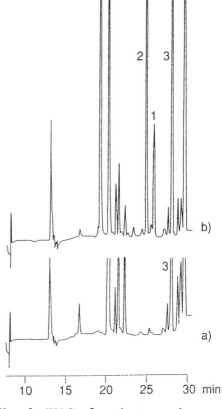

Fig. 5. HPLC of human plasma: a) blank, **b**), spiked with 20 ng/ml of sumarotene (peak **1**) and Z-isomer, Ro 18-6776 (**2**) and 100 ng/ml etarotene (**3**) as i.s. **D2** set at 303 nm.

Fig. 6. HPLC of patient samples, with the highly sensitive method for acitretin: **a**), pre-dose; **b**), 24 h after 50 mg/day orally for 3 weeks. Observed ng/ml: acitretin, 5.77 (**1**); 13-*cis*-acitretin, 56.6 (**2**); Ro 11-6738, 20 (**3**; spiked-in i.s.).

an even higher proportion of patient plasma samples showed this effect. The isomerization became relevant only after the plasma samples had been stored for >10 h in the autosampler - precluding overnight injections. Isomerization was not preventable by adding reagents (e.g. p-OH-mercuribenzoate, L-cys, iodoacetate, GSH); these, moreover, accelerated pre-column clogging.

Finally we found that the isomerization occurred only in the presence of acetonitrile, and that undiluted plasma samples were completely stable for >20 h in the autosampler. Accordingly, in the method adopted (peak pattern as in Fig. 6) the acetonitrile requisite for adequate recovery was added, on-line, by means of a T-piece (**T**, Fig. 2) [3]. After initial i.s. addition (in 5 µl acetonitrile) and centrifugation, 1 ml plasma was injected; 1% ammonium acetate/acetonitrile was the mobile phase

(M1A; 100:2) and the on-line diluent (M1B; 6:4). For both analytes ~90% recoveries were thereby obtained, with mean inter-assay precision 3.8% and LOQ 0.3 ng/ml. Fig. 6 shows chromatograms for pre- and post-dose patient plasma samples. A single pre-column could tolerate ~30 injections, fewer than after acetonitrile addition prior to injection (mean 77.5 injections, corresponding to 64.5 ml plasma) [11]. However, use of the tandem pre-column selector and pressure monitor enabled full automation with overnight injections.

DETERMINATION OF RETINOIDS IN SKIN AND OTHER TISSUES

Because retinoids are used in dermatology and oncology, a knowledge of their concentration in skin and papilloma tissue is very important. Moreover, liver and fat, as possible sites of retinoid storage, are also major matrices for retinoid determinations. The column-switching technique was applied to the determination of isotretinoin, acitretin, etretinate and sumarotene in all these tissues. At the outset, methods had to be developed for human skin biopsies with only small sample amounts, and a sample work-up procedure was validated for 1-30 mg of skin. In some preliminary studies, higher amounts (up to 80 mg of liver or papilloma) were also assayed. The tissue sample was homogenized in a glass homogenizer in 0.6 ml of ethanol (containing the i.s.). Water (0.3 or 0.4 ml) was added and, after another brief homogenization and centrifugation, 0.5 ml was directly injected into the HPLC column-switching system. The validated skin method for acitretin showed the following characteristics (20 mg samples):- the linear range was 20 to 40,000 ng/g at least, and the LOQ, which was dependent on the amount of tissue, was 20 ng/g. Recoveries for acitretin and 13-*cis*-acitretin were 98.3% and 86% respectively, and intra-assay precision was ~6% (range 40 to 20,000 ng/g).

The solvent volumes for homogenization, which were identical to those in the plasma method for etretinate, suited well for skin analysis. For liver, additional interferences were prone to occur, and the run-time had to be prolonged to elute late-running peaks prior to the next injection. During analysis of fat samples, a decrease of peak heights was observed after repeated injections, especially for the most lipophilic compounds (e.g. etretinate). This appeared to be due to coating of the pre-column by lipids, which became insoluble in the homogenizing medium after the water-addition step. This peak-height decrease was no longer seen when the ethanolic solution was injected directly and water was introduced, on-line, by the T-piece (T, Fig. 2). Without water addition, there would be some breakthrough of the retinoids during pre-concentration on the pre-column. This improved procedure will henceforth be used for all tissues. The on-line addition of water was also applied to *in vitro* studies with incubated liver 12,000 **g** supernatants, where up to 20 ml of ethanol-containing solutions of acitretin and etretinate could be injected and pre-concentrated [12].

CONCLUSIONS

A fully automated system is described for the determination of first-, second- and third-generation retinoids, using on-line SPE with automated column switching and RP-HPLC with UV detection. This technique is especially suitable for retinoid determinations because of the sensitivity of these compounds to photoisomerization and oxidation. However, special injection conditions had to be used to overcome strong, high-affinity protein binding of the retinoids. To this end, plasma samples were either directly injected after acetonitrile addition, or deproteinized with ethanol. Tissue samples were homogenized, centrifuged, and the supernatant injected directly.

High sensitivity was obtained by injecting up to 1 ml of plasma under routine conditions, with automated pre-column replacement when a pre-column became clogged. Thereby, with simple UV detection, LOQ's of 0.3-0.5 ng/ml were obtained for acitretin, isotretinoin, sumarotene and their metabolites. The 3-valve column switching system has several advantages over conventional assays for retinoids [2] or other column-switching methods for isotretinoin [13, 14]. In particular, this system allowed forward- and back-purging of the pre-column, purging of capillaries to prevent memory effects, and on-line addition of acetonitrile or water to improve the recoveries, avoiding any isomerization. Omitting liquid extraction and solvent evaporation minimized oxidation of the retinoids.

Acknowledgements

The authors thank Mr H. Suter for the drawings and, **for** correcting the manuscript, Dr D. Dell and Mr B. Haurer.

References

1. Packer, L., ed. (1990) *Meth. Enzymol. 189 & 190 (in toto).*
2. Wyss, R. (1990) *J. Chromatog. 531,* 481-508.
3. Wyss, R. & Bucheli, F. (1992) *J. Chromatog. 593,* 55-62.
4. Roth, W., Beschke, K., Jauch, R., Zimmer, A. & Koss, F.W. (1981) *J. Chromatog. 222,* 13-22.
5. Wyss, R. & Bucheli, F. (1988) *J. Chromatog. 456,* 33-43.
6. Wyss, R. & Bucheli, F. (1988) *J. Chromatog. 424,* 303-314.
7. Wyss, R. (1990) *Meth. Enzymol. 189,* 146-155.
8. Wyss, R. & Bucheli, F. (1988) *J. Chromatog. 431,* 297-307.

9. Wyss, R. & Bucheli, F. (1988) in *Bioanalysis of Drugs and Metabolites, especially Anti-inflammatory and Cardiovascular* [Vol. 18, this series] (Reid, E., Robinson, J.D. & Wilson, I.D., eds.), Plenum, New York, pp. 271-272.
10. Wyss, R. & Bucheli, F. (1992) *J. Chromatog.* *576*, 111-120.
11. Wyss, R. & Bucheli, F. (1990) *J. Pharm. Biomed. Anal. 8*, 1033-1037.
12. Chou, R.C., Wyss, R., Huselton, C.A. & Wiegand, U.W. (1991) *Life Sci. 49*, PL-168- PL-172.
13. Creech Kraft, J., Eckhoff, C., Kuhnz, W., Löfberg, B. & Nau, H. (1988) *J. Liq. Chromatog. 11*, 2051-2069.
14. Eckhoff, C. & Nau, H. (1990) *J. Lipid Res. 31*, 1445-1454.

Added by Ed.- Ref. 14 concerns the identification in human plasma, as well as assay, of retinoic acids (vitamin A metabolites) - all-*trans* (tretinoin), 13-*cis* and 13-*cis*-oxo. Use was made of three RP-HPLC systems and, for examining eluted compounds as methyl esters, GC-MS.

#B-6

THE ROLE OF ANTIBODIES IN DRUG-DEVELOPMENT BIOANALYSIS

J.D. Robinson

Bioanalytical Department, Hoechst U.K. Ltd.,
Walton, Milton Keynes, Bucks. MK7 7AJ, U.K.

For meeting different bioanalytical requirements in the various stages of drug development, three bioanalytical approaches are particularly suitable: GC, HPLC and RIA*. In respects such as specificity and throughput, RIA and also EIA are advantageous if sufficient time is available for setting up the assay. Consideration is given to achieving specificity in IA's, e.g. by performing HPLC initially and assaying analyte-containing fractions. Ab's of appropriate specificity can also serve for immunoaffinity extractions.

During the course of drug development, large numbers of samples are analyzed. These samples arise from a variety of studies, ranging from initial toxicology studies to later studies in patients (clinical trials - Phases 2, 3 & 4). Drug concentrations have to be measured to ascertain the drug's behaviour (absorption, metabolism and elimination), to define its pharmacokinetics, to compare different formulations and to provide information for the regulatory authorities. A typical development programme can generate ~45,000 samples for analysis. Although this work-load can be spread over 10 years or so, most pharmaceutical companies have many drugs going through the process at any one time. Thus, a typical bioanalytical laboratory can expect to receive 30,000-40,000 samples per year. Furthermore, it is often desirable to have the results from one trial to be able to design the next and subsequent trials. Thus a rapid turn-round of samples is essential.

In order to meet these demands the analytical technique must be chosen with care in relation to requirements such as sensitivity, specificity, accuracy and reproducibility whilst being cost-effective and sufficiently robust to be widely applicable. Of the major bioanalytical procedures available, three would seem to satisfy most of these requirements -

*Abbreviations.- Ab, antibody; IA, immunoassay (EIA, enzyme; RIA, radio); ELISA, enzyme-linked immunosorbent assay.

GC, HPLC and RIA.   Table 1 compares these three techniques with respect to the various parameters.   It can be concluded that RIA offers many advantages over the chromatographic techniques in respects such as sensitivity, sample throughput and wideness of application.

Table 1.   Comparison of the major analytical procedures.

Feature	GC	HPLC	IA
Extraction step	Yes	Yes	No
Derivatization step	Yes	Sometimes	No
Throughput, samples/week	200	200	800
Specificity	Good	Good	Varies
Difficulty of operation	Moderate	Moderate	Slight
Sensitivity	ng/ml	ng/ml	pg/ml
Development time, months	2-3	2-3	3-12

RADIOIMMUNOASSAY

IA has long been a method of choice for the rapid screening of samples in the clinical chemistry laboratory and has gained widespread acceptance as a method for detecting drugs of abuse.   Since the first drug RIA's were described in 1969, the technique has been used to analyze drugs in all therapeutic categories.   The diversity of the IA approach is illustrated by the range of different procedures that have been developed for the analysis of, e.g., ACE inhibitors.   Both EIA's and RIA's have been used for this group of closely related drugs [1].   The EIA's are illustrated in the work of Tanaka and co-authors [2] who developed an ELISA for cilazapril.   This compound is a prodrug that is converted to its active diacid metabolite *in vivo*.   By raising antisera to both the parent compound and the active metabolite, specific ELISA's were developed for the two compounds with sensitivities of 30 pg/ml using a 10 µl plasma sample [2].

Another EIA approach, highlighting the specificity of antisera, is a reported assay for captopril [3].   The Ab's were raised against an immunogen consisting of captopril coupled to IgG using 4-(maleimidoethyl)cyclohexane carboxylic acid (MCC). It was found that the samples had to be reacted with MCC to derivatize the captopril at the outset so that it would be recognized by the Ab.   This illustrates the so-called 'bridge-recognition' problem – that the antiserum recognizes not only the drug itself, but also the bridging group between the drug and the carrier protein. In this example the recognition was highly specific and pre-assay derivatization was needed.

In an RIA for ramipril and its active diacid metabolite ramiprilat [4], a lysine derivative of ramiprilat was coupled to the carrier protein to produce the immunogen.   The resulting assay was specific for the diacid metabolite and did not recognize the parent compound. However, it was possible to hydrolyze ramipril by the addition of esterase, thereby allowing the analysis of both compounds using the same antiserum. This assay was interesting in that it could be used to analyze ramipril and its metabolite and also the sister compound trandolopril and its metabolite trandoloprilat, using the same procedures. The assays, however, showed no cross-reaction with any of the other ACE inhibitors.

These examples serve to illustrate the application of IA techniques to the analysis of drugs in a sample without any pre-treatment. They also confirm that the procedure can be sensitive, precise, simple and capable of handling large numbers of samples in one analytical run. The application of IA in drug analysis has been excellently reviewed by Chamberlain [5].

Often the specificity of an IA may be questioned, particularly when the results obtained by, e.g., a chromatographic method do not agree with those obtained for the same sample when analyzed by IA. Problems caused by cross-reacting materials can frequently be overcome by the use of a simple solvent extraction procedure prior to analysis of the samples. This procedure is often used for the analysis of steroid drugs, or in the IA of actual endogenous steroids. Another remedy for problems of specificity in an IA is use of a combination of techniques, e.g. combined HPLC and RIA as now considered.

HPLC-IMMUNOASSAY

The HPLC-IA combination can produce a technique that has the selectivity of a chromatographic method coupled with the sensitivity of IA. This approach is illustrated by reported assays for the β-agonists clenbuterol, salbutamol and cimaterol in urine [6]. The sample is firstly cleaned up by means of a simple solid-phase extraction procedure on a C-18 cartridge; the eluate is evaporated to dryness and the residue resuspended in EIA buffer for analysis by the EIA procedure. Any samples which give a positive response in the EIA are then subjected to a combined procedure.- The sample is again cleaned up on a cartridge and the residue suspended in water. An aliquot is then injected onto an HPLC column (LiChrospher RP-Select B). Under these conditions the drugs are concentrated on top of the column and can then be sequentially eluted using a step gradient incorporating 3% acetonitrile for 5 min, followed by 20% acetonitrile for 7 min and finally 100% acetonitrile for 4 min. From the column eluate, 0.5 ml fractions are collected and then analyzed directly by EIA; thereby pg/ml and ng/ml levels can be measured.

In Vol. 12 the author and colleagues described identification of plasma metabolites of loprazolam with the aid of HPLC-RIA [7].   Two antisera with different specificities for loprazolam and its metabolites were available, together with several less sensitive chromatographic methods. After careful optimization of an HPLC procedure to gain the best separation of the known metabolites, using a mobile phase that was least detrimental to the RIA, it was possible to analyze fractions of the HPLC eluate directly without the need for solvent evaporation and resuspension. The procedure was used to show the presence of loprazolam metabolites in the plasma of volunteers after administration of the drug. Analysis of the samples by RIA alone was not specific enough, nor was the HPLC sensitive enough, to confirm the presence of a metabolite. However, the combination of techniques proved to be both sensitive and specific.

Exemplifying a twin-approach variation, Farjam & co-authors [8] used Ab's as an on-line clean-up procedure for their samples prior to HPLC analysis. In their immuno-affinity system the Ab's were immobilized on a solid phase through which the sample was passed, and retained the compounds of interest. Drug desorption can be either immunoselective, by means of a solute that has a high affinity for the Ab and so displaces the analyte from its binding sites, or non-selective, the analyte being eluted from the Ab by means of an organic modifier such as methanol. Both these procedures can be done on-line using a series of column-switching valves. The immunoaffinity columns can be used to concentrate the analyte from relatively large volumes of sample material, and to clean up the sample prior to the chromatographic step. This approach uses the selectivity of the Ab's to extract the analyte from a large volume of starting material in the same way that one would use a solid-phase extraction cartridge, with the advantage that it is much more specific.

Elsewhere in this book (art. #D-6) Statham and co-authors describe a similar procedure for extraction of salbutamol from biofluids to aid the extraction and purification of material prior to MS analysis. This work further exemplifies the usefulness of Ab's in the analysis and purification of drugs and metabolites.

CONCLUDING COMMENTS

The use of Ab's in drug development has now extended beyond the simple direct IA procedures that have been developed in the past. The Ab's that form the basis of the IA are now used in combination with other techniques to provide sensitive and specific assays (HPLC-RIA, HPLC-EIA), as sample pre-treatment prior to on-line analysis (immunoaffinity-HPLC) and in immunoaffinity columns to extract samples prior to MS analysis.

If Ab's and IA's in their various guises are to be even more successful in the drug development process, it is important that the groundwork of raising Ab's be started as early as possible in the programme. In many cases the initial work of producing immunogens, immunizing animals and screening antisera can be carried out in parallel with the development of a chromatographic method. The Ab's can then be used in an IA which can be validated by the chromatographic procedure and then used in its place for the analysis of the large numbers of samples that arise during the drug development programme. These same Ab's can be used as sample preparation reagents to aid metabolite identification, in the same way that solid-phase extraction cartridges are used except that they are more specific.

References

1. Robinson, J.D. & Lewis, S. (1987) in *Bioanalysis of Drugs and Metabolites, especially Anti-inflammatory and Cardiovascular* [Vol. 18, this series] (Reid, E., Robinson, J.D. & Wilson, I.D., eds.), Plenum, New York, pp. 143-148.
2. Tanaka, H., Yoneyama, Y., Sugawara, M., Limeda, I. & Ohta, Y. (1987) *J. Pharm. Sci. 76*, 224-227.
3. Kinoshita, H., Nakamura, R., Tanaka, S., Tohira, Y. & Sawada, M. (1986) *J. Pharm. Sci. 75*, 711-713.
4. Eckert, H-G., Munscher, G., Oekonomopoulos, R., Urbach, H. & Wissman, H. (1985) *Arz. Forsch./Drug Res. 35*, 1251-1256.
5. Chamberlain, J. (1985) *Analysis of Drugs in Biological Fluids*, CRC Press, Boca Raton, FL.- See pp. 115-136.
6. Meyer, H.H.D., Rinke, L. & Durch, I. (1991) *J. Chromatog. 564*, 551-556.
7. Robinson, J.D., Wilson, I.D., Bevan, C.D. & Chamberlain, J. (1983) in *Drug Metabolite Isolation and Determination* [Vol. 12, this series] (Reid, E. & Leppard, J.P., eds.), Plenum, New York, pp. 111-118.
8. Farjam, A., Lingeman, H., Timmerman, P., Soldaat, A., Brugman, A., van de Merbel, N., de Jong, G.J., Frei, R.W. & Brinkman, U.A.Th. (1990) in *Analysis for Drugs and Metabolites including Anti-infective Agents* [Vol. 20, this Series] (Reid, E. & Wilson, I.D., eds.), Royal Society of Chemistry, Cambridge, pp. 365-370.

#B-7

STABILITY OF ANALYTES, INCLUDING ANTI-ASTHMATICS, DURING BIOLOGICAL SAMPLE HANDLING

R.J. Simmonds, S.A. Wood and C.A. James

Upjohn Laboratories - Europe, Fleming Way,
Crawley, West Sussex RH10 2NJ, U.K.

The accurate and reproducible measurement of drugs and drug metabolites becomes much more difficult when these are unstable in any of the stages that an analytical sample goes through from biological matrix to (e.g.) introduction into a column. Instability may be due to enzyme hydrolysis in vivo or in vitro, or to chemical degradation (e.g. extraction), and may be a design feature of the compound itself, e.g. a prodrug. Instability of any analyte must be investigated in biological samples, during storage and extraction, in dilute solution, and even during analysis itself. Sources of instability must be identified and ameliorated during method development, before an assay can be validated and used to support regulatory studies. This article gives examples of different types of analyte instability that we have seen, and strategies adopted to stabilize samples.

Any bioanalytical assay must be adequately, and formally, validated before it is used in support of regulatory studies. It is near-impossible to validate an unreliable assay to meet current FDA deadlines [1]. A common source of unreliability is instability of analytes, whether in the biological matrix, or during storage, or in any of the method steps, e.g. thawing and extraction, keeping as extracts or dilute solutions, or chromatography. In our experience this problem is increasing, perhaps because assays are becoming more sensitive and discriminating, such that any minor breakdown of analyte that would have gone unnoticed in the past may now rather obviously affect the reproducibility of an otherwise adequate method. D. Dell & co-authors [2, 3] have provided useful guidelines for assessing analyte stability. It is, then, essential that any analyte instabilities be investigated and at least controlled before a method is formally validated.

Here we present some types of unstable drugs and metabolites that we have encountered, and the strategies adopted to render them stable during the assay so that biological samples can

*Note by Ed.- In earlier vols., the Index entry 'Lability' is pertinent.

be collected, stored, extracted and analyzed reproducibly.
Causes of instability are legion. Thus, **in the biological
matrix** the compound may (a) be unstable at physiological pH
(pH 7.3–7.5), (b) be rapidly hydrolyzed by esterases *in vivo* or
in vitro, or (c) react with other components, e.g. in urine;
during extraction the compound may (d) be unstable at the
preferred extraction pH, (e) be unstable when adsorbed onto
solid phases, or (f) react with solvents,[θ] e.g. aldehydes reacting
with methanol to form hemi-acetals; **in dilute solution** the compound may undergo (g) hydrolysis, (h) oxidation, (i) photodecomposition, or (j) microbial attack.

Metabolites and their conjugates tend to be more unstable
than the parent drug. However, with the ever-increasing
sophistication of drug (or prodrug) design, instability in
the actual biological matrix may be a desired feature of the
drug itself. Some examples are now considered.

ATRACURIUM

This neuromuscular blocking agent is unstable at physiological pH and, *in vivo* and *in vitro*, undergoes rapid Hofmann
elimination and ester hydrolysis. Effective inhibition of
ester hydrolysis was obtained by collecting blood samples
with EDTA, which also served as anticoagulant. After immediate
centrifugation, 0.2 ml plasma was diluted to 1 ml with 15 mM
sulphuric acid. This enabled samples to be stored safely,
but did put constraints on the extraction method. In view
of the pH instability, an acidic mobile phase suited best
for HPLC [4].

PRODRUGS AND THEIR METABOLITES

Prodrugs are increasingly used to improve a drug's solubility, compatibility with blood, or bioavailability. By design,
prodrugs are rapidly transformed *in vivo*, e.g. by enzymic
breakdown which may continue *in vitro* in blood and also in
plasma which contains many enzymes, notably carboxyesterases.
To complicate matters further, there are species and individual
differences in enzyme types and levels.

Bunaprolast (U-66,858), an anti-asthmatic, is hydrolyzed
to U-68244, an active metabolite, both *in vivo* and, rapidly,
in rat plasma *in vitro* (Fig. 1A). Only a reduction, not
complete inhibition, was obtainable by adding potent esterase
inhibitors (Fig. 1B), including phenylmethylsulphonyl chloride
(not shown). In contrast, with dog plasma the esterase
action was effectively stopped by all inhibitors tried. In
fact, as a general rule, the smaller the animal the more
active do esterases and other enzymes appear to be; thus
any treatment that inhibits rat enzymes will be effective
for other species including man, although the converse may
not hold.

[θ] See end of refs. list.– *Ed.*

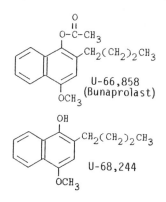

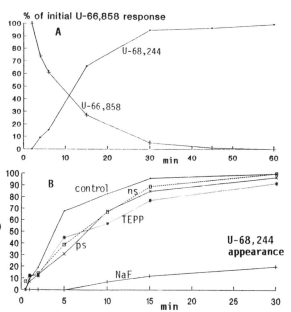

Fig. 1. Stability of U–66,858 (7 µg/ml) in rat plasma: **A**, tested alone; **B**, with added esterase inhibitor (~100 µg/ml) – ns, neostigmine; ps, physostigmine; TEPP, triethyl pyrophosphate. For the effect of fluoride, see text.

An encouraging point here is that NaF (5 mg/ml) was more effective than the more toxic, not to say lethal, options tried (Fig. 1B), but with the drawback that the concentrations used could cause red-cell lysis. U–68,244, which we were also measuring by HPLC (Fig. 1), was likewise unstable and was readily oxidized to unknown products. Degradation occurred in whole blood, during extraction, and in solution awaiting analysis.

In order to obtain reliable values for both bunaprolast and U–68244, plasma had to be rapidly separated from the red cells after sampling (with the useful aid of a high-speed mini-centrifuge). All subsequent steps in sample work-up (protein precipitation and solid-phase extraction) were carried out in the presence of 50 mM ascorbic acid, as also used to prime the extraction cartridges. Ascorbic acid was chosen in preference to, e.g., sodium metabisulphite, because the latter would attack HPLC column packings and impair the chromatography. With these precautions, the assay proved reliable and robust in routine use [5].

Methylprednisolone prodrugs, Solu–Medrol and ProMedrol, were designed to be unstable *in vivo,* and so had to be investigated for stability in plasma *in vitro* (Fig. 2). Solu–Medrol was stable, even in rat whole blood, with EDTA as anticoagulant, but ProMedrol required the addition of 5 mg/ml NaF (Fig. 2B). However, despite having apparently stabilized

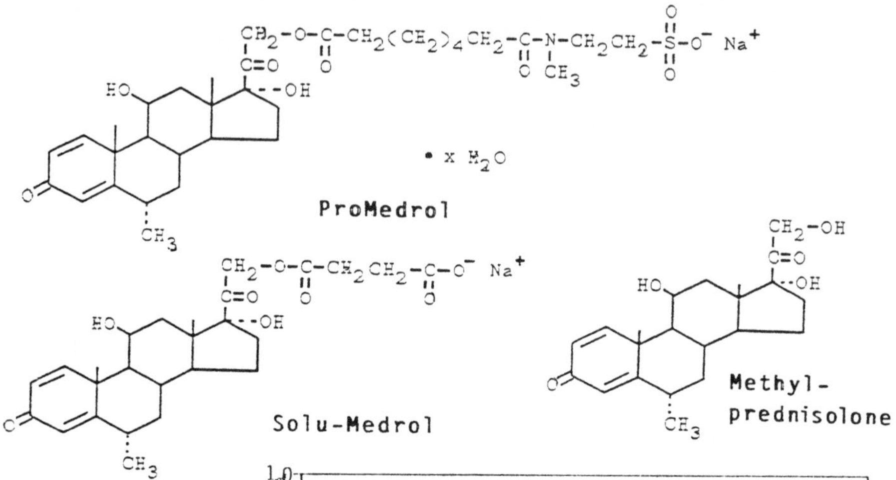

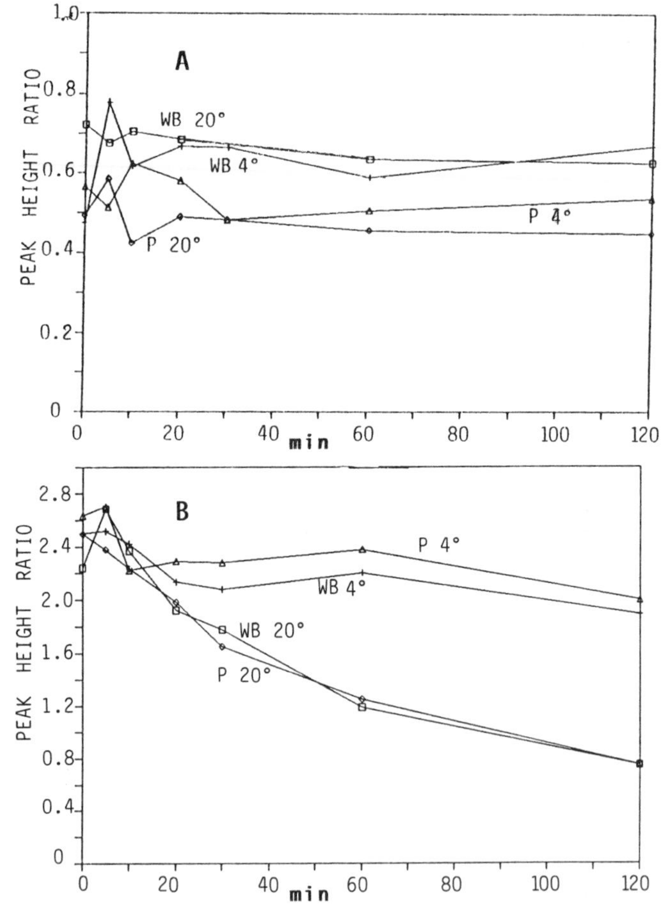

Fig. 2. Formulae of methyl-prednisolone prodrugs, and their stability in rat whole blood (WB) and plasma (P). A, Solu-Medrol; B, ProMedrol. EDTA present as anticoagulant.

ProMedrol, anomalous concentrations of methylprednisolone derived from ProMedrol were found in practice. The stability of ProMedrol was therefore investigated in all assay stages:

- Prime C-2 Bond Elut cartridge with 2 × 1 ml acetonitrile,
2 × 1 ml water.
- Load 250 µl plasma, diluted to 1 ml with water, + intl. std.
- Wash with 2 × 1 ml water, 2 × 1 ml 10% (v/v) aq. methanol.
- Elute with 600 µl 50% aq. acetonitrile/0.2% trifluoroacetic
acid (TFA).
- Reduce eluate to 200 µl, add 20 µl satd. sodium citrate,
spin, and apply 100 µl to HPLC column.

Somewhat unexpectedly, ProMedrol was unstable in the
eluting solvent. The addition of 20 µl of saturated sodium
citrate to eluates brought their pH to 5.5 (the pH of ProMedrol
formulations), and with this modification extracts could be
stored in autosampler vials for at least 24 h, and the assay
performed reliably. A typical chromatogram is shown in Fig. 3.

OTHER INSTABILITY EXAMPLES

Sometimes the cause of instability is the very feature
of the molecule that makes it therapeutically effective, as
can be the case with antioxidants or 'free-radical scavengers'.
Whilst photodecomposition, commonly encountered with such
compounds, can be ameliorated by the use of amber glassware
and a subdued-light environment, oxidation in whole blood
and in solution is more difficult to control. For example,
oxidation to give a quinone can occur even in frozen plasma
with vitamin E and its analogues, catecholamines, and the
following classical example of a drug [6].

Apomorphine gives
rise to a quinone:-

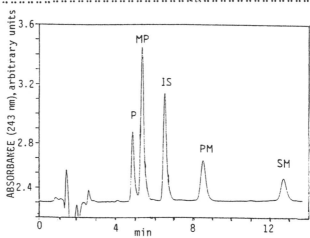

Fig. 3. Assay of
methylprednisolone
(MP) and its pro-
drugs (SM, Solu-
Medrol; PM, Pro-
Medrol) by HPLC.
P, methylpredni-
sone (MP metabolite).

IS, internal std.
Column (run at 45°):
Spherisorb ODS 2
Excel, 250 × 4.6 mm.
Eluent (1.5 ml/min):
25% (v/v) acetonitrile
/0.3% TFA/0.4% di-
methyloctylamine.

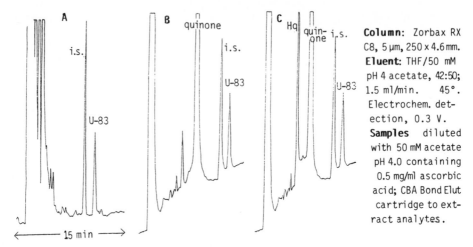

Column: Zorbax RX C8, 5 μm, 250 x 4.6 mm.
Eluent: THF/50 mM pH 4 acetate, 42:50; 1.5 ml/min.   45°. Electrochem. detection, 0.3 V.
Samples diluted with 50 mM acetate pH 4.0 containing 0.5 mg/ml ascorbic acid; CBA Bond Elut cartridge to extract analytes.

Fig. 4. Chromatograms for plasma spiked with U-83 (to 1 μg/ml), i.s. and (*not* in **A**) the quinone *(see text)* with ascorbic acid present. **A & B**: sample injected immediately (no degradation of U-83 to the quinone seen in **B**). **C**: **B** reinjected after 1-2 h (note reduction of quinone to hydroquinone, Hq).

This quinone formation may occur during sampling, storage, freezing/thawing, extraction (particularly solid-phase), or even chromatography. In general, where an assay is required for a parent drug and an active metabolite that appears through oxidation of the drug, there is then the possibility that in preventing the oxidation of the parent drug, e.g. by adding glutathione or butylated hydroxytoluene, the metabolite becomes destabilized. With apomorphine, in aqueous solution or in plasma, stabilization can be achieved by adding ascorbic acid. However, the effect on other analytes needs to be checked; the quinone may become reduced [6].

U-83, a recent Upjohn anti-asthmatic compound, presents problems that are illustrated by the HPLC patterns in Fig. 4. It is oxidized in solution, or in plasma and tissues, to the corresponding quinone. Ascorbic acid (0.5 mg/ml) was used to stabilize U-83 in the routine assay, which involved solid-phase extraction and RP-HPLC with electrochemical detection, and gave repeatable results for the parent drug. The methods for extraction and chromatography were effective for the quinone, a possible active metabolite, but if ascorbic acid were present the observed levels were variable, even in calibration samples. In the presence of U-83, but without added stabilizer, quinone levels were variable, sometimes high. In the absence of ascorbic acid, U-83 was being oxidized; with ascorbic acid the quinone was variably reduced to the hydroquinone. No consistent way could be found to stabilize

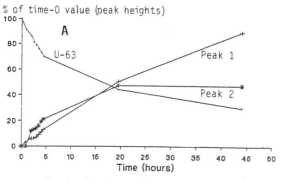

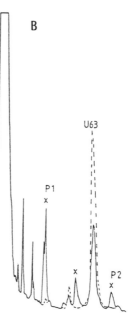

Fig. 5. Stability of trospectomycin (U-63) in human urine incubated at 37°. **A** shows the time courses, and **B** the HPLC patterns: ----, initial; ———, after 28 h (assay [7] with post-column reaction). For P1 see text; P2 and other peaks that appeared are of unknown identity.

both analytes together during sample collection, storage and extraction, or in cartridge eluates. In such cases it may be better to avoid solid-phase extraction, since oxidation during extraction in the presence of a reducing agent can hardly be avoided: adsorption takes place over a large surface area, and it is difficult to avoid drawing air through cartridges in the rinsing and elution steps.

Trospectomycin (U-63) was considered in our article [7] concerning assay in plasma, with expectation of applicability to other biofluid samples. It is in fact an example of compounds that appear to be more stable in plasma, possibly because of protein binding, than in dilute aqueous solution, and even less stable in urine, due to reaction with other solutes in this very complex matrix. Trospectomycin is stable in plasma but degrades in urine at body temperature in a matter of hours (Fig. 5). The possible major product, propylactinospectinoic acid (peak P1), is also unstable in urine, even when stored frozen. No way of stabilizing urine samples has been found, nor can the breakdown in the bladder be compensated for, since this is highly variable and depends upon the actual composition and pH of the urine. Buffering of control urine samples to pH 4 (at which trospectomycin is at its most stable) enabled stable calibration and quality-control samples to be prepared, but even immediate acidification of clinical samples did not prevent up to 40% loss of drug during extended storage at -20° before analysis.

It is apparent that trospectomycin is reacting with some of the many components in urine, and degrading to largely unknown products.   It is difficult to know quite what to do in such situations. The bladder contents can be acidified *in vivo* by oral administration of, e.g. sodium citrate, and urine samples can be extracted immediately after collection and extracts stored; but neither of these procedures is more than a partial answer to the problem, and in practice they may not be convenient or even feasible.

CONCLUDING COMMENT

Even though analyte stability during the ultimate assay procedure will have been established during assay development and validation, the stability in real samples from actual studies may differ from that in spiked control biological matrices. For this reason it is wise to re-assay representative real samples after a period of time to confirm that there is not a problem.

References

1. Shah, V.P. (1987) *Clin. Res. Prac. Drug Reg. Affairs 5*, 51-60 [& see pp. 89-90 in ref. 3].
2. U. Timm, M. Wall & D. Dell (1985) *J. Pharm. Sci. 74*, 972-977.
3. Dell, D. (1990) in *Analysis for Drugs and Metabolites including Anti-infective Agents* [Vol. 20, this series] (Reid, E. & Wilson, I.D., eds.), Royal Society of Chemistry, Cambridge, pp. 9-22.
4. Simmonds, R.J. (1985) *J. Chromatog. 343*, 431-436.
5. Wood, S.A., Rees, S.A. & Simmonds, R.J. (1990) *J. Liq. Chromatog. 13*, 3809-3824.
6. Smith, R.V., Wilcox, R.E. & Humphreys, D.W. (1980) *Res. Comm. Chem. Path. Pharmacol. 27*, 183-186.
7. Wood, S.A. & Simmonds, R.J. (1990) *as for* 3., pp. 103-108.

Reminder by Editor: past vols. are worth consulting.

Adverse effects of solvents, or of impurities therein, have featured in earlier vols. (Index entry 'Solvents'), e.g. pp. 263-264 in Vol. 12 (Plenum; ed. E. Reid & J.P. Leppard). Informative arts. (A.H. Beckett; A.C. Moffat) and discussions appeared in: *The Poisoned Patient - Ciba Found. Symp. 26* (1974); see also a review (1976) by E. Reid, *Analyst 101*, 1-18.
Problems in achieving valid results feature particularly in ref. #3 arts. and in Vol. 10 (out-of-print; Horwood, as for Vol. 7). Vols. 12, 16 & 18 are pertinent too (Plenum; ed. E. Reid *et al.*).

#B-8

COMPARISON OF TWO METHODS OF CALIBRATING LINEAR HPLC ASSAYS (EVERYTHING COMES TO THE ANALYST WHO WEIGHTS)

G.S. Land, W.J. Leavens[*] and B.C. Weatherley

Department of Bioanalytical Sciences, Wellcome Research
Laboratories, Langley Court, Beckenham, Kent BR3 3BS, U.K.

An attempt is now made to identify some key factors which need to be assessed in order to derive the most appropriate standard curve from s set of calibration data. Analysis of the relationship between variance and the concentration of calculating standard is performed to determine the appropriate weighting factor. Three bioanalytical methods were selected, and each calibration data set was analyzed using both unweighted and weighted least-squares linear regression. Confidence intervals (95%) were established for each individual calibration line, and its relative weights were calculated from the pooled data set. The described weighting scheme does not suffer from the distortions associated with weighting on the basis of $1/(conc'n)^2$. Because valuable bioanalytical information may be lost through the use of inappropriate calculation procedures, the present scheme is strongly advocated.

The pharmacokinetic profile of a drug is used to describe its bioavailability. It is therefore important, especially in bioequivalence studies, that bioanalytical results be free from bias and sufficiently precise over the desired concentration range so that additional analytical random error does not affect the statistical power of the study.

HPLC methods, commonly using UV and fluorescence detectors, mostly give a linear response with increasing concentration of analyte and can be used over wide dynamic ranges. To obtain analytical data from such assays correctly, the method by which the linear calibration parameters are estimated can be of paramount importance. Here we attempt to identify some key factors which need to be assessed to derive the most appropriate standard curve from a set of calibration data.

METHODS

Three bioanalytical methods were selected:
- Case 1: a plasma assay of an anti-epileptic compound using liquid-liquid extraction, followed by normal-phase HPLC and UV detection with peak area ratio measurement *vs.* an internal standard;
- Case 2: a urine assay of an antiviral compound using ASTED sample preparation, followed by reversed-phase HPLC and UV detection with simple peak area measurement.

[*]Any correspondence should be sent to Mr Leavens.

- Case 3: a plasma assay of a Hb-O_2 affinity modifier, using solid-phase extraction, RP-HPLC and UV detection, with peak-area ratio measurement; the dynamic range was greater than in Case 1. The measured responses (peak areas or ratios, as appropriate) from calibration data pooled from 5-10 analyses were examined to assess empirically whether errors showed constant variance (square of S.D.) at each concentration. If the data did not support this essential requirement, then a weighted scheme or an appropriate transformation was considered [1].

Assuming that the variance of the response was smoothly related to the concentration, by a power function (P) and proportionality parameter (C) augmented by a constant variance (C_0), weighted factors were estimated by non-linear least squares curve-fitting of the pooled variance data to the equation:-

$$Y_i = C_0 + C.(X_i)^P$$ where X_i was the concentration of the

calibrating standard and Y_i the measured variance of pooled response. C_0, C and P were all positive parameters iteratively computer-estimated [2] after logarithmically transforming the above equation. Relative weights for each *individual* calibration line in the 3 'Cases' were calculated from the parameters thus determined for the *pooled* data:- $$W_i = 1/[C_0 + C.(X_i)^P]$$

where W_i is the relative weight at concentration X_i. Each calibration data set was analyzed using both weighted and unweighted least-squares linear regression of the form $Y = m.X + c$. The analysis also displayed 95% confidence intervals (CI) about the line [3]. The concentrations of all low, medium and high standards were read from each line (X_{est}), and the % bias from the mean of the known values (X_{std}) was calculated by:-   Bias (%) = $100.[(X_{est} - X_{std})/X_{std}]$.

RESULTS

Variance of response *vs.* standard concentration from pooled data is illustrated graphically for each 'Case' (Fig. 1). The estimated parameters subsequently used to calculate weighting factors are depicted on each graph.

Overall results for accuracy and bias at three standard concentrations are shown in Table 1, in anticipation of Figs. 2-4. These Figs. show confidence intervals (95%) for both weighted and unweighted calibration lines, following least-squares linear regression analysis, all depicted around the line for two examples selected from each 'Case'. It is common practice to weight on the basis of 1/C or $1/C^2$. Calculations based on a $1/C^2$ as well as the present scheme appear in Figs. 2-4 and in Table 1, and comment appears in the DISCUSSION.

[Continuation (DISCUSSION) *on* p. 109

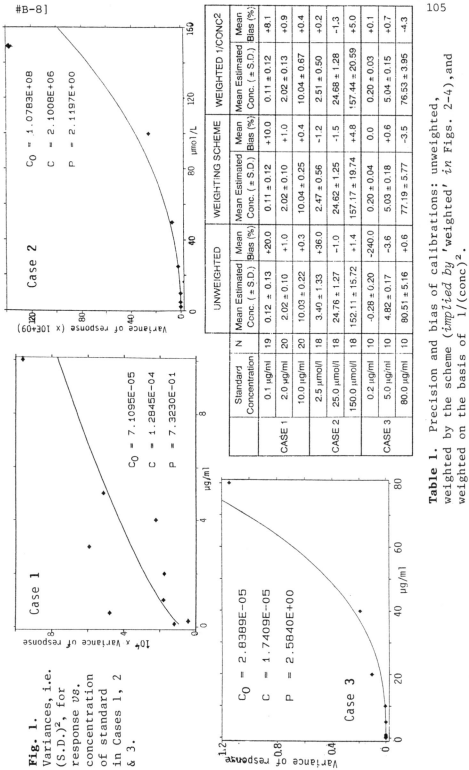

	Standard Concentration	N	UNWEIGHTED		WEIGHTING SCHEME		WEIGHTED 1/CONC2	
			Mean Estimated Conc. (±S.D.)	Mean Bias (%)	Mean Estimated Conc. (±S.D.)	Mean Bias (%)	Mean Estimated Conc. (±S.D.)	Mean Bias (%)
CASE 1	0.1 µg/ml	19	0.12 ± 0.13	+20.0	0.11 ± 0.12	+10.0	0.11 ± 0.12	+8.1
	2.0 µg/ml	20	2.02 ± 0.10	+1.0	2.02 ± 0.10	+1.0	2.02 ± 0.13	+0.9
	10.0 µg/ml	20	10.03 ± 0.22	+0.3	10.04 ± 0.25	+0.4	10.04 ± 0.67	+0.4
CASE 2	2.5 µmol/l	18	3.40 ± 1.33	+36.0	2.47 ± 0.56	-1.2	2.51 ± 0.50	+0.2
	25.0 µmol/l	18	24.76 ± 1.27	-1.0	24.62 ± 1.25	-1.5	24.68 ± 1.28	-1.3
	150.0 µmol/l	18	152.11 ± 15.72	+1.4	157.17 ± 19.74	+4.8	157.44 ± 20.59	+5.0
CASE 3	0.2 µg/ml	10	-0.28 ± 0.20	-240.0	0.20 ± 0.04	0.0	0.20 ± 0.03	+0.1
	5.0 µg/ml	10	4.82 ± 0.17	-3.6	5.03 ± 0.18	+0.6	5.04 ± 0.15	+0.7
	80.0 µg/ml	10	80.51 ± 5.16	+0.6	77.19 ± 5.77	-3.5	76.53 ± 3.95	-4.3

Table 1. Precision and bias of calibrations: unweighted, weighted by the scheme (*implied by 'weighted' in* Figs. 2-4), and weighted on the basis of $1/(conc)^2$.

Fig. 1. Variances, i.e. $(S.D.)^2$, for response *vs.* concentration of standard in Cases 1, 2 & 3.

Case 1

C_O = 7.1095E-05
C = 1.2845E-04
P = 7.3230E-01

Case 2

C_O = .1078E+08
C = 2.1008E+06
P = 2.1197E+00

Case 3

C_O = 2.8389E-05
C = 1.7409E-05
P = 2.5840E+00

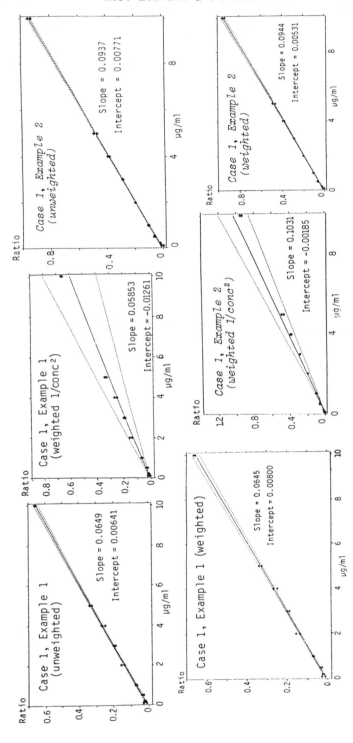

Fig. 2. Confidence limits (95%) in Case 1, Example 1 & *Example 2.*
In Figs. 2–4 the unqualified term 'weighted' implies the present scheme.

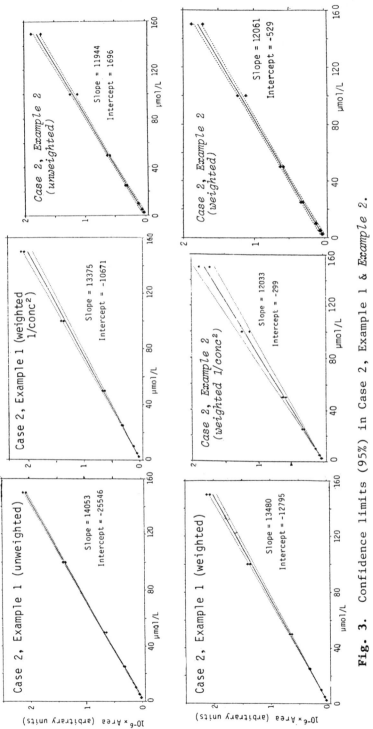

Fig. 3. Confidence limits (95%) in Case 2, Example 1 & Example 2.

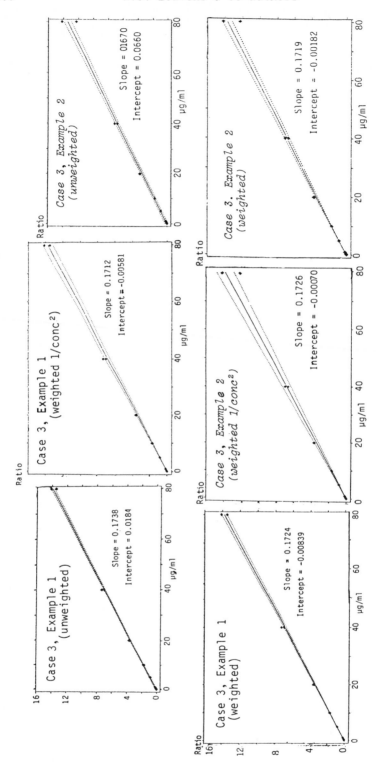

Fig. 4. Confidence limits (95%) in Case 3, Example 1 & *Example 2*.

DISCUSSION

Case 1 exemplifies an assay where the variance did not change enormously with concentration (Fig. 1). Over the 0.1-10 µg/ml range the variance changed by a factor of 10, an S.D. range of 3. As shown in Fig. 2, there was little difference between weighted and unweighted slopes and intercepts, but the weighted line shows the statistically valid widening of confidence interval (CI) at the high end of the range. The narrowing of the CI at the lower end for unweighted regression results from the common bioanalytical practice of distributing more standards towards this end.

Case 2 demonstrated an assay where the variance depended very strongly upon concentration (Fig. 1), with a 700-fold difference (27-fold for S.D.; unweighted values in Table 1). Consequently both unweighted lines (especially in Example 2) were erroneously high at low concentrations, with intervals much wider than the duplicate ranges.

Case 3 had variance which similarly showed a strong dependency on concentration, with a 50,000-fold change (S.D. change 230-fold) over the concentration range 0.2-80 µg/ml. Fig. 4 demonstrates a random 'perfect' example of near-coincidence of unweighted and weighted lines, only the unweighted CI at the low end being quite unrealistic. A more typical situation was seen in the other example where the unweighted fit was biased for all 5 lowest standards, requiring the analyst to limit the quantification range or to reject the entire assay. In contrast, the line from the present weighting scheme was clearly acceptable over the entire range.

Table 1 indicates that, in all three 'Cases', at medium and high concentrations there was little difference between weighted and unweighted methods of constructing calibration lines. Unweighted calibrations consistently showed better accuracy at highest concentrations, but the weighted method still produced acceptable accuracy (mean bias <5%). In contrast, at low concentrations the bias shown by the unweighted regression method was virtually abolished by weighted regression, even for 'Case 1' where variance was most weakly dependent upon concentration (P value only 0.7).

Where the variance was strongly concentration-dependent, in 'Case 1' (P 2.1) and 'Case 2' (P 2.6), the extreme bias produced by unweighted calibration (36% and 240% respectively) seriously limited the lower limit of quantification.

For each 'Case' that we happened to select, weighting by $1/C^2$ gave bias and precision results (Table 1) similar to those obtained using our weighting scheme. However, our scheme will hold for any calibration data, unlike $1/C$ and $1/C^2$ as more commonly used to weight calibration data.

In fact the $1/C^2$ basis in our examples showed extreme widening of the 95% CI's at the high end and absurd narrowing at the low end. Our scheme does not suffer such shortcomings, and through calculation of concentration by reverse interpolation will give valid CI's. Weighting based on $1/C$ or $1/C^2$ cannot help producing near-infinite weighting as C tends towards zero, and therefore an impossibly narrow CI at the low end of a large dynamic range. This may also become apparent by inspection of the standardized residuals after weighting with our scheme compared with the alternatives.

CONCLUSIONS

Unweighted least-squares calibration should be used only when variances change imperceptibly with concentration. Calibration bias, produced by inappropriately weighted calibrations, artificially raises the lower limit of quantification. Weighting the calibration using a scheme such as the one described here is to be strongly recommended because valuable bioanalytical information may be lost or distorted through using inappropriate data-handling methodology. We are not aware of any other publications where a dynamic variance:concentration model was used to produce a three-parameter weighting scheme.

Acknowledgements

We thank H.J. Merrett and F.R. House, Management Services Division Business Team, for their help and advice in preparing this material, and also J. Ingram, Dept. of Bioanalytical Sciences, Pharmacokinetic Division, for help in its presentation.

References

1. Sharaf, M.A., Illman, D.L. & Kowalski, B.R. (1986) *Chemometrics*, Interscience/Wiley, New York, p. 119.
2. Chance, E.M., Curtis, A.R., Jones, I.P. & Kirby, C.R. (1981) *FACSIMILE: A computer program for flow and chemistry simulation, and general initial value problems.* Computer Sciences & Systems Division, AERE, Harwell, Oxon., U.K.
3. PROGRAM: *Genstat 5 Release 1.2 (IBM/CMS)*, Rothampstead Experimental Station, Harpenden, Herts., U.K.

#B-9

CHOICE AND OPTIMIZATION OF CALIBRATION FUNCTIONS

[1]H.M. Hill, [1]A.G. Causey, [2]D. Lessard, [3]K. Selinger
and [4]J. Herman

[1]Hazleton UK, Otley Road, Harrogate, N. Yorks. HG3 1PY, U.K.

[2]Phoenix International Life Sciences,
St. Laurent (Montreal), Québec, Canada H4R 9Z7

[3]Glaxo Research Laboratories,
Research Triangle Park, NC 27709, U.S.A.

[4]LabLogic America, Montreal-West, Québec, Canada H4X 1Y8

This article discusses the pivotal role of calibration functions in ensuring the accuracy of data derived from bioanalytical methods using chromatographic procedures, specifically HPLC and HR-GC*. In order to derive the most appropriate calibration function the analyst is faced with a number of choices. He must decide on the following: #the calibration range, including the LLOQ; #the number and spacing of calibration points used to construct the curve; #which type of calibration model is most appropriate; #the acceptability of the curve based on goodness of fit to the model.

Although a number of publications have attempted to provide some general guidelines on evaluating calibration functions [1-5], these have not been widely accepted, and until recently these parameters have been subjectively evaluated[θ]. Recently, however (Dec. 1990), over 500 industrial scientists, regulators and academics met in Washington, DC, to reach a consensus on analytical methods validation. The main sponsors of the meeting were the AAPS and the FDA. The consensus report of this meeting has appeared recently in several journals including *Journal of Pharmaceutical Sciences* [6].

Some of the more specific recommendations relating to calibration functions are detailed here and should be regarded as minimal requirements rather than as definitive. The following points from the consensus are quoted from the section *Principles*

---

*Abbreviations.- AAPS, American Association of Pharmaceutical Scientists; FDA, Food and Drug Administration; HR, high-resolution (GC); LLOQ, lower limit of quantification; ULL, upper limit of linearity.

[θ]See Vol. 20 [3], based on the 1991 Forum: Dell discussed question-naire results (p. 21).- *Ed.*

for Establishing a Valid Method.
#"The concentration range over which the analyte will be
determined must be defined in the method on the basis of
evaluation of actual standard samples over the range (including
their statistical variation). This procedure defines the
standard curve."

#"It is necessary to use a sufficient number of standards
to define adequately the relationship between concentration
and response. The relationship between response and concentration must be demonstrated to be continuous and reproducible.
The number of standards to be used will be a function
of the dynamic range and the concentration-response relationship. In many cases, five to eight concentrations (excluding
blank values) may define the standard curve. More standard
concentrations may be necessary for non-linear relationships
than would be for linear relationships."

Similarly the section *Application to Routine Drug Analysis* states:
#"A standard curve should be generated for each analytical
run for each analyte in the unknown samples assayed with
that run. It is important to use a standard curve that will
cover the entire range of concentrations in the unknown
samples. Estimation of unknowns by extrapolation of standard
curves below the lowest standard or above the high standard
is not recommended."

CALIBRATION RANGE, NUMBER AND DISTRIBUTION OF CALIBRATION STANDARDS

The number of points to be used in constructing a
calibration curve recommended in the consensus report [6]
is 5-8 non-zero points. However, no recommendation is provided
as to their distribution; for a linear function over short
calibration ranges (1 up to 10) the points should be weighted
towards the lower concentration, whereas for wider concentration
ranges even spacing is ideally preferred, although for practical
purposes this may not be desirable. For a range between
LLOQ (= 1) and top standard (=100) it is now suggested that
the distribution be:
    0, LLOQ, 2 × LLOQ, 5, 10, 20, 40, 80, 100.

This distribution provides sufficient points to define a non-
linear function. At the same time, inclusion of a standard
close to the LLOQ ensures that another calibrant will be
available if the LLOQ standard is compromised, thus minimizing
the need to re-assay unknowns close to the LLOQ. Similarly,
because extrapolations beyond the top standard are disfavoured,
it is essential to ensure that while this point covers most
(90%) of the unknown concentrations, it does not compromise
the calibration model.

The choice of the LLOQ is defined in the consensus report [6] as that concentration for which the bias is no greater than ±20% and the precision (C.V.) is not greater than 20%.

While use of duplicate assays is recommended for methods with C.V.'s >15%, for chromatographic methods single calibration standards generally appear appropriate. Significant increases in the number of calibration standards may seriously decrease the number of unknowns per batch.

Again these are the minimal limits, and 'tighter' criteria can clearly be specified within these guidelines.

CHOICE OF APPROPRIATE MODEL

The mathematical models used to define concentration-response relationships may be divided into two groups, linear models and non-linear models. Data are fitted to these models using the least-squares procedure. The limitations, assumptions and advantages of this procedure have been discussed elsewhere [3, 5]. The most widely used linear models are either of the $y = mx + c$ form or are log-log transformations. Where assay variance increases with diminishing concentration, weighted regression may be carried out [6]. The most appropriate weighting function is based on the actual variance at each concentration.

The advantages of taking account of variances at different concentrations are demonstrated in the preceding article (#B-7; Land, Leavens & Weatherley). While these authors have developed a sophisticated approach to determining the appropriate weighting for each standard, it is necessary to generate sufficient data to provide a definitive indication of the variance at each concentration prior to running the assay. In fact their data presentation is useful in confirming that the approximate weighting using the reciprocal of the concentration or the square thereof closely matches that from their own weighting scheme, whilst they reckon theirs to be better in certain circumstances.

The requirements for weighting may be determined by plotting, for calibration standards, the residual values, i.e. (difference between the nominal and the back-calculated concentration) *vs*. concentration [3]. Those models which provide residuals that have a constant error should therefore be regarded as the most appropriate. If, however, plotting of residuals from a linear model gives skewed distribution of residuals [3], then a non-linear model may be more appropriate, e.g. log linear, log quadratic or quadratic.

The log quadratic function was first used by Wilkinson [7] to define the concentration/response for an assay of ethanol using headspace analysis and GC-FID. This model provides a number of advantages in that the log format provides a weighting that ensures a good fit at lower concentrations. In addition it has universal applicability in that it will reduce to a log-log format.

A more limited but perhaps more widely used format is the quadratic function, $y = ax^2 + bx + c$. In this case it is essential to use weighting to ensure a good fit at lower concentrations (Figs. 1A & 1B). A further problem associated with quadratic functions is the ULL, where:

$$dy/dx = 0 = 2x + b, \text{ hence } x = 0.4(b/a)$$

However, this is a gross over-estimate and may be better defined in the same way as the lower quantitation (LLOQ), i.e. the concentration at which the assay variability (C.V.) is ±20%. It does, however, provide guidance as to the usable limits of the concentration range. An alternative to the use of the quadratic function is to split the calibration function into two linear ranges. However, this produces ambiguity in the area of overlap for the split curves. With the wider use of computerized curve-fitting programs, one can use continuous non-linear functions which eliminate this problem.

The choice of the most appropriate calibration model should be made during the method-development and validation phase. Once this choice is established, criteria for the acceptability of each curve should be applied prior to quantifying each batch of unknowns.

ACCEPTABILITY OF CALIBRATION FUNCTION

Having chosen the most appropriate model based on an examination of residuals, it is necessary to evaluate the acceptability of the curve for each batch of samples. The criterion most widely used to evaluate goodness of fit is the correlation coefficient. While this is a useful first indication, it is of limited value [3]. A better indication of goodness of fit is a comparison of the differences between the nominal calibration standard concentration and the back-calculated standard. A number of ways of evaluating this were discussed in the previous volume [3]. One parameter whose virtues have been extolled by Knecht & Stork [4] is the 'Guete coefficient' (g) or goodness-of-fit coefficient

$$g = \frac{\sqrt{\Sigma(\% \ difference)}}{n - 1} \quad \text{where } n = \text{no. of calibration stds.}$$

$$\% \ difference = \frac{back\text{-}calculated - nominal}{nominal} \times 100$$

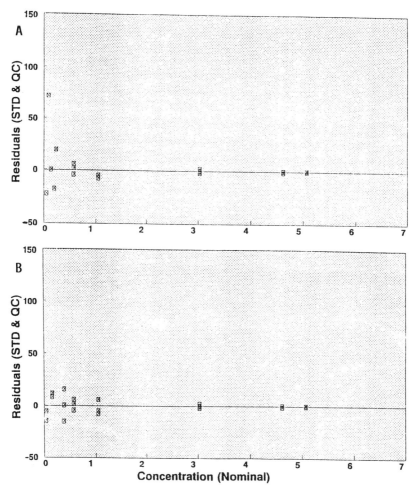

Fig. 1. Comparison of residuals: **A,** an unweighted quadratic calibration function; **B,** the same curve weighted with the reciprocal of the concentration.

Not only is it possible to compare the goodness of fit for specific curves, but by comparing the goodness of fit for a range of different models it is possible to determine the most appropriate model. Table 1 provides a comparison of g for a range of different weightings for a linear model and compares it with a non-parametric robust regression analysis.

It may be possible to improve the goodness of fit by excluding a 'spurious' value in calculating the calibration curve. However, this can only be done by objectively applying rejection criteria. One such criterion is to eliminate from the calculation those standards for which the back-calculated value is ±20% of the nominal. However, in considering

Table 1. Use of 'quality' parameter to measure fit: effect of different procedures for calculating a linear calibration curve on the goodness of fit parameter (g).

For Table 2 below
Actual concn.
$\mu g/ml$:

	Nominal	Linear	1/c	1/c^2	Robust
STD H	60.00	58.80	59.10	59.80	59.62
STD G	47.20	45.00	47.40	48.00	47.83
STD F	30.00	29.70	29.70	30.10	30.00
STD E	15.00	13.90	13.90	14.00	13.99
STD D	7.50	7.51	7.38	7.45	7.42
STD C	1.00	1.19	1.02	1.00	1.00
STD B	0.50	0.70	0.52	0.50	0.50
(g) %Quality		18.02	2.05	2.30	2.90

59.80
48.00
30.10
14.00
7.45
1.00
0.50

	Actual	+1SD	+2SD	+3SD	+4SD	+5SD
1/c	0.9984	0.9956	0.9914	0.9860	0.9795	0.9722
1/c^2	0.9982	0.9959	0.9925	0.9880	0.9826	0.9764

Table 2. Influence of outliers on linear regression (*see above* for actual concentrations). The calculated concentrations were from a linear calibration curve using a weighting of 1/c^2 (Table 1). The next higher standard was decreased in incrementals of 1 S.D., i.e. 10%, in order to evaluate the effect on the correlation coefficient. These changes have no effect on the data generated by 'robust regression analysis'.

this approach there is no reason to suppose that such values are outliers. Table 2 compares the effect of distorting the next highest standard by increments of 1 S.D., assuming that an acceptable S.D. would be 10%. An inspection of Table 2 shows that 2 S.D.'s is the limit for acceptable correlation coefficients, i.e. better than 0.99.

A more appropriate procedure for calculating the calibration function is the non-parametric regression procedure of Theil, the 'incomplete' method as discussed by Miller & Miller [5]. The method is used to fit linear models of the form $y = mx + c$. Pairs of data sets (concentration/response) are ranked in order; where the number of pairs is odd the median is rejected. The slope is calculated for pairs of points and the median value from the calculated slopes is used to describe the curve. Using this slope, values of c are calculated for each point by substitution in the equation. The median value is then chosen as the best estimate of the intercept.

The major advantage of this 'robust regression analysis' is that it is unaffected by 'outliers'. While it is possible to calculate a goodness-of-fit for curves using this procedure, it is not possible to derive a correlation coefficient. This may account for its lack of acceptance since conventional wisdom seems to require calculation of a correlation coefficient. In our experience (cf. Fig. 1B), robust correlation analysis provides a fit similar to that of weighted linear regression analysis without the inherent problem of acceptance or rejection of spurious calibration values.

Despite the advantages of robust regression analysis, calculation of linear calibration curves by the least-squares procedure appears to be the most widely accepted procedure. It is therefore necessary to provide general guidance on the acceptability of a calibration curve. We have developed the following empirical rules to guide analysts in their acceptance or rejection of a specific calibration curve.

- A calibration curve should consist of at least 5 evaluable standards, and no more than 2 standards, which must not be adjacent, may be excluded from the calculation.
- Standards may be excluded from calculation of the calibration curve if the back-calculated value differs by >20% of the nominal value.
- The correlation coefficient must be at least 0.990.
- The Goodness-of-Fit Coefficient should be no more than 10%.
- If there is an 'interference' equivalent to or greater than 20% of the LLOQ, then this value must be discarded, and the next lowest standard for which the interference constitutes <20% should be used as the LLOQ.

SUMMARY

The recent AAPS/FDA consensus report [6] on validation of bioanalytical methodology provides minimal standards of acceptability. While still adhering to the criteria therein, we have attempted to provide some practical guidelines for choosing the most appropriate mathematical model to define the calibration response relationship and provide acceptance criteria for quantifying unknown samples.

References

1. Bonate, P.L. (1990) *J. Chromatog. Sci. 28*, 559-562.
2. Karnes, T.H., Shiu, G. & Shah, V.P. (1991) *Pharmaceutical Res. 8*, 421-426.
3. Phillips, L.J., Alexandra, J. & Hill, H.M. (1989) in *Analysis for Drugs and Metabolites, including Anti-infective Agents* [Vol. 20, this series] (Reid, E. & Wilson, I.D., eds.), Royal Society of Chemistry, Cambridge, pp. 23-36.

4. Knecht, J. & Stork, G. (1974) *Fresenius Z. Anal. Chem. 270*, 97-99.
5. Miller, J.C. & Miller, J.N. (1988) *Statistics for Analytical Chemists*, Horwood, Chichester (2nd edn.), pp. 154-156.
6. Shah, V.P., Midha, K.K., Dighe, S., McGilveray, I.J., Skelly, J.P., Yacobi, A., Layloff, T., Viswanathan, C.T., Cooke, C.E., McDowall, R.D., Pittman, K.A. & Spector, S. (1992) *J. Pharm. Sci. 81*, 309-312.
7. Wilkinson, P.K., Wagner, J.G. & Sedman, A.J. (1975) *Anal. Chem. 47*, 1506-1510.

#B-10

HPLC ANALYSIS OF TISSUES FOR ANTIBIOTIC DRUGS: PRINCIPLES AND VALIDATION

P. Heizmann

F. Hoffmann-La Roche & Co., Ltd.,
CH-4002 Basel, Switzerland

*Therapeutic studies with antimicrobial drugs entail ascertaining the drug concentration at the infection site. To appraise the closeness of the measured concentrations to the actual ones, the validation of the analytical method should focus on the recovery in the extraction (sample preparation) step, on calibration procedures and on assay reproducibility. For recovery experiments, spiked tissue samples do not in any case reflect actual in vivo conditions. To mirror these and to evaluate optimal extraction procedures, animal tissues from experiments with ^{14}C-labelled drugs can be used. Calibration procedures should reflect as closely as possible the recovery conditions from in vivo samples. However the calibration is carried out, the analyst should know how realistic his results are. Reproducibility of the analytical procedure is another important criterion for the quality of analytical results. Strategies for analytical validation in the field of tissue analysis are now discussed, with particular examples.**

Bacterial infections are normally located not in the vascular compartment but in the tissues. The activity of an antimicrobial drug therefore depends on its concentration at the infection site (tissue). Antibiotic tissue concentrations may be evaluated from the (free) plasma levels of the drug [1] or can be measured from tissue fluids or from whole tissues. Since tissue fluids are not always available, different experimental models have been developed to obtain a biological fluid similar to the interstitial tissue fluid. These approaches and techniques, as well as general principles of antibiotic tissue penetration, have been discussed elsewhere [1-7]. Methodological problems associated with the HPLC analysis of antibiotics in whole tissues are now described, together with consideration of problems of interpretation.

METHODOLOGICAL ASPECTS

To interpret measured levels correctly, the analyst should have an idea of how close the found concentrations are to the real ones, and the extent to which the results might

*In previous vols., see 'Tissues' in Index for pertinent items.- *Ed.*

be influenced by methodological problems. Among the numerous publications dealing with antibiotic tissue analysis (including bioassays), some authors discuss possible methodological influences on their analytical results [8-15]. Thus, in Fong's publication [13] ciprofloxacin levels in bone and muscles were calculated in two ways: method 'A' had a single extraction and standards were prepared with control tissue, and 'B' had multiple extractions and standards prepared in buffer. In the case of muscle samples, method 'B' gave values only 58% of those from 'A'.

This example indicates that, as for drug analysis of body fluids, analysis of whole tissues needs a carefully validated assay. During validation, the main emphasis should be on the following:- sample preparation, recovery studies, calibration procedure, evaluation of assay precision, and correct description of the assay.

Sample handling

Superficial blood traces should be removed from the freshly taken tissue, e.g. by washing with physiological saline and gently drying with gauze. During storage, weight losses by liquid evaporation must be taken into account for small tissue pieces, lest there be substantial over-estimation of the antibiotic concentration [8].

Sample preparation

Procedures for sample preparation depend on the type of tissue, the physicochemical characteristics of the drug, the detection mode and the sensitivity and selectivity needed. UV detection, for instance, is generally less sensitive and less selective than fluorescence detection, and normally needs a more extensive sample clean-up. Soft tissues (lung, liver) are much easier to handle than hard tissues (bone) or fatty tissues. Homogenization is the first step in the sample preparation procedure. An aliquot of the sample is suspended in an appropriate solvent and then finely disintegrated (mechanically or enzymatically [10]). Aqueous or water-miscible solvents are normally used (Table 1A).

After homogenization, drug isolation from the homogenate can be done by the same analytical procedures as for biofluid analysis (Table 1B). The homogenate may merely be centrifuged and put direct into the column [11-16]. Extractive procedures include liquid-liquid [24, 26, 27] and liquid-solid [12, 17] partition. Back-extraction procedures may be used for further purification of the extract [19, 28, 29, 31].

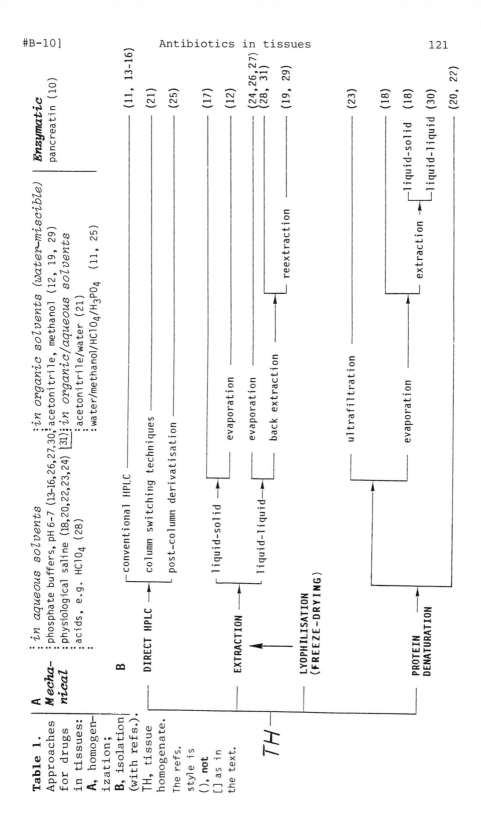

Table 1. Approaches for drugs in tissues: **A,** homogenization; **B,** isolation (with refs.). TH, tissue homogenate. The refs. style is (), **not** [] as in the text.

A **Mechanical** | **Enzymatic** pancreatin (10)

: *in aqueous solvents*
:: phosphate buffers, pH 6-7 (13-16,26,27,30;
:: physiological saline (18,20,22,23,24)
:: acids, e.g. HClO₄ (28)

: *in organic solvents (water-miscible)*
:: acetonitrile, methanol (12, 19, 29)

: *in organic/aqueous solvents*
:: acetonitrile/water (21)
:: water/methanol/HClO₄/H₃PO₄ (11, 25)

B

DIRECT HPLC → conventional HPLC ——— (11, 13-16)
     column switching techniques ——— (21)
     post-column derivatisation ——— (25)

EXTRACTION → liquid-solid → evaporation ——— (17)
     liquid-liquid → evaporation ——— (12)
           → back extraction → reextraction ——— (24,26,27) (28,31)
                    ——— (19, 29)

LYOPHILISATION (FREEZE-DRYING)

PROTEIN DENATURATION → ultrafiltration ——— (23)
     → evaporation → extraction → liquid-solid ——— (18)
                       liquid-solid ——— (18)
                       liquid-liquid (30)
                       ——— (20, 22)

TH

Some authors have added protein-denaturing solvents such as acetonitrile, methanol, ethanol [18, 23, 30] or acids [20, 22] to the homogenate suspension, as a means of further purification or to improve the extraction of the drug from the tissue. More recently, direct injection [20, 22], ultrafiltration [23] or additional extractive procedures [18, 30] have been performed. Column switching [21] for sample clean-up, or post-column derivatization for fluorescence enhancement [25], may be applied. UV-detection has been used, besides fluorescence or electrochemical detection [19].

Some of these procedures are very simple, others are more complicated. Whenever possible, the analyst should try to develop the simplest procedure. In practice, however, methodological requirements such as selectivity, sensitivity, precision and recovery may strongly influence the sample preparation procedure and make it more or less complicated.

Recovery

It is vital to investigate the recovery of the drug from the tissue. Procedures with low recoveries may lead to the under-estimation of the drug content in the tissue, unless the calibration procedure compensates for this. Tissues from experiments *in vivo*, as well as tissue samples spiked *in vitro*, should be used for recovery studies.

Experiments with 'radiolabelled' tissues (tissues of animals dosed with the labelled drug) are optimal for drug-recovery evaluations, provided that metabolites are absent, or present in negligible amounts. If the tissues contain metabolites, recovery studies with radiolabelled materials need to be much more subtle. This was not the case in the Table 2 example, where tissue samples from dosed rats were combusted to determine total ^{14}C and, in parallel, processed for HPLC determination of the unchanged drug, allowing recovery comparisons.

'Cold' *in vivo* tissues (non-radiolabelled drug dosed) may be used to ascertain whether, after repeating the sample preparation procedure on the same specimen, the amount of extracted drug remains constant or increases [13-15, 31]. From these experiments, it will be ascertained whether single or repeated sample processing steps should be performed, or whether the sample preparation procedure has to be modified in order to improve recovery. Absolute recoveries cannot be determined from 'cold' *in vivo* samples.

The extraction of the drugs from the tissue may be a time-dependent process. Recovery may then depend on the elapsed time following the homogenization of the sample in an eluting solvent. In the experiment illustrated in Fig. 1,

Table 2. Concentrations of a fluoroquinolone in rat tissues. Total ^{14}C ascertained by combustion is expressed as concentration equivalents of the unchanged drug. Recovery of the oral dose (10 mg/kg) = ratio of value from HPLC (tissue homogenate in pH 7.5 phosphate buffer extracted with DCM/isoPrOH, 7:3; dried down under N_2) to ^{14}C value.
Courtesy of E. Weidemann.

TISSUE	^{14}C µg equiv/g	H P L C µg/g	RECOVERY %
Plasma	3.20	2.92	91.3
Lung	4.55	4.59	100.9
Heart	7.25	8.21	113.2
Kidney	13.0	13.3	102.3
Fat	0.54	0.55	101.9
Muscle	6.19	6.33	102.9
Ovary	3.97	4.08	102.8
Eye	1.50	1.34	89.3
Uterus	6.13	5.67	92.5
Tube	5.42	4.47	82.5
Cartilage	4.04	3.57	88.4
Bone	4.28	3.55	82.9

Fig. 1. Assay recoveries (as peak heights) of fluoroquinolones from *in vivo*-exposed mouse skin tissues *vs*. time after homogenization (blendor; room temp.) in 67 mM phosphate. For pre-HPLC extraction, see Table 2 legend.

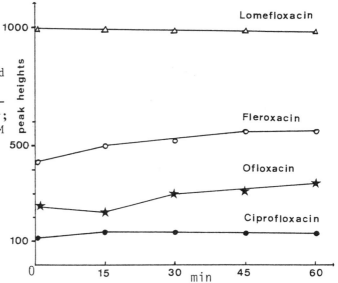

in-vivo tissue samples were homogenized. Aliquots were then submitted to the assay procedures at different times after homogenization. For two of the compounds (fleroxacin, ofloxacin), recoveries increased with time. Accordingly, in the final assay procedure, the homogenates were allowed to stand, for 0.5 h at room temperature, prior to extraction.

Spiked tissue homogenates should also be used in recovery experiments. However, it is unlikely that they reflect the *in-vivo* drug binding situation. Assuming poorer *in-vitro*

Fig. 2.[*] Procedure
entailing addition of
standard: calibration
lines for the determi-
nation of a β-lactam,
cefetamet, in human
tonsilitic tissues
(3 samples; one line
for each). Blendor
homogenate (~3 ml 0.5 M
$HClO_4$/g) spiked and
centrifuged; see text.

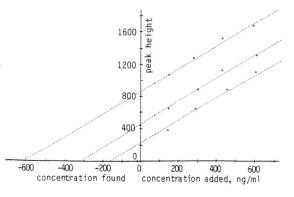

binding of drug to the homogenized tissue (e.g. where large
tissue particles are present after insufficient homogenization),
higher recoveries from the spiked tissues are to be expected
as compared with the real samples. However, it is also
the case that, through a homogenization procedure, tissue
barriers may break down and expose the drug to structures
to which it normally does not bind [8]. Thus, binding to
the *in vitro* tissue could be more extensive or stronger as
compared to that for the *in vivo* sample.

Calibration procedure

The importance of preparing the calibration standard
properly cannot be over-emphasized. Different results, depending on the type of standard used, have been reported [8,
13]. Ideally, the sample serving as standard should contain
the antibiotic bound and distributed exactly as in the *in-vivo*
tissue, thus giving identical recoveries. In reality, this
is difficult both to investigate and to fulfil.

The recovery experiments discussed above are vital for
an appropriate preparation of calibration standards. Most
authors use tissue homogenates [13, 14, 17, 19, 22, 28, 30]
for standard preparation, but others use serum [15, 21] or
aqueous solutions of the drug [18, 25]. Spiking homogenized
drug-free tissues appears to be the best procedure for preparing
calibration standards. Adding the drug to the tissue before
homogenizing has been suggested [8]. The degree of dilution
in standards and in samples should be identical, since dilution
reduces the degree of tissue binding [8].

In human experiments, obtaining sufficient quantities of
drug-free pre-dose tissues is usually precluded. Yet if
sufficient *in vivo* tissue material is obtained, the procedure
entailing addition of standard may be used to prepare a
calibration line. This is exemplified in Fig. 2. Each
tissue sample was homogenized, and aliquots of the homogenate

Courtesy of J. Kneer.

Table 3. Comparison of the recoveries (expressed as HPLC peak heights) of various fluoroquinolones from spiked skin-tissue homogenates (mouse) and from spiked plasma. Spike levels: 2 µg/ml plasma and 2 µg/g tissue.

	PEAK HEIGHTS			HEIGHT RATIO		
	Plasma	Tissue	% of Plasma	Plasma	Tissue	% of Plasma
Fleroxacin	446	383	86	5.41	5.31	98
Ciprofloxacin	184	113	61	1.57	1.24	79
Ciprofloxacin*	173	160	92	1.50	1.29	86
Ofloxacin	257	196	76	2.84	3.06	107
Lomefloxacin	334	250	75	2.98	2.66	94

*2 extractions

were then spiked with different amounts of the drug and submitted to the assay procedure. For each of the unknown samples, a separate calibration line was obtained, the x-intercepts representing the drug content of the unknown samples.

If no drug-free tissue is available and/or if the sample amount is limited, a hopefully valid alternative for the preparation of the calibration standard should be sought, e.g. spiked plasma if recovery from tissue is in the same range as that from plasma. To evaluate this, additional recovery experiments have to be performed. Table 3 shows the results of such an investigation. Drugs were spiked in parallel into drug-free plasma and into drug-free tissue. The spiked samples were then submitted to sample preparation. The recoveries (from measured peak heights) were lower from the tissue than from plasma, such that the use of plasma standards would lead to under-estimation of drug content in the tissues, especially with ciprofloxacin. In such a situation, the analyst has to decide whether or not the inaccuracy is acceptable. If not, additional experiments have to be performed. In the actual example, the recovery of ciprofloxacin could be improved by extracting a second time (Table 3). Furthermore, the use of an internal standard gave some improvement of the results, as shown ('RATIO' values).

Precision

Assay precision - the relative S.D. (%) for independent replicates on the same sample - is shown in Table 4 for an extractive tissue-assay method:- it was calculated from replicate analysis of, firstly, spiked tissue homogenates and, secondly, *'in-vivo'* samples. Especially for ciprofloxacin

Table 4. Precision (% C.V.) of an analytical assay (extraction) for the determination of fluoroquinolones in mouse skin tissues: a) from spiked tissue homogenates, each of the N replicates being a spiked homogenate; b), from *in-vivo* samples: C.V. from individual (N = 12) C.V.'s.

	Ciprofloxacin	Lomefloxacin	Ofloxacin	Fleroxacin
a) CV %	7.4 (N = 13)	6.2 (N = 7)	3.0 (N = 8)	4.4 (N = 7)
b) $\overline{CV}$ %	29	8.9	11.6	6.1
	11.2 *			

* improved method

and ofloxacin, the data obtained from the spiked tissues did not reflect the *in-vivo* situation. For ciprofloxacin, precision could be improved by modifying the assay: instead of analyzing small amounts of the tissue separately, a larger sample was homogenized, and aliquots of the resulting pool were submitted to the sample preparation procedure.

More realistic information on precision is generally obtained from replicate analysis of *in vivo* samples.

Description of the assay

Many procedures in reports or publications are not described well enough for the reader to be able to assess and compare the analytical results. The description should include sample preparation, recovery experiments and precision data.

ADDITIONAL FACTORS AFFECTING ANALYTICAL INTERPRETATIONS

Blood contamination.- Organs such as lung, liver and kidney contain varying amounts of blood, reported [8, 20] to be 2-10% for lung. This could contribute to the measured tissue levels. Some authors have corrected for the blood content of the sample [8, 10, 14, 16, 20, 30, 31], and others have regarded this as unimportant [9, 32] - hardly the case if the drug concentration in blood is high compared with that in tissue [8]. With kidney there could also be an error due to urine, but no method for ascertaining urine content seems to have been described.

Drug distribution within the tissue.- Tissue schematically comprises three compartments (Fig. 3): the vascular (blood delivery) system, the extracellular (interstitial) tissue fluid (which is normally the site of infection and, for muscle, is reckoned to be 7-16% [6]), and the cell mass. The unbound (free) drug equilibrates between these compartments. The

VASCULAR   EXTRAVASCULAR

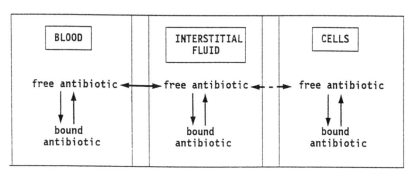

Fig. 3 (above).
Schematic description
of a tissue.

Fig. 4 (right).
Mean concentrations
of cefuroxime in
rabbit samples after
injection of 25 mg/
kg, i.m.
Figs. 3 & 4 are from
ref. [6], by
permission.

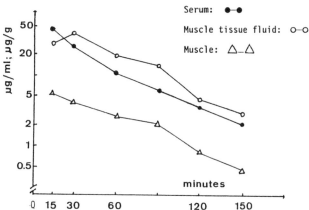

total amount of extravascular antibiotic comprises an extracellular portion in the interstitial fluid and (in some cases)
an intracellular portion.   Drug levels measured in whole
tissue may thus over- or under-estimate the actual levels
in the extracellular fluid, usually the most relevant ones
since this fluid is the site of most bacterial infections.
With β-lactam antibiotics, for example, penetration into cells
is relatively poor [8].   In animal experiments, cefuroxime
was measured in serum, tissue fluid and whole tissue (Fig. 4).
Evidently the drug within the tissue was localized mainly
in the tissue fluid.

CONCLUSION

Clearly, from the points discussed, antibiotic concentrations measured in whole tissues need careful interpretation
for evaluation of their therapeutic significance. Methodological
influences, besides factors which might influence drug distribution within the tissue, should be ascertained and possibly
taken into account. The more thoroughly these aspects are
evaluated, the more valuable is the information.

128   P. Heizmann

References (Amb = Antimicrob.; Ctr = Chemother.)

1. Ryan, D.M., Cars, O. & Hoffstedt, B. (1986) *Scand. J. Infect. Dis. 18*, 381-388.
2. Carbon, C. (1990) *Eur. J. Clin. Microbiol. Infect. Dis. 9*, 510-516.
3. Mazzei, T. & Periti, P. (1989) *J. Ctr 1*, 75-79.
4. Wise, R. (1986) *Rev. Infect. Dis. 8, Suppl. 3*, 325-332.
5. Barza, M. & Cuchural, G. (1985) *J. Amb Ctr 15, Suppl. A*, 59-75.
6. Ryan, D.M. & Cars, O. (1980) *Scand. J. Infect. Dis. 12*, 307-309.
7. Bergan, T., Angeset, A. & Olzewski, W. (1987) *Rev. Infect. Dis. 9*, 713-718.
8. Cars, O. & Ögren, S. (1985) *Scand. J. Infect. Dis., Suppl. 44*, 7-15.
9. Kaplan, J.M., McCracken, Jr., G.H. & Snyder, E. (1973) *Amb Agents Ctr 3*, 143-146.
10. Loebis, L.H. (1985) *J. Amb Ctr 16*, 757-761.
11. Kees, F., Naber, H., Schumacher, H. & Grobecker, H. (1988) *Ctr 34*, 437-443.
12. Lagana, A., Curini, R., D'Ascenzo, G., Marino, A. & Rotatori, M. (1987) *J. Chromatog. 417*, 135-142.
13. Fong, J.W., Ledbetter, W.H., Vandenbroucke, A.C., Simbul, M. & Rahm, V. (1986) *Amb Agents Ctr 29*, 405-408.
14. Fong, J.W., Rittenhouse, B.R., Simbul, M. & Vandenbroucke, A.C. (1988) *Amb Agents Ctr 32*, 834-837.
15. Wittmann, D.H. & Kotthaus, E. (1986) *Infection 14, Suppl. 4*, 270-273.
16. Perea, E.J., Ayarra, J., Iglesias, M.C.G., Luque, J.G. & Loscertales, J. (1988) *Ctr 34*, 1-7.
17. Frijs, J.M. & Lakings, D.B. (1986) *J. Chromatog. 382*, 399-404.
18. Marunaka, T., Maniwa, M., Matsuhima, E. & Minami, Y. (1988) *J. Chromatog. 431*, 87-101.
19. Foulds, G., Shephard, R.M. & Johnson, R.B. (1990) *J. Amb Ctr 25, Suppl. A*, 73-82.
20. Wijnards, W.J.A., Vree, T.B., Baars, A.M., Hafkenscheid, J.C.M., Kohler, B.E.M. & van Herwaarden, C.L.A. (1988) *J. Amb Ctr 22,*
21. Allen, H.H., Khalil, M.W., Vachon, D. & [*Suppl. C*, 85-89. Glasier, M.A. (1988) *J. Amb Ctr 22, Suppl. B*, 111-116.
22. Wijnands, W.J.A., Vree, T.B., Baars, A.M. & van Herwaarden, C.L.A. (1988) *J. Amb Ctr 22, Suppl. B*, 67-77.
23. Grellet, J., Couraud, L., Saux, M.C. & Roche, G. (1989) *La Presse Médicale 18*, 1589-1592.
24. Viitanen, J., Auvinen, O., Tunturi, T., Männistö, P. & Haataja, H. (1984) *Ctr 30*, 211-215.   [330.
25. Scholl, H., Schmidt, K. & Weber, B. (1987) *J. Chromatog. 416*, 321-
26. Kusajima, H., Ishikawa, N., Machiada, M., Uchiada, H. & Irikura, T. (1986) *Amb Agents Ctr 30*, 304-309.
27. Montay, G. & Tassel, J.P. (1985) *J. Chromatog. 339*, 214-218.
28. Forchetti, C., Flammini, D., Carlucci, G. & Cavicchio, G. (1984) *J. Chromatog. 309*, 177-182.
29. Shepard, R.M. & Falkner, F.C. (1990) *J. Amb Ctr 25, Suppl. A*, 49-60.
30. Bawdon, R.E. & Madsen, P.O. (1986) *Amb Agents Ctr 30*, 231-233.
31. Dellamonica, P., Bernard, E., Etesse, H. & Garaffo, R. (1986) *J. Amb Ctr 17, Suppl. B*, 93-102.

#ncB

NOTES and COMMENTS relating to

ASSAY STRATEGIES FOR VARIOUS DRUGS

#ncB-1

A Note on

ANALYSIS OF PLASMA AND EDIBLE TISSUES FOR
ESTABLISHING SULPHONAMIDE WITHDRAWAL PERIODS

L.A. Meijer, K.G.F. Ceyssens, R. Siemons, R. Kramer
and J. Verstegen

DOPHARMA RESEARCH, 4941 VX Raamsdonksveer, The Netherlands

For determining sulphadiazine in plasma, muscle and kidney, a validated HPLC method (based on [1] & [2]) has been developed. It has given reliable values for low levels in samples from studies to establish pharmacokinetics and withdrawal time (W) — the time needed for an animal to clear its body of a drug and metabolites down to an accepted or pre-determined tolerance level (C_t). Generally during the terminal elimination phase the ratio (R) of tissue to plasma concentrations is constant and plasma concentration-time curves can be extrapolated to the desired C_t in tissue. Then W can be estimated [3]:

$$W = T_{\frac{1}{2}} \cdot (\ln R.C_0 - \ln C_t)/\ln 2$$

Sulphadiazine residues are highest in kidney (**R** ~2.75) [4].

Analysis (Fig. 1) employed a C-18 column and acetonitrile/10 mM sodium acetate acidified to pH 4.6 with acetic acid. Extracts prepared from muscle and kidney (e.g. pig) with dichloromethane, and cleaned up with Florisil columns. Plasma samples were deproteinized with acetonitrile; after centrifugation the supernatant was evaporated to dryness and the residue dissolved in mobile phase. Based on at least 6 points, calibration curves were fitted by least-squares linear regression.

Validation was given special attention because of the need for reliability at low concentrations. To determine the accuracy, drug-free plasma or, after homogenizing, tissue was spiked with drug, over the concentration range of interest. For true *vs.* measured concentrations the regression-line slope should be unity and the intercept zero if there are no significant errors. Precision ('repeatability') was determined by analysis of different concentrations of analyte under the same conditions. The linearity of the method was defined as the calibration-curve slope (an approach sometimes termed 'sensitivity'). The lower limit of the range of the method was the LOQ (Limit of quantification; cf. LOD, detection). The LOD and LOQ were determined by means of the 95% Confidence Interval and the 95% Dispersion (Prediction) Band, as in [5]. The formula was as follows for LOQ (and was the same for LOD except that the [1 +] term was omitted), where $\beta = \alpha = 0.95$, $b_1 = b_0$ estimate:

$$LOQ = X_{UB} + T_{(1-\beta)} \, (s_{yx}/b_1) \sqrt{([1+]\frac{1}{n} + (2 X_{UB} - X_{avg})^2/\Sigma xx)} \quad \begin{array}{l} \theta\,'UB' = \text{its} \\ \text{upper limit} \end{array}$$

Fig. 1. HPLC patterns for **(A)** plasma, **(B)** tissues (pig) spiked with the drug.

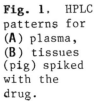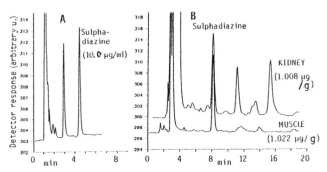

The LOQ was optimized through decreasing the calibration curve range, and was calculated (LOD also) each time a curve was established. Table 1 summarizes the validation data.

Table 1. Summary of assay parameters (64 sets of pig samples).

	Muscle tissue	Kidney tissue	Plasma
Accuracy	No errors (α=0.05)	No errors (α=0.05)	No errors (α=0.05)
LOD	0.015 µg/g	0.023 µg/g	0.047 µg/ml
LOQ	0.027 µg/g	0.042 µg/g	0.079 µg/ml
Linearity	Evaluated	Evaluated	Evaluated
Repeatability (CV)	better than 13.6 %	better than 8.3 %	better than 4.5 %
Range	0.027 to 0.204 µg/g	0.042 to 0.202 µg/g	0.079 to 21.4 µg/ml
Specificity	Evaluated	Evaluated	Evaluated

Application.- A pharmacokinetic study in pigs after a single oral [or i.v.] dose gave: AUC, 25042 ±2957 [37529 ±8746] min × µg/ml, $T_{\frac{1}{2}}$ 187±47 [327 ±161] min, C_0 354 ±224 µg/ml, C_1 3.33 ±0.4 [2.13±0.6] µl/g × min, and Vdβ 0.88 ±0.20 [0.96 ±0.39] ml/g (n = 4; ± = S.D.). Using the oral dose parameters with C_t 0.1 µg/g (safety factor 100), W was estimated as:

$$W_{kidney}: [3.9 \times (\ln(2.75 \times 579.1) - \ln(0.001)]/0.693/24 = 3.4 \text{ days.}$$

To confirm the calculated W, a residue study was performed in 4 ~40 kg pigs fed sulphadiazine for 5 days (60 mg/kg per day) and slaughtered 4 days later. For kidney and muscle respectively, analyzed in triplicate, the ng/g values ranged from 1.9 to 6.1 and from 1.3 to 2.4. Since no residues exceeding the C_t were found, evidently plasma kinetics had correctly established the W value.

References

1. Petz, M. (1983) *Lebensm. Unters. Forsch. 176*, 289-293.
2. Haagsma, N. & Water, C. van de (1985) *J. Chromatog. 333*, 256-261.
3. Mercer, H.D., Baggot, J.D. & Sams, R.A. (1977) *J. Toxicol. Environ. Health 2*, 787-801.
4. Woolley, J.L. & Sigel, C.W. (1982) *J. Vet. Res. 43*, 768-774.
5. Wernimont, G.G. (1985) in *Use of Statistics to Develop and Evaluate Analytical Methods*, A.O.A.C., Arlington, VA, pp. 64-82.

#ncB-2

A Note on

MODIFICATION OF AN ENZYME INHIBITION SCREENING ASSAY FOR BIOANALYSIS OF A PEPTIDE

N.K. Burton and R.J. Simmonds

Upjohn Laboratories - Europe,
Fleming Way, Crawley RH10 2NJ, U.K.

In today's commercial and regulatory environment, assays to support animal studies and human clinical trials are required ever earlier in the development of a new drug. Where the new compound has characteristics that facilitate the development of HPLC or GC methods, this is no great problem. However, for some of the new generation of drugs there is no easy solution to method development problems due, for example, to the need to assay enantiomers or to determine sub-ng amounts.

U-77 is a small, modified peptide (~1000 Da) in the RIP* class. Initial studies required a bioanalytical assay which was -
(a) sensitive: the compound would probably be administered at low doses, and the assay should at least be able to measure 10 ng/ml in serum;
(b) robust and routine, as needed for clinical studies where large numbers of samples would be analyzed, such that the assay had to be robust and repeatable;
(c) GLP-compliant, capable of validation to an acceptable standard of accuracy and variability;
(d) needed immediately, since initial animal studies were about to start.

The following assay methods were considered and evaluated with respect to the time needed for development, selectivity, sensitivity and possible routine application.
HPLC.- Chromatographic characteristics of U-77 were poor in terms of robustness, repeatability and selectivity; sensitivity was poor with UV detection. Fluorescence and electrochemical detection were not possible and, because of modifications made to the peptide to confer metabolic stability, few reactive groups were available for derivatization to enhance selectivity and sensitivity of detection.

Abbreviations.- ANG I, angiotensin I; GLP, Good Laboratory Practice; RIA, radioimmunoassay; RIP, renin-inhibitory peptide.

RIA.- The molecule itself was too small to be immunogenic, and therefore had to be coupled to a carrier protein. Again, being a modified peptide, few groups were available for conjugation.   RIA development schemes need at least 6-9 months, and this time-scale was considered too long.

GC is unsuitable for the analysis of peptides.

Screening assay.- In order to screen a large number of RIP's *in vitro*,   an assay had been developed during the drug discovery process and used by colleagues in Upjohn Laboratories, Kalamazoo.   This assay, although successful in the discovery context, was reported to have a sensitivity of 100 ng/ml; whilst we assumed that this could be improved, there was no information as to whether it was suitable for the routine assay of biological samples, and no formal validation had been carried out.

However, as none of the above options nor TLC appeared viable, at least within the 2-3 months' time frame allowed, the screening assay was investigated to see whether it could be modified to produce acceptable characteristics and used to support large clinical trials, at least in the short term.

The assay was originally developed to determine the inhibition constants of a large number of RIP's.   The problem, therefore, was whether a screening assay could be formalized and validated well enough to become a routine analytical tool.   The basis of the assay was a commercially available kit for the determination of renin activity in serum by assessing the conversion of angiotensinogen to ANG I (Baxter Healthcare Ltd., Compton, Berks.). Scheme 1 shows diagrammatically the basic method.   This was set up in our laboratories and assessed by the addition of various amounts of U-77. Inhibition was indeed obtained, but sensitivity with respect to U-77 was inadequate (100 ng/ml). This could be improved to 20 ng/ml by decreasing the concentrations of renin and substrate (containing angiotensinogen) present in the incubation mixture, and further improved (to 5 ng/ml) by pre-incubation of renin and U-77 before addition of the substrate.   The effect of these procedures on the robustness of the assay was not ascertained.

A problem which had to be overcome was the interference caused by endogenous renin present in biological samples. Methods for extracting U-77 or inactivation of endogenous renin before assay were therefore necessary. Solid-phase extraction and plasma protein precipitation with organic solvents, whilst quite efficient, produced their own interferences in the assay, resulting in high 'blank' readings reducing sensitivity to 10-20 ng/ml.   It was found that the best method for inactivating endogenous renin was by immersing the diluted biological sample in boiling water for 5 min.   Fortunately U-77 was stable during this procedure, and it was incorporated in the assay, formal validation of which was therefore attempted.

Serum and renin and angiotensinogen substrate kept 1 h at 37°
↓
Incubation mixture aliquots added to tubes coated with anti-ANG I;
^{125}I-labelled ANG I added; kept overnight at 37°
↓
Centrifuge, and aspirate supernatant
↓
Concentration of bound ^{125}I assessed by gamma-counting for 60 sec

Scheme 1. Outline of the method.[*] The ratio bound/free ANG I
(B/B_0) is inversely proportional to the U-77 concentration
in the original incubation mixture.

....

METHOD

Calibration standards were prepared in control rat serum
at U-77 concentrations of 1.6, 6.3, 12.5, 25 and 100 ng/ml.
Quality control samples were prepared in bulk at 5, 10 and
20 ng/ml. Standards and QC's were assayed in triplicate.

Rat serum (20 µl) was diluted with 120 µl Milli-Q water
(Millipore, Watford) in polypropylene test tubes and immersed
in a boiling water bath for 5 min. After cooling to room
temperature, 130 µl pH 6.0 buffer (150 mM phosphate, 3 mM
EDTA, 1 mg/ml bovine serum albumin), 20 µl renin (0.2 µg/ml)
and 10 µl phenylmethylsulphonyl fluoride were added. After
incubation for 30 min at 37°, 50 µl of substrate was added
and the mixture further incubated for 1 h at 37°. The tubes
were then put on ice for 10 min, and then 100 µl was transferred
to an anti-ANG I coated tube, whereafter 1 ml of ^{125}I-ANG I
was added. The tubes were kept for 3 h at room temperature,
then the liquid was aspirated from the tubes and the bound
radioactivity measured in a Minigamma counter (Pharmacia-LKB).

RESULTS

Fig. 1 shows a typical standard curve for U-77. When
fitted to a spline function, using RIACalc (Pharmacia-LKB),
the working range of the curve was ~5-20 ng/ml with respect
to U-77. Within-assay and *(italicized) between-assay* QC values
were as follows (n = 36):
- for 5 ng/ml, observed mean = 4.6; C.V. 11.1%, *18.6%;*
- for 10 ng/ml, observed mean = 9.1; C.V. 7.9%, *18.4%;*
- for 20 ng/ml, observed mean = 21.7; C.V. 11.1%, *13.2%.*

Evidently accuracy was reasonable, though precision was
poor throughout the range. This variability was probably
due to the boiling step of the assay, where some evaporation
of sample is inevitable. Alternatively, inactivation of endogenous renin may not be complete and repeatable. In an initial
test of the practicality of the assay, the plasma profile
of a Sprague Dawley rat receiving 16 µg/min U-77 by infusion

[*]Company term for the RIA Kit: 'Gamma Coat [^{125}I] Plasma Renin
Activity'.

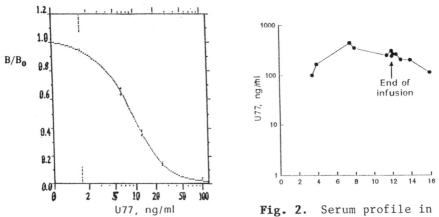

Fig. 1. A typical standard curve.

Fig. 2. Serum profile in a rat receiving U-77 i.v.

was determined (Fig. 2). Drug concentrations found compared favourably with those expected from pharmacokinetic modelling.

Due to discontinuation of the project, the assay was not developed further, nor could sources of variability be investigated. However, at this stage the assay appeared at least adequate for initial animal experiments.

CONCLUSIONS

During the initial development of an assay the problem of assay sensitivity often causes significant delays in the initiation of pharmacokinetic studies. Screening assays developed during drug discovery have the potential for use in initial pharmacokinetic studies while further assay development is undertaken. A screening assay such as the present one, based on enzyme inhibition, could if properly optimized and validated for bioanalysis also become the definitive assay for clinical development.

#ncB-3

A Note on

DETERMINATION OF CLARITHROMYCIN AND A METABOLITE IN
SERUM AND URINE BY RP-HPLC WITH COULOMETRIC DETECTION

[1]K. Borner, [1]H. Hartwig and [2]H. Lode

[1]Institute for Clinical Chemistry   [2]Pneumological Dept. 1,
   and Clinical Biochemistry,   Hospital Zehlendorf-
   Klinikum Steglitz, Hindenburgdamm 30, Heckeshorn, Free
   D-1000 Berlin 45, Germany   University, Berlin 39

Require- *An HPLC assay method for*
ment   *CLM* and its active metabol-*
          ite OH-CLM in biofluids from
          volunteers and patients,
          needing high selectivity
          and adequate sensitivity.

End-step *RP-HPLC (semi-micro); 2-cell*
          EC detection; 30 min run.

Sample   *Serum (+ i.s.) or urine made*
handling *alkaline and solvent-extracted; dried*
          down, and residue dissolved in mobile phase.

Comments *The detector is very specific and sensitive (LOQ*
          ~50 µg/L), but needs meticulous attention, notably to sol-
          vent purity. Use of a column of smaller i.d. than
          standard columns reduces the exposure of the electrode
          surface to matrix material and also gives better
          separation. HPLC and bioassay give concordant values
          for CLM + OH-CLM (OH-CLM acts against various pathogens).

CLM* is a recently developed macrolide antibiotic active
towards a wide range of organisms [1], as is OH-CLM which,
amongst the diverse metabolites [2], is the major one. HPLC
assay of CLM and OH-CLM in human biofluids is now described.

MATERIALS AND METHODS

Reagents and chemicals.- Abbott (Wiesbaden) supplied CLM
(A-56268, lot #91-763-AR; active agent 97.6%) and, used as
i.s. (44 mg/L in acetonitrile), erythromycin-9-O-methyloxime
(A-41039, lot #37-265-AX). E. Merck (Darmstadt) supplied the
reagents and solvents, either 'for HPLC' or 'Suprapur' grade,

_____

*Abbreviations.- CLM, clarithromycin; OH-CLM, (R)-14-hydroxy-
CLM; EC, electrochemical; i.s., internal standard; LOQ, limit
of quantification; RP, reversed phase.

including water except where this was re-purified using a
Milli-Q filtration system (Millipore). For spiking into the
drug-free human serum (Boehringer Mannheim), the CLM and OH-CLM
(from Abbott Labs.) were dissolved in acetonitrile.

Samples were stored at -80° immediately after collection
without any additives; the analytes were stable for at least
6 months.   **Urine** aliquots (0.2 ml) were mixed with 0.1 ml
0.1 M Na_2CO_3 and extracted with 3 ml ethyl acetate/hexane
(1:1 by vol.).   After centrifugation 2 ml supernatant was
taken to dryness in a vacuum concentrator (Fröbel, Lindau);
the residue was re-dissolved in 1 ml mobile phase.   **Serum**
aliquots (0.5 ml), after addition of i.s. (30 µl) and 0.2 ml
Na_2CO_3, were extracted as above, and the final residue dissolved
in 0.5 ml mobile phase.

HPLC (mostly with Perkin Elmer modules) and **detector.-**
The mobile phase comprised 8.2 g anhydrous sodium acetate
in 300 ml re-purified water, 370 ml acetonitrile and 78 ml
methanol, made up to 1 L with water after adjusting to pH 8.8
with ~0.3 ml glacial acetic acid.   After passage through
a guard column and the main column, the eluate fed into
two EC cells in series, each connected to a coulometer.
The EC detector (model 5100A; Environmental Sciences Ass.,
Bedford, MA) had a guard cell (5020; 0.85 V) preceding the
column guard cell, and a high-sensitivity dual carbon cell
(5020).   In the final method the conditions were as in the
legend to Fig. 1, with a Nucleosil packing (Macherey & Nagel,
Düren; guard column 15 × 3 mm).

Microbiological assay (agar plate diffusion) was performed
in triplicate using, as test organism, *Micrococcus Luteus* 8340.

RESULTS

Separations.- Several column packings (Nucleosil 5 C-8,
Lichrosorb 5 RP2) yielded fair separations of the parent drug
from the i.s. with ternary mixtures of aqueous buffer, acetonitrile and methanol. A better separation including the metabolite OH-CLM was achieved with a Nucleosil semi-micro column
as specified in the legend to Fig. 1, which shows separation
examples.   Re-cycling of the 1 L of mobile phase for 2-7 days
was feasible.

Detection (Fig. 2).- The coulometric detector was operated
in the oxidative screen mode.   Potentials over 0.85 V lead
to high background current; the guard cell located just after
the pump proved essential for keeping this low.

Validation.- For calibration curves, see Fig. 3 (which
shows linear ranges). *Italicized* values that follow denote *urine
- CLM [or OH-CLM]*.   Only serum-level plots are ratios, *vs.* i.s.

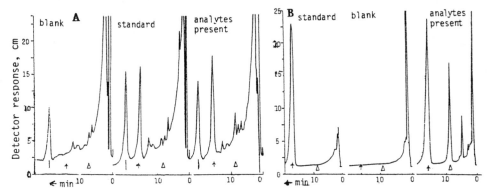

Fig. 1. HPLC of CLM (position denoted ✝) and OH-CLM (Δ):– serum (i.s.at |) and, *top right*, urine, each spiked with CLM, 1.82 & 51.3 mg/L respectively, and OH-CLM, 0.54 & 21.8 mg/L; pure CLM std., 1.5 & 80 mg/L respectively; also (serum only) i.s. Column: Nucleosil 120–3 C–8, 125 × 3 mm; mobile phase: *see text*; run at 30°, 0.7 ml/min; 12 µl injected. Detector (cf. Fig. 2) amplification 100×55; response time 2 or 10 sec.

Fig. 2, *right*. Hydrodynamic voltammogram for CLM; detector twin-cell. Setting +0.50 V for detector #1; oxidation potentials denoted ▲.

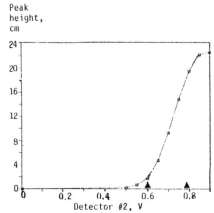

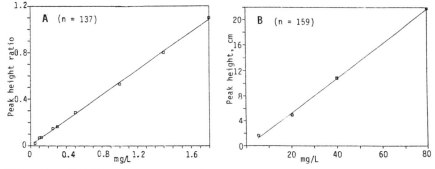

Fig. 3. Calibration lines: A, serum; B, *urine* (r = 0.933, *0.999*).

#Detection limit/LOQ, mg CLM/L: 0.05/0.10, *1.25/1.25*
#C.V.'s, % (with mg/L value).– **Within**-series: 8.0 (0.48), 5.3 (1.46), each n = 12; *4.1 (9.0), 4.0 (39.6) [7.2 (10.8), 4.8 (13.8)], each* n = *10*. **Between**-series: 8.0 (0.40), n = 11; 9.7 (3.26), n = 8; *4.0 (43.8), n = 6; [5.5 (17.0), n = 5].**
#Recovery, % (with mg/L value).– 113.1, C.V. 16.4% (1.0) and 109, C.V. 11.2% (1.0), each n = 10; *105.5 (100) [102.5 (100)], each* n = *1.**
*The *[ ]* values are for OH-CLM, *urine only*.

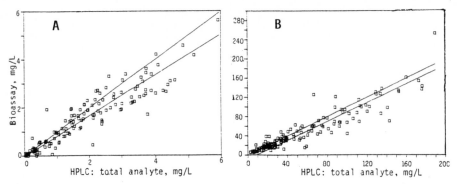

Fig. 4. Method comparisons: bioassay values *vs.* values (CLM + OH–CLM) by HPLC; each *upper* line represents a $y = x$ relationship. **A,** serum (n = 137): the *lower* line represents the equation $c_{bioassay} = 0.844 \times c_{HPLC} - 0.02$ (cf. HPLC for CLM, excluding OH–CLM: $1.096 \times c_{HPLC} + 0.15$). **B,** urine (n = 159): the *lower* line corresponds to $c_{bioassay} = 0.977 \times c_{HPLC} - 8.2$ (cf. HPLC for CLM, excluding OH–CLM: $1.401 \times c_{HPLC} + 1.7$). There was no exclusion of 'outliers'.

As shown in Fig. 4, there was satisfactory accord between bioassay and (for CLM + OH–CLM) HPLC. For serum a detection limit of 0.01 mg/L was achievable with a larger column (Nucleosil 5 C-8, 125 × 4 mm; 50 µl injected), but electrodes became passivated more rapidly and had to be cleaned up more frequently. Pre-dose samples from 20 human volunteers showed no interferences.

DISCUSSION

RP-HPLC with EC detection is the preferred analytical method for most macrolide antibiotics and their metabolites. For CLM, OH–CLM in the sample strongly influences the microbiological assay value (Fig. 4). HPLC of CLM needs sample clean-up and analyte enrichment. Operation of the EC detector requires extremely pure reagents and solvents, high cleanliness in all manipulations and a very dedicated technician. The present method permits CLM and its main metabolite to be determined in serum and urine up to 12 h after administration of a single oral dose of 250 mg. Thus far, the method has been successfully applied to several studies on pharmacokinetics in human volunteers.

References

1. Fernandes, P.B., Bailer, R., Swanson, R., Hanson, C.W., McDonald, E., Ramer, N., Hardy, D., Shipkowitz, N., Bower, R.R. & Gade, E. (1986) *Antimicrob. Agents Chemother. 30*, 865-873.
2. **Ferr**ero, J.L., Kopp, B.A., Marsh, K.C., Quigley, S.C., Johnson, D.J., Anderson, D.J., Lamm, J.E., Tolman, K.G., Sanders, S.W., Cavenaugh, J.H. & Sonders, R.C. (1990) *Drug Metab. Dispos. 18*, 441-446.

#ncB-4

A Note on

DETERMINATION OF VALPROIC ACID IN PLASMA: ADAPTATION OF A PACKED-COLUMN GC METHOD TO CAPILLARY COLUMNS

[1]H.M. Hill, [2]D. Lessard and [3]B.K. Martin

[1]Hazleton UK, Otley Road, Harrogate, N. Yorks, HG3 1PY, U.K.

[2]Phoenix International Life Sciences, St. Laurent (Montreal), Quebec, Canada H4R 9Z7

[3]Bios Consultancy and Contract Research, Bagshot GU19 5ER, U.K.

The assay of valproic acid does not present any problems in terms of sensitivity, since the desired lower limit of quantification (LLOQ), 1 µg/ml, is easily attainable by GC. Its volatility, however, necessitates procedures to minimize loss of both the analyte and the internal standard. The methods we have developed entail use of the DuPont Prep System with solid-phase extraction (SPE; Type W cartridges) and of GC-FID, with packed columns in the first instance. Because high sensitivity was not required, a limitation to use of SPE in sample preparation for GC did not apply; SPE with cartridges is applicable in GC methods (with NPD, ECD or FID) only for sensitivities down to 0.1 µg/ml, due to the presence of impurities that become extracted from the plastic cartridge.

The extraction procedure for packed-column GC is shown on the *right* of Scheme 1. After loading under acidic conditions and washing with deionized water, the eluate obtained with methanolic KOH was dried down; then the solution was kept in a tightly sealed vial from which an aliquot was taken for GC. The GC stage was as shown on the *right* of Table 1.

RENDERING THE METHOD SUITABLE FOR CAPILLARY-GC

With conversion of conventional GC systems to capillary systems, attempts to run the original assay method by capillary GC encountered chromatographic problems. The cause was the high concentration of salts present in the extract, which affected both the injection system and the column.

The problem was overcome by using a modified extraction procedure, as shown on the *left* of Scheme 1. The drug was eluted from the cartridge directly into a receptacle containing

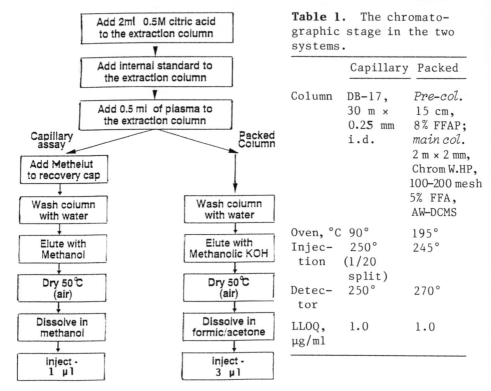

Scheme 1. Extraction procedures.

Table 1. The chromatographic stage in the two systems.

	Capillary	Packed
Column	DB-17, 30 m × 0.25 mm i.d.	*Pre-col.* 15 cm, 8% FFAP; *main col.* 2 m × 2 mm, Chrom W.HP, 100–200 mesh 5% FFA, AW-DCMS
Oven, °C	90°	195°
Injection	250° (1/20 split)	245°
Detector	250°	270°
LLOQ, µg/ml	1.0	1.0

'Methelut', whereby the analyte was derivatized. Thereby loss through volatilization was minimized and, moreover, the chromatographic characteristics were improved.

COMMENTS

Whatever the GC methodology, with valproic acid it was essential to minimize loss of a very volatile analyte. This was achieved for the packed-column method by formation of the potassium salt; but injection of excess salts as an inherent consequence impaired chromatographic performance in the capillary system. This was remedied by use of a modified extraction procedure entailing derivatization. The valproic acid assay illustrates that problems may arise in method transfer to different instrumentation.

    One of the significant advantages of the DuPont Prep System is the constant flow rate it applies to each cartridge. This is difficult to ensure in a batch basis using standard vacuum manifold systems. Since the Prep System is no longer widely available, care must be taken to ensure methodological reproducibility when using manifold systems, with inclusion of additional QC/standard samples for each 'manifold batch'.

Comments on #**B**, P.F. Carey – SURVEY ('Preface')
              #**B-1**, P.H. Degen – ASPECTS OF METHOD DEVELOPMENT
              #**B-2**, B. Law – HPLC OF BASIC DRUGS

Carey agreed with **remarks by H. de Bree** and by **B. Law** that method development often suffers because of deficient chemical knowledge amongst laboratory staff, whose background is often biological. **Sue Rees remarked** that the ever-increasing potency of drugs calls for more sensitive and specific assays even in 'discovery research': some derivatization may prove necessary, but should be regarded as a last resort.

P.H. Degen, answering S. Westwood.- For levoprotiline glucu- ronide, high-temperature acid hydrolysis can't be avoided in view of resistance to β-glucuronidase.

I.D. Wilson asked B. Law about the merit of the PC-based system for doing molecular orbital calculations etc. as compared with the 'VAX'. **Reply.**- As yet it can't do all the things that I would like to do, but it does look to be a promising approach.

S. Westwood asked Law whether his study of strongly basic analytes had been extended to columns not based on silica. **Reply:** some have been evaluated but have not proven useful.

Comments on #**B-4**, C.A. James – HPLC OPTIMIZATION BY SOFTWARE
              #**B-6**, J.D. Robinson – ANTIBODY USEFULNESS
              #**B-7**, S.A. Wood – ANALYTE STABILITY

James, answering D. Stevenson.- In future, although not at present, expert systems may be able to call on experience such as a bioanalyst can bring to bear on developing a method (example: HPLC conditions for a related compound). **S. Westwood remarked** that it would seem best to develop a separation using the actual biofluid for which the method is required.- Can this be done using peak tracking by peak areas? **Reply (to P. Heizmann also).**- Such approaches have proven unsuccessful, especially for small peaks where the area is not reproducible; use of 'information-rich' detection might help. Tracking small peaks (2-5 mm) of endogenous origin is difficult. Information- rich detectors such as diode array usually have insufficient sensitivity to detect these peaks and give spectral information. If peaks can be detected by two different high-sensitivity detectors, such as UV + fluorescence or fluorescence + ECD, then area ratios might help. *(See also foot of* p. 146.*)*

Robinson, answering D. Chapman.- In eluting the compound of interest from immunoaffinity columns, using methanol, no problems have been encountered.

[**To Wood:**] **R. Whelpton.**- Use of stabilizers calls for care: e.g. ascorbic acid will reduce amine oxides. Catechols may be stable in acid medium, so anti-oxidants may not be needed if the pH is right. **Response (from N.K. Burton,** of the Upjohn group).- Our 'U83' actually oxidized to the quinone in acid conditions and dropped out of solution in an alkaline environment. So the remedy does not lie in changing the pH; the position is more complicated.

Comments on #**B-9**, H.M. Hill - CALIBRATION FUNCTIONS
             #**B-10**, P. Heizmann - ANTIBIOTICS IN TISSUES
             #**ncB-3**, K. Borner - CLARITHROMYCIN ASSAY

Comment to Hill by B. Law.- The 'quality factor' which you cite, g, is merely a normalized RSM deviation.

Remarks to Heizmann by D. Perrett.- Additional factors in performing tissue analysis need consideration.- (1) Sample collection: metabolism may have to be stopped by freeze-clamping. (2) How results should be expressed: wet wt., dry wt., intra-cellular water, or per unit of protein or DNA. (3) Hb estimation to correct for blood in tissues. (4) Need for protein estimation (e.g. Lowry method) to check efficiency of protein removal.

P. Degen, to Borner.- For external quality assurance to ensure 'portability' of methods between laboratories, I suggest provision of spiked material on which the originating laboratory does a primary assay.

Citations contributed by Senior Editor
- *most centered on techniques (as in Sect.* #**D**) *rather than drugs*
- J Chr *denotes* Journal of Chromatography

HPLC packings.- (1) Silica, adsorption properties/ remedies: Nawrocki, J. (1991) Chromatographia **31**, 177-192 & 193-205 [cf. Law's art., # B-2]. (2) Basic properties of silica gel and carbon packings: Berek, D. & Novak, I. (1990) *same* **30**, 582-590. (3) Evaluation of CN (aminopropyl) packings for very polar bioactive analytes (of environmental interest): Lafleur, A.L. et al. (1988) J. Chromatog. Sci. **26**, 337-344.

Obviating protein removal.- (1) **ISRP columns** (examples: serum anticonvulsants): Pinkerton, T.C. (1988) Am. Lab. **20**, 70-76. [Pioneer paper from same lab. (1985): Anal. Chem. **57**, 1757-1763. Application to oligonucleotides in culture media: Pompon, A. et al. (1992) Biochem. Pharmacol. **43**, 1769-1775.] (2) **Triethylammoniun acetate** in NP mobile phase (examples: serum β-blockers/analgesics; pH 4): Bui, K.H. & French, S.B. (1989) J. Liq. Chromatog. **12**, 861-873.
On-line protein removal.- (1) Electrodialysis ('EDIST'): Debets, A.J.J. et al. (1990) Chromatographia **30**, 361-366.

(2) **Protein—coated pre—column** (RP-HPLC) with switching: Morita, I. & Yoshida, H. (1990) J Chr **527**, 127-133.

Protein—removal stratagems (plasma theophylline as an example).- (1) **'Syringe'** mini-columns containing Celite 545 (eluate HPLC-loaded): Homma, M. et al. (1989) Anal. Chem. **61**, 784-787. (2) Zn^{2+} + **solvent**, micro-scale: Lam, S. & Malikin, G. (1989) J. Liq. Chrom. **12**, 1851-1872.

Detection/difficult chromatograms.- (1) **EC**, especially voltammetric: Buchberger, W. (1990) Chromatographia **30**, 577-581. (2) Peak **overlaps**: Crilly, P.B. (1990) Chemometrics **4**, 51-59. (3) **'Indirect'** detection where mobile phase has a responsive component (RP-HPLC; principles): Schill, G. & Arvidsson, E. (1989) J Chr **492**, 299-318. (4) Mininizing **plasma—blank interferences** (C-18 & CN columns, 254 nm) by 'modifier optimization': Kelly, M.T. et al. (1990) J Chr **528**, 277-292. (5) **'Analytical Artifacts'.**- Middleditch, B.S. (1989), Elsevier, 1033 pp.: 'encyclopoedia' of contaminants and other **interference sources** in GC, HPLC, MS & TLC.

Normal—phase HPLC.- (1) [cf. Law's art., #B-2] A range of **basic drugs**, with initial on-line SPE: Kelly, M.T. et al. (1989) Analyst **114**, 1377-1380. (2) **LC—MS with DLI** (direct liquid injection), exemplified by derivatized **retinoids** [cf. Wyss, #B-5]: Garland, W.A. et al. (1991) Trends Anal. Chem. **10**, 177-184.

TLC densitometry including FID and radioscanners (review): Poole, C.F. & Poole, S.K. (1989) J Chr **492**, 539-584.

HPLC optimization.- (1) Column length, eluent composition & flow rate, wavelength, i.s. amount: Hayashi, Y. et al. (1990) Chromatographia **30**, 85-90. (2) **'ECAT'** (Expert Chromatographic Assistance Team); combination of expert-system approach and conventional programming reckoned to be generally best [cf. James, #B-4]: Williams, S.S. (1990) Trends Anal. Chem. **9**, 63-65. (3) **'ESCA'** (Expert Systems in Chemical Analysis), design and evaluation, and aspects of **method validation**: Buydens, L.M.C. et al. (1990) Trends Anal. Chem. **9**, 58-62. (4) The **'ruggedness expert system'** (RES): Van Leeuwen, J.A. et al. (1991) Chemom. & Intell. Lab. Systems **11**, 37-75.

CALIBRATION AND ASSAY VALIDATION: an in-depth publication - Cowens, J.W., Papp, G., Parsons, J. & Greco, W.R. (1990) J. Liq. Chromatog. **13**, 887-911.

The RP-HPLC assay considered (detection at 287 nm) was for determining, in biofluids, an anti-cancer agent — a thiazolidinedione derivative with two diastereomeric forms. Statistical procedures were developed for constructing a linear calibration curve (over a 1000-fold range), for estimating unknowns, for validating the assay and for generating QC charts.

[Continued overleaf

Comments on the foregoing article - *an appreciated response by* **H.M. Hill** *(cf. art. #B-9) to Editor's "Please comment" request*

[The article's] major limitation is that it does not provide practical guidance on choice of calibration concentrations to define the curve. Indeed with only 5 points for 3 orders of magnitude I feel justification may be necessary. Thus for one analyte there were only the following calibration standards:- 0.1, 0.25, 1.0, 10 and 100 ng/ml. The authors aimed to avoid having to produce 'split plots' based on 2 calibration curves, one for the low concentration range (0.1-10) and a second for the 10-100 range. This was accomplished by using a log-transformed calibration function with empirical weighting (weight reciprocal of the variance estimate derived during validation).

Significantly, they stressed the importance of validating the calibration function as well as the method itself. The appropriateness of each calibration curve from each assay batch to be used for evaluating unknown data was based on two major parameters used as QC criteria:
- error mean square of the curve;
- g: precision is poor when the magnitude of the slope is small and/or if the slope is poorly defined. [Ref. to g: Draper, N. & Smith, H. (1981) *Applied Regression Analysis*, 2nd edn., Wiley, New York.]

It is possible for the error mean square to be acceptable while the other parameters vary significantly, and *vice versa*. While plotting of trends in these variables may provide clues to problems associated with the assay method, the authors did not provide definitive prospective guidelines on criteria for acceptability of calibration functions. Waiting until the end of a study in order to evaluate overall data quality is commendable, but it may delay any corrective action.

Weighting *[cf. #B-8 & #B 9]: earlier book material of relevance*

The question of dependence of variability on concentration and the usefulness of weighting approaches has been considered in past volumes (ed. E. Reid......), notably in Sect. #A of Vol. 20, particularly pp. 15-18 (D. Dell), 21 (Questionnaire), 28 (H.M. Hill), 59 (R.J.N. Tanner), 83 and 88. On p. 86 there is a précis of a pertinent art. by E. Doyle [pp. 534-540 in: J.W. Gorrod *et al.*, eds. (1986), *Development of Drugs and Modern Medicines*, Horwood/VCH, Chichester]. Dell's art. cites a pioneer art. (1981) by J.P. Leppard in Vol. 10 (Horwood; out-of-print) [for Corrigenda see Vol. 12 (Plenum) or consult E. Reid]. For **outliers** see p. 87 in Vol. 20.

HPLC prediction *[cf. #B-4; Forum discussion - W. Pacha/P. Heizmann]:* Experiences with the 'Drylab' (I & G) system could be informative. Unpredictable biomatrix peaks are a notable problem.

#ABC

COMPENDIUM by E. Reid (Guildford)

RECENT PAPERS ON DRUG ASSAY OR METABOLITE DELINEATION
mostly using HPLC or GC; listed in therapeutic groups

1. ANTI-INFLAMMATORIES/ANALGESICS/EICOSANOID-RELATED AGENTS
2. ANTI-ASTHMATICS/BRONCHODILATORS/ANTI-ALLERGICS
3. (ANTI)ADRENERGICS/Ca^{2+} CHANNEL BLOCKERS, etc. [overlaps **2, 4** & **5**]
4. CARDIOVASCULAR AGENTS
5. NEUROACTIVE AGENTS incl.(ANTI)CHOLINERGICS
6. ANTI-INFECTIVE/ANTI-CANCER AGENTS
7. UNCLASSIFIED AGENTS, e.g. anti-diabetics, relaxants
Amplified titles, with book-series cross-refs., appear in the 'text'

This compendium (1988-early 1992) may help if a procedure for a particular drug has to be set up. UP-DATING SEARCH:- for a paper now listed, check (by first author) in *Citation Index* for newer papers that cite it; look up the drug in post-**574** vols. of *J. Chromatog.* (look for Cumulative Indices).

Method descriptions (maybe including RIA) besides those now listed can be located in the present series [list on final p.] by Analyte Index consultation (Cumulative version herein and in Vol. 12). A few 'therapeutic type' entries (e.g. 'Cytotoxic agents') appear in each General Index, which also lists non-chromatographic approaches (e.g. 'RRA', 'ELISA').

The present compilation was made without 'quality control'. Some papers have deficient presentation, e.g. unexplained use of NP rather than RP columns. There are clues in some entries to meritorious features, e.g. procedural comparisons (* *alerts*) or to an 'assay slant' - e.g. automation of an existing method, trial of ISRP or other novel columns, or design for 'ther(apeutic) monitoring' where high sensitivity may not be paramount.

The term **chiral** implies distinction between **enantiomers.**

Metabolite identification/profiling (not merely assay; may include species comparisons) is implied by the term **invstg.** (investigation) coupled with **mtbs** (metabolites). The focus is on the analytical methodology, as distinct from delineation of metabolic pathways.

Metabolite type(s) that feature in particular papers are broadly indicated by a 'superscript alerting system' ([1] [1] [2]; SEE OVERLEAF) as also used in each Analyte Index. The alerting to conjugates ([2]) may be omitted if a hydrolysis step (hydrol.) was included in the assay. For general guidance, consult entries in each General Index:- 'Conjugates' (see particularly pp. 267-268 in Vol. 12), also 'Metabolites'. *See also final p. of the Compendium.*

KEY TO ABBREVIATIONS

Prefix p = a prodrug
mtb(s) = metabolite(s) looked at:
invstg. = investigation, not
merely assay; *superscripts:*
Phase I, [1] (**bold** [1] if including
N–des......; oxid. = some oxidative metabolism); Phase II, [2]:–
conjugates, maybe N/A if cleaved
(hydrol.) by acid or enzyme
SAMPLE: human plasma or serum
unless, e.g., (whole) blood stated.

• Automation/robotics *denoted* auto;
  swtch = switching
• deriv = derivative prepared

MATERIALS (water disregarded; grad. =
gradient) include: acet., acetate;
MeCN, acetonitrile; MeOH, methanol;
phos., phosph$^{ate}_{oric}$; TBA, Bu$_4$ammonium;
TEA, Et$_3$N; THF, tetrahydrofuran
•Precipitation (protein): pptn.
•Extraction: LE, liquid–liquid; SPE,
  solid–phase; →dry = dried down
•GC detection: ECD = electron capture
• HPLC col. types: C-18, CN, AGP (α_1-acid
  glycoprotein), ion–exch(ange), ISRP, etc.
•DETECTION: nm value implies UV;
  fluor = fluorescence; EC = electrochemical; radio = radioactivity
•sens. = sensitivity (LOD or LOQ)

References: ('91) = 1991, etc.; AnMcAgCh = *Antimicrob. Agents Chemother.*;
Chr'ia = *Chromatographia*; Dr Rs = *Arzneim. Forsch./Drug Res.*; J Chr = *J.
Chromatog.*; J Lq Chr = *J. Liq. Chromatog.*; J Ph Bm An = *J. Pharm. Biomed.
Anal.*; J Ph Sc = *J. Pharm. Sci.* MS, *Mass Spectrom.* |*means especially
worth scrutiny

#1. ANTI–INFLAMMATORIES and relevant ANALGESICS [others in #7];
[at end of #1] EICOSANOID–RELATED AGENTS
Previous coverage: Vol. 18 (compendium: p. 218)

5–Aminosalicylic acid & [1]**mtbs**, some new (urine & faeces also):- for
plasma, pptn. (MeCN; LE to remove); NP col. (with pre-injector 'saturation'
col.), 34°; MeOH/phos./hexadecyl-triMe-ammon. Br⁻ pH 6.5; fluor; sens.
0.002-0.5 ng/ml. Agasøster, T.J.B. et al. ('91) J Chr **570**, 99-107.

Antipyrine & oxid. [1]**mtbs** (urine), as MODEL system for assessing monooxygen-
ase status:- LE (pH 3 & then, for 3-carboxy **mtb**, 0.9); C-8 col.; MeCN/acet./n-
decylamine pH 3.8; 242 nm. Palette, C. et al. ('91) J Chr **563**, 103-113.

Aurothiolates, incl. protein-bound, & [1]**mtbs**;(speciation studied):- FPLC,
cross-linked agarose col.; 0.2 M NH$_4$HCO$_3$; 254 nm for protein, atomic absorption for Au. Rayner, M.H. et al. ('89) Int. J. Clin. Pharmacol. **9**, 377-383.

Brifentanil (rat):- SPE; C-18; MeCN/phos. pH 4.0; 226 nm & radio; sens.
0.1 ng/ml. Kvalo, L.T. et al. ('91) J Chr **565**, 391-399.

Diclofenac:- pptn. (MeCN), →dry; C-8 col.; aq. MeCN pH 3.3; 280 nm;
sens. 25 ng/ml. El-Sayed, Y.M. et al. ('88) J. Pharm. Pharmacol. **40**, 727-729.

Diclofenac & oxid. [1]**mtbs:**- (1) (urine also):- LE, →dry; C-18, 30°; MeCN/MeOH
/THF/phos.; 282 nm; sens. 50 ng/ml. Lansdorp, D. et al. ('90) J Chr **528**, 487-494.
(**2**) SPE; C-8 col.; MeCN/acet. pH 3; EC; sens. <1 ng/ml. Zecca, L. et al.
('91) J Chr **567**, 425-432.

* **Diclofenac.**- (1): pptn. (MeCN); C-18; phos./MeCN pH 6.6; deriv (photo),
then fluor; sens. 6 ng/ml. Wiese, B. et al. ('91) J Chr **567**, 175-183.
(**2**; auto): SPE; C-18; MeOH/phos. pH 7.2; 282 nm; sens. 10 ng/ml. Sioufi, A.
et al. ('91) J Chr **565**, 401-407. (**3**; drug applied to skin)(urine also): cap.
GC-MS; sens. ~1 ng/ml. Sioufi, A. et al. ('91) J Chr **571**, 87-100.

Diflunisal & [2]**mtbs**, some prone to rearrange (usually no hydrol.; urine):-
C-18; MeOH/phos. pH 2.5, grad.; fluor. Loewen, G.R. et al. ('89) J Ph Sc
78, 250-255.

Diflunisal/Indomethacin/Naproxen/Piroxicam/Sulindac (NSAID's):- SPE,→dry; C-18; MeOH/phos. pH 3; EC; sens. <20 ng/ml. Kazemifard, A.G. et al. ('90) J Chr **533**, 125-132.

Dimethindene & **¹mtb** (urine):- LE, →dry; AGP **chiral** col., 30°; phos./MeCN/PrOH; *254 nm; sens.<3 ng/ml. Vessels deactivated. Radler, S. et al. ('91) J Chr **567**, 229-239.

Dipyrone & **¹mtbs** (incl. urine):- pptn. (PCA); LE, →dry; C-18; MeCN/MeOH /buffer pH 2.8; 265 nm; sens. 0.1 μg/ml. Damm, D. ('89) Dr Rs **39**, 1415-1417.

Doxofylline (**Doxophylline**):- SPE; C-18; phos./MeOH/MeCN pH 3.0; UV det'n; sens. 30 ng/ml. Lagana, A. et al. ('90) Biomed. Chr. **4**, 205-208.

Etodolac:- LE; C-18; MeCN/phos.; fluor. Dey, M. et al. ('89) Int. J. Pharm. **49**, 121-128 [≈ UV method of Cosyns et al. ('83) J PhSc **72**, 275-277].

Etodolac (possibly hydrol.; urine also):- LE, →dry; **chiral** deriv, →dry; NP col.; hexane/Et acet./isoPrOH; 280 nm; <0.2 μg/ml. Jamali, F. et al. ('88) J PhSc **77**, 963-966.

Eugenol (incl. bile & urine):- LE, →dry; C-18; phos./MeCN/MeOH pH 2; 220 nm; sens. 10 ng/ml. Fischer, I.U. et al. ('90) J Chr **525**, 369-377.

Feprazone:- LE, →dry; GC-FID. Berry, D.J. et al. ('88) J Ph BmAn **6**, 493-501.

Flurbiprofen:- LE, →dry; **chiral** (AGP) col.; iPrOH/dimethyloctylamine/phos. pH 6.5; 246 nm; sens. 50 ng enantiomer/ml. Geisslinger, G. et al. ('92) J Chr **572**, 163-167.

Ibuprofen **¹mtbs** (urine):- LE, →dry; **chiral** deriv; C-8 col.; MeOH/BuOH; 232 nm; sens. 6 μg/ml. Rudy, A. et al. ('90) J Chr **528**, 395-405.

Ibuprofen.- (1): LE; **chiral** (AGP) col.; PrOH/Me₂octylamine/phos.; 227 nm; sens. 0.2 μg/ml. Pettersson, K-J. et al. ('91) J Chr **563**, 414-418. (2): pptn. (MeCN; ammon. sulphate to salt out); C-18; MeCN/phos. acid (pH 2.2); 220 nm; sens. 25 ng/ml. Rustum, A.M. ('91) J. Chr. Sci. **29**, 16-20.

* **Ibuprofen** or **Mefenamic acid** [MODELS for ion-pair HPLC of acidic drugs] (auto, swtch):- LE, →dry; C-18 *cols.*, #1→#2 → #3; [TBA Br⁻, *cols. 1 & 2*]phos./MeCN pH 3.5-7, depending on analyte & col.; 220 nm; sens. <0.5 ng/ml. Yamashita, K. et al. ('91) J Chr **570**, 329-338.

Indomethacin: see **Diflunisal**.   **Iloprost, Ketorolac:** see end of #1.

* **Ketoprofen** (rat):- direct onto protein-coated CN-Pr pre-col.; **chiral** col. (ovomucoid; swtch); MeOH, pH 3.2; 265 nm. Tamai, G. et al. ('91) Anal. Sci. **7**, 29-32.

Ketoprofen.- (1; **Probenecid** also) (rat; urine also), & (2): LE, →dry; **chiral** deriv, →dry. (1) C-18, MeCN/phos./TEA; (2) NP col., isoPrOH/heptane, 275 nm. Sens.: 0.5 (1) or <0.1 (2) μg enantiomer/ml. (1) Palylyk E. et al., (2) Hayball, P.J. et al.: ('91) J Chr (1) **568**, 187-196; (2) **570**, 446-452.

Loxtidine:- LE, →dry; cap. GC-ECD; sens. 2.5 ng/ml. Martinez, G. et al. ('90) J Chr **533**, 235-240.

Mefenamic acid: see **Ibuprofen**.

Naproxen:- pptn. (MeCN); C-18; MeCN/phos. pH 2.5; fluor; sens. 0.1 μg/ml. Wanwimolruk, S. ('90) J Lq Chr **13**, 1611-1625. See also **Diflunisal**.

Niflumic acid (urine also):- pptn. (MeOH); C-18; MeCN/acet. pH 6.4; 279 nm; sens. 0.1 μg/ml. Avgerinos, A. & Malamataris, S. ('90) J Chr **533**, 271-274.

Pentazocine:- LE; C-18; MeCN/phos. acid; fluor; sens. 4 ng/ml. Moeller, N. et al. ('90) J Chr **530**, 200-205.

Piroxicam [& see **Diflunisal**], auto:- SPE, swtch; C-18; phos./MeCN/TEA pH 3.0; 331 nm; sens. 0.2 μg/ml (20 μl sample). Saeed, K. et al. ('91) J Chr **567**, 185-193.

Pirprofen & **¹mtbs** (incl. urine, hydrol.):- SPE (NP for urine); C-18; MeCN/phos. pH 5.5, grad.; nm varied; sens. 0.2 μg/ml. Sioufi, A. et al. ('89) J Chr **495**, 195-203.

Pranprofen:- LE, →dry; C-18; MeCN/acetic acid; 254 nm; sens. 50 ng/ml. Kajiyama, H. et al. ('91) Int. J. Clin. Pharmacol. Res. **11**, 123-127.

Sulindac: see **Diflunisal.**

Tebufelone (rat, incl. paw tissue):- LE, ⇥dry; GC-MS(EI); sens. 2 ng/ml. Eichhold, T.H. & Doyle, M.J. ('90) Biomed. Environ. MS **19**, 230-234.

Tripelennamine and several **other basic drugs** (spiked into plasma) as *MODEL cpds. for NP/eluent optimization:- CN pellicular or C-18 col. (to concentrate), then (swtch) NP; MeOH/ammon. nitrate pH 8; 254 nm. Kelly, M.T. et al. ('89) Analyst **114**, 1377-1380.

Ximoprofen & ¹**mtbs** (urine):- LE, ⇥dry; C-18; MeCN/phos. pH 7.4; 220 nm; sens. 1 µg/ml. Morris, G.R. et al. ('90) JChr **530**, 377-385.

Continuation of #1: EICOSANOID-RELATED AGENTS
See p.225 for a metabolic scheme; LT, leukotriene; TX, thromboxane

LT's (incl. bronchial fluids from asthmatics):- deriv (giving dinitroben-zoates, EC-responsive); C-18; MeOH/HFBA/TEA pH 3.0; EC; sens. 8 pg onto col. Steffenrud, S. & Salari, H. ('89) Biomed. Chr. **3**, 5-9.

LTD₄ antagonists.- (1; MK 0571): pptn.; C-18; 282 nm: Berna, R.A. et al. ('91) JChr **563**, 458-465. (2): pptn. (MeOH); *either* C-8 at 50°, THF/PrOH/phos. acid, *or* C-18 at 50°, MeOH/phos. pH 7.2; UV *and* fluor; sens. 0.5 µg/ml. Bedard, P.R. & Cotton, M.L. ('89) JLqChr **12**, 307-322. (3): SPE, ⇥dry; chiral deriv, ⇥dry [**chiral** amide-formation requisite even with:] **chiral** col. (phenylurea type showed batch diff.); TEA/pentane/iPrOH/MeCN; fluor; sens. 0.05 µg enantiomer/ml. Robinett, R.S.R. et al. ('91) JChr **570**, 157-165.

Lipoxygenase inhibitors.- (1) (E & Z isomers; some stereoselectivity):-pptn. (MeCN); C-18; MeCN/Bu₄ammon. OH⁻/acet./MeOH pH 6.9; EC; sens. 10 ng/ml. Matuszewski, B.K. et al. ('88) JPhSc **77**, 880-884. (2): LE, ⇥dry; C-18; MeOH/acet. pH 4.5; EC; sens. 1 ng/ml: Kuwahara, S.K. et al. ('90) JChr **534**, 260-266.

TXB₂.-(1): SPE, ⇥dry; deriv, ⇥dry; GC-MS(NCI); sens. 3 pg/ml. Herold, D.A. et al. ('89) JChr **496**, 180-188 [similar method for **PGE₂**]. (2), incl. *keto-PGF₁α & nor derivs.; rat also): immunoaffinity extraction; deriv, ⇥dry; GC-MS. Chiabrando, C. et al. ('89) JChr **495**, 1-11.

11-Dehydrothromboxane B₂.- (1): Assay studies (incubates & urine; for other eicosanoids also) and (of general use) simple **preparation of stable-isotope stds.:** Leis, H.J., et al., e.g. ('90) JChr **526**, 169-173, also papers in Biomed. Environ. MS. (2, urine): immunoaffinity extraction [RIA-relevant Ab to acyclic product from the lactone; vague]; GC-MS; sens. 25 ng/ml: Watanabe, K. et al. ('89) JChr **468**, 383-394 [int'l std.: Ohyama, Y. et al. ('89) Chem. Pharm. Bull. **37**, 3171-3173]. (3), pSK&F 105809, an imidazole with a >S=O group which becomes >S in the active ¹mtb SK&F 105561 (**cyclooxygenase** also inhibited; rat/mouse):- SPE, ⇥dry; C-18; MeCN/ammon. acet., grad.; 270 nm; sens. 0.2 or (**mtb**) 0.05 µg/ml. Marshall, P.J. et al. ('91) Biochem. Pharmacol. **42**, 813-824 (cf. 825-831).

A **TXA₂synthetase inhibitor** (imidazole):- Iwata, T. et al. ('90) JPhSc **79**, 295-300. [**TXA₂** *background:* Cross, P.E. & Dickinson, R.P. ('91) Chem. Br. **27**, 911.]

Consult # the introduction; #p.148 *for Journal-title & other* | * means scrutiny
ABBREVIATIONS. *They include:-* | warranted

p = prodrug
mtb(s) = metabolite(s): **invstg.**
  if investigated; Phase I,
  ¹, or ᴵ if incl. *N*-des (oxid. =
  oxidative metabolism); Phase
  II, ² (hydrol. = hydrolyzed)

auto = automated system
deriv = derivatization
LE = liq.-liq. extraction
⇥dry = drying-down
pptn. = precipitation
swtch = switching

MeCN = acetonitrile
HPLC cols.: e.g. C-18;
nm value = UV detection
sens. = sensitivity
SAMPLE usually
human plasma/serum

* **Ciprostene** (dog, rabbit):- affinity SPE (immobilized Ab; preparation described), →dry; deriv; GC-MS; sens. 50 pg/ml. Komatsu, S. et al. ('91) JChr
* **568**, 460-466. Another **prostacyclin**-like agent: **Iloprost**:- affinity-SPE (Ab); GC-MS; sens. <0.1 ng/ml. Krause, W. et al. ('85) Prostaglan. Leukotr. Med. **17**, 167.- Cited by Hildebrand, M. et al. ('90) Int. J. Clin. Pharmacol. Res. **10**, 285-292.

   Enprostil Acid (**PG** analogue):- SPE (CN); deriv; col. *sequence:* CN, C-18
* & C-18; MeOH/acetic acid; **laser fluor**; sens. 10 pg/ml. Kiang, C-H. et al. ('91) JChr **567**, 195-212.

   Ketorolac tromethamine (**cyclooxygenase** inhibitor[†]) & oxid. active [1]**mtb** (urine also):- LE, →dry; C-18; MeOH/phos./TBA-phos. pH 6;   312 nm; sens. <50 or (**mtb**) <10 ng/ml. Wu, A.T. et al. ('90) JChr **534**, 241-246.

   Ketorolac trometamol (rat; hydrol.):- LE; **chiral** deriv; C-18; MeCN/phos./TEA; 280 nm;   sens. <0.1 µg/ml. Jamali, F. et al. ('89) JLqChr **12**, 1835-1850.
                                                        [†likewise **Terbufelone**, OPPOSITE.

#2. ANTI-ASTHMATICS/BRONCHODILATORS [some in #3 or, if eicosanoid-related, in #1]; ANTI-ALLERGICS/relevant ANTIHISTAMINICS [some H₁-receptor antagonists, e.g. anti-ulcer, are in #7]
                  [#2 complements Sect. #C of the present book]

   p**Bambuterol** & active [1]**mtb, Terbutaline** (urine also):- SPE; deriv; GC-MS(CI); sens. 0.25 or (urine) 2 ng/ml. Lindberg, C. et al. ('90) Biomed. Environ. MS **19**, 218-224. *Amplification* (**chiral**; *other refs.) in art.* #C-3.

   Cetirizine (urine):- LE; C-18; MeOH/THF/ TBA-phos.; 230 nm; sens. 20 ng/ml. Accord with earlier cap.-GC assay. Rosseel, M.T. et al. ('91) JChr **565**, 504-510.

   Chlorpheniramine & [1]**mtbs, invstg.** - by TLC, MS & GC-MS (urine; rat also):- LE, →dry; ASSAY: C-8 col.; MeCN/(e.g., depending on the **mtb**) TBA-OH; sens. 265 nm. Kasuya, F. et al.('91) Xenobiotica **21**, 97-109.

* **Chlorprenaline**:- LE, →dry; **chiral** (ovomucoid) polymer col. [better than silica]; EtOH/phos.; UV (vague); sens. 25 ng enantiomer/ml. Miwa, T. et al. ('91) JChr **566**, 163-171.

   Clemastine: see Tripelennamine in Group #1.

   Clenbuterol (urine).-   (1): LE; deriv to cyclic boronate; cap. GC-MS; sens. 0.5 ng/ml. Polettini, A. et al. ('91) JChr **564**, 529-535. (2) (plasma & urine):- LE, →dry; deriv; cap. GC-MS(neg.-ion CI); sens. 5 pg/ml. Girault, J. et al. ('90) Biomed. Environ. MS **19**, 80-88.

   Cromoglycate (rabbit, after drug inhalation):- ultrafilter; C-18; phos. /MeOH/TBA-hydrox. pH 7.4; 238 nm; sens. <10 ng/ml. Williams, A.S. et al. ('90) J. Pharm. Pharmacol. **42**(Suppl.), 14P. *See also art.* #C-3.

   Cromolyn ≡ Cromoglyate.

   Dimetindene oxid. [1]**mtb, invstg.**:- C-18; MeCN/phos. pH 2.5, grad.; UV & radio. GC-MS(EI) also. De Graeve, J. et al. ('89) DrRs **39**, 551-555.

   Doxazosin:- SPE; CN col.; MeOH/PCA/heptanesulphonate; fluor; sens. 1 ng/ml. Jackman, G.P. et al. ('91) JChr **566**, 234-238.

   Famotidine (urine also):- SPE; C-8 col.; MeCN/phos. acid; 267 nm; sens. 5 ng/ml. Cvitkovic, L. et al. ('91) JPhBmAn **9**, 207-210.

   Ketanserin & [1]**mtb** incl. oxid. (rat):- pptn. (MeOH); C-18; MeCN/acet. pH 7.0; fluor; sens. (25 µl sample) <200 ng/ml. Wong, Y.W.J. et al.('91) JChr **571**, 318-323.

   Methadone:- LE, →dry; **chiral** (AGP) col.; MeCN/phos. pH 6.6; 212 nm; sens. 10 ng/ml. Glassware silanized. Beck, O. et al. ('91) JChr **570**, 198-202.

   Oxatomide (swtch):- ISRP & 2 × C-18 cols. (50°); MeOH/ammon. acet.; fluor; sens. 0.5 ng/ml. Fujii, J. et al. ('90) JChr **530**, 469-473.

Picumast & **^{1}mtbs** incl. oxid./active, **invstg.** (dog, incl. urine):- C-18; MeOH/phos.; sens. 2 ng/ml. TLC also. Besenfelder, E. et al. ('89) DrRs **39**, 1317-1320.†

Temelastine & oxid. ^{1}mtb:- LE, phenyl col.; MeCN/phos. pH 4.5, grad.; 280 nm; sens. 1 ng/ml [cf. Rogers et al. ('86) JPhSc **75**, 813-814]. Doyle, E. et al. ('90) JPhSc **79**, 524-526. [*In vitro* **mtbs invstg.**: Oldham, H.G. et al. ('90) Drug Metab. Disp. **18**, 146-152.]

Terbutaline: see *above*, Bambuterol *(Alternative assignment: Group #3)*.

Terfenadine.- (1) + oxid. ^{1}**mtb** (auto feasible):- SPE, →dry; CN col., 35°; MeCN/MeOH/acet.; fluor; sens. 8 ng/ml. Coutant, J.E. et al. ('91) JChr **570**, 139-148. (2) + oxid. ^{1}**mtb** (urine):- LE, →dry; **chiral** cols.: for **drug,** cyclodextrin & MeOH/perchlorate; for **mtb,** BSA & isoPrOH/**either** phos. **or** formate pH 8; 210 or 254 nm; sens. unclear. Chan, K.Y. et al. ('91) JChr **571**, 291-297.

Tibenelast (incl. urine):- pptn. (TCA); C-18; MeCN/acetic acid; 313 nm; sens. 50 ng/ml. Peyton, A.L. & Ziege, E.A. ('90) JChr **529**, 377-387.

Zaprinast (incl. rat; auto):- SPE; C-8 col.; MeCN/acet. pH 4.0; 275 nm; sens. 1 µg/ml. Picot, V.S. et al. ('90) JChr **527**, 454-460.

[†Also (1991) Biomed. Chr. 5, 32-37.

#3. ADRENERGIC ANTAGONISTS, esp. β-blockers (& some agonists)/Ca^{2+} CHANNEL BLOCKERS; ACE/ALDOSE REDUCTASE INHIBITORS – *therapeutic overlap with #2, #4 & #5; Vol. 18 has earlier refs., esp. on p. 219*

[An illuminating 'Commentary': *'Evolving Concepts of Partial Agonism: The β-Adrenergic Receptor as a Paradigm'.-* Jasper, J.R. & Insel, P.A. ('92) Biochem. Pharmacol. **43**, 119-130; Fig. 7 shows antagonist/agonist/partial agonist binding concepts. Vols. 13 & 21, this series, feature receptors.]

An **ACE inhibitor** (*p*; a thiazepine), CS-622, & active ^{1}**mtb**:- inhibitor-*binding assay*; estimation of product, as hippuric acid: LE; C-18; MeCN/acetic acid; 228 nm. Shioya, H. et al. ('91) JChr **568**, 309-314 [cf. ('89) **496**, 129-135: GC-MS assay of ^{1}**mtb**, sens. 0.5 or (urine) 5 ng/ml].

β-Blockers (strategy; 13, incl. **Labetalol** & **Nadalol**:- SPE(CN); CN col.; MeCN/phos. pH 3.0; UV or fluor as appropriate. Musch, J. et al. ('89) JPhBmAn **7**, 483-497.

Atenolol:- LE, →dry; **chiral** deriv; C-18; MeCN/acet. pH 7; fluor; sens. (100 µl plasma) 10 µg enantiomer/ml. Rosseel, M.T. et al. ('91) JChr **567**, 239-245.

Belfosdil [a phosphonate, stable in plasma] (urine also):- LE, →dry; GC-NPD; sens. 0.025 ng/ml. Glassware silanized! Mulvana, D. et al. ('90) JChr **527**, 343-350.

Benidipine & oxid. ^{1}**mtbs, invstg.** (rats, dogs; urine & bile too):- pptn.; C-18, grad.; UV & radio; TLC on fractions. Kobayashi, H. et al. ('88) DrRs **38**, 1753-1756.

Benzimidazole derivative, **'S-I'** (β-adrenergic antagonist) – MODEL for studying fate of ^{11}**C-labelled drugs by PET** (positron emission tomography) (urine & lung also; **mtbs** near-lacking):- if plasma: pptn., →dry [lung: dil. HCl homogenate centrifuged]; SPE, →dry; C-18; acetic acid/iPrOH; labelled fractions (→dry) re-run:- C-18; TFA/MeCN, grad.; 254 nm & radio (also for ^{3}H). Also (post-SPE) OPTLC studies, with radio-scan of plate. Jones, H.A. et al. ('91) JChr **570**, 361-370.

Bunolol (& oxid. ^{1}**mtb**, aqueous humour only):- tricky deriv. steps (protein removed; discussed); cap. GC-neg. ion MS; sens. <0.1 µg/ml. Watson, D.G. et al. ('91) JChr **570**, 101-108.

Captopril (incl. CSF; Bu$_3$phosphine to reduce mixed disulphides to free form):- LE; deriv; C-18; MeCN/acetic acid; 260 nm; sens. 5 ng/ml. TLC also. Colin, P. et al. ('89) JLqChr **12**, 625-643.

*p***Chlorambucil t-butyl ester** & active **mtbs** (chlorambucil &, by β-oxidation, phenylacetic mustard; incl. tissues):- LE; C-18; AcOH/MeCN, grad.; 254 nm; sens. <100 ng/ml. Greig, N.H. et al. ('90) JChr **534**, 279-286.

* **Chlorprenaline:-** LE, ⇒dry; **chiral** col. (polymer, ovomucoid), 30°; phos. /EtOH; sens. 25 ng/ml. Miwa, T. et al.('91) J Chr **566**, 163-171.

Cicloprolol:- LE, ⇒dry; C-18; MeCN/phos. pH 2.5; fluor; sens. 5 µg/ml (suits for ther. monitoring; GC alternative). Guinebault, P. et al. ('90) J Chr **525**, 359-367.

Cinnarizine & other Ca²⁺ channel blockers (rat):- SPE (cation-exch.), ⇒dry; MeCN/phos. pH 5.5; 210 nm; sens. <100 ng/ml. Waki, H. & Ando, S. ('89) J Chr **494**, 408-412. **Flunarizine** also studied.

Diltiazem [& see Tripelennamine, Group #1].- (1),+ ¹**mtbs**: LE, ⇒dry; C-18; MeOH/acet./MeCN/TEA pH 7.3; 237 nm; sens. <10 ng/ml: Yeung, P.K.F. et al. ('89) J PhSc **78**, 592-597. (2): LE, ⇒dry; C-18; phos./MeCN; 239 nm; sens. 2 ng/ml: Parissi-Poulou, M. et al. ('90) Int. J. Pharm. **62**, R13-R16. * (3),+ ¹**mtbs**, ther. monitoring: SPE (surpasses LE [Boulieu, R. et al. ('90) J Chr **528**, 542-546]); C-18; MeCN/phos./TEA pH 5.9; 237 nm; sens. 25 ng/ml: Bonnefous, J.L. et al. ('90) J Lq Chr **13**, 3799-3807 (cf. 291-301). (4) (dog): LE; C-18, MeCN/TFA; then **chiral** col. (Chiracel OC); EtOH/DEA; 240 nm; sens. 2 (L-) or 3 (D-) ng/ml: Ishii, K. et al. ('91) J Chr **564**, 338-345.

Diproteverine & ¹**mtbs** (auto):- pptn. (MeCN); SPE; C-18; MeCN/MeOH/THF/phos. pH 7, step-grad.; 248 nm; sens. 7 ng/ml. Metcalf, A.C. et al. ('90) Anal. Proc. **27**, 24-25.

Enalapril, Perindopril & **Ramipril** (di-acid ACE inhibitors) & cyclized lactam ¹**mtbs**, invstg. (urine):- SPE, freeze-drying; deriv; GC-MS(EI). Drummer, O.H. et al. ('88) Dr Rs **38**, 647-650.

Esmolol & **mtb**, near-inactive (pig blood):- LE [blocks blood esterase]; CN col.; MeOH/phos./TEA pH 3.15; 221 nm; sens. 5 ng/ml. Fan, C.D. et al.('91) J Chr **570**, 217-223.

Felodipine, dosed as 'pseudo-racemate' with ²H-label in S-enantiomer, & oxid.¹**mtb** (dogs):- LE, ⇒dry; GC-MS **chiral** study. Eriksson, U.G. et al. ('91) Xenobiotica **21**, 75-84; cf. Ahnoff, D.M. et al. ('87) J Chr **394**, 419-427.

Felodipine & **mtbs**, invstg.:- LE & SPE; deriv; cap. GC-MS; sens. 0.2 ng/ml. Nishioka, A.R. et al. ('91) J Chr **565**, 237-246.   [**Flunarizine:** see Cinn...

Fenoterol (racemic; urine also):- RIA; sens. 20 pg/ml. Rominger, K.L. et al. ('90) Dr Rs **40**, 887-895.

Gallopamil & ¹**mtbs**, invstg.:- LE, ⇒dry; **drug:** packed or cap. col. GC-NPD; **mtbs:** C-18, fluor, & TLC. Weymann, J. et al. ('89) Dr Rs **39**, 605-607.

Imirestat (lens also):- SPE; deriv; cap.GC-ECD; sens. 2.5 ng/ml. McCue, B. et al. ('91) J Chr **565**, 255-264.

Labetalol (incl. sheep foetal plasma etc. - placental-transfer study):- * LE; microbore C-18; MeCN/phos. pH 3.1; fluor; sens. 1.6 ng/ml. Yeleswaram, K. et al. ('90) J Chr **565**, 383-390.

Lacidipine & ¹**mtbs:-** pptn. (MeCN); SPE, ⇒dry; C-18, 40°; MeCN/MeOH, 300 nm, to collect peaks for **RIA;** sens. (3 ml sample) 20 pg/ml. Pellegatti, M. et al. ('92) J Chr **572**, 105-111. *Cf. art. by colleagues,* #ncA-1.

Manidipine & ¹**mtb** (swtch):- LE, ⇒dry; C-18 cols.; MeCN/phos./[1st col: nonane-sulphonate] pH 3.0; 230 nm; sens. 0.1 ng/ml. Miyabayashi, T. et al. ('89) J Chr **494**, 209-218.

Consult # the introduction; # p. 148 for Journal-title & other | *means scrutiny
ABBREVIATIONS. *They include:-* | warranted

p = prodrug	auto = automated system	MeCN = acetonitrile
mtb(s) = metabolite(s): **invstg.**	deriv = derivatization	HPLC cols.: e.g. C-18;
if investigated; Phase I,	LE = liq.-liq. extraction	nm value = UV detection
¹, or ¹ if incl. N-des (oxid. =	⇒dry = drying-down	sens. = sensitivity
oxidative metabolism); Phase	pptn. = precipitation	*SAMPLE usually*
II, ² (hydrol. = hydrolyzed)	swtch = switching	*human plasma/serum*

Mepirodipine:-   LE, ⇥dry; cap. GC-MS(NCI); sens. 30 pg/ml. Teramura, T. et al. ('90) J Chr **528**, 191-198.

Metoprolol [rat/human studies; pseudoracemate (one enantiomer having $^2H_2$ label)], & oxid. **[1]mtbs**, enantioselectivity **invstg.** (microsomal incubates & urine): routinely LE, ⇥dry.   *Approaches* for different **mtbs** [complexity of presentation, as of actual pathways].- (*a*) C-18 with MeCN/TEA pH 3 or (for acidic **mtb**) phos./MeOH pH 5.5, then fluor. (*b*) If enantiomers to be separated, **chiral** deriv; ⇥dry; 2 C-18 cols. with MeCN/MeOH/phos. pH 7.0; 2 collected peaks (diastereoisomers) ⇥dry, deriv. and examined by GC-MS. (*c*) GC-MS following deriv. Murthy, S.S. et al. ('90) Biochem. Pharmacol. **40**, 1637-1644.

Metoprolol.- (1): SPE, ⇥dry; **Chiracel** C-18 col.; hexane/EtOH/DEA; fluor; sens. 4 ng/ml. Herring, V.L. et al. ('91) J Chr **567**, 221-227. (2): LE, ⇥dry;
* **chiral** col. (cellulose carbamate on silica); hexane/isoPrOH/octylamine; fluor; sens. 5 ng/ml. Straka, R.J. et al. ('90) J Chr **530**, 83-93.

Nicardipine.- (1), + **[1]mtb** (rat blood, urine, bile; human urine): LE, ⇥dry; cap. GC-ECD; sens. 5 ng/ml. Watari, N. et al. ('90) J Chr **530**, 438-446. (2): LE; cation-exch. col.; MeOH/ammon. perchlorate; EC; sens. 5 ng/ml. Eastwood, R.J. et al. ('90) J Chr **530**, 463-468.

Nicergoline & **[1]mtbs**, some oxid. (urine also):- LE, ⇥dry; C-18-MS(API); MeCN/ammon. acet., grad.; sens. 2 or (urine) 10 µg/ml. Banno, K. et al. ('91) J Chr **568**, 375-384.

* **Nifedipine** & **[1]mtb**, **invstg.** (hamster):- LE, ⇥dry; GC; radio [apparatus: J Chr **382**, 31-38]. Akira, K. et al. ('88) Chem. Pharm. Bull. **36**, 3000-3007.

Nifedipine (auto):- SPE, ⇥dry; C-8; aq. MeOH; 235 nm; sens. 5 ng/ml. Sheridan, M.E. et al. ('89) J Ph BmAn **7**, 519-522.

* **Nifedipine** & oxid. **[1]mtbs** (blood & urine also):- LE (selective); C-18; cetrimide; 275 nm; sens. <0.5 ng/ml. Rose, M.C.M. et al. ('91) J Chr **565**, 516-522.

Nimodipine:- LE, ⇥dry; cap. GC-NPD; sens. 0.5 ng/ml. Rosseel, M.T. et al. ('90) J Chr **533**, 224-228.

Nisoldipine & 18 **[1]mtbs** (some oxid.; **invstg.**; animals & man, incl. bile & pre-dried urine):- LE, ⇥dry; isolation by TLC & C-18; deriv for GC-FID or -ECD, + radio. NMR also. Scherling, D. et al. ('88) Dr Rs **38**, 1105-1110.

Nitrendipine **[1]mtbs**, **invstg.**:- SPE; C-8 col.; MeCN/phos. acid; 234 & 345 nm. Böcker, R.H. et al. ('90) J Chr **530**, 206-211. (Microsomal incubates.)

Oxodipine (rat):- LE, ⇥dry; C-18; MeCN/phos./THF pH 7; EC; sens. 1 ng/ml. Egros, F. et al. ('90) J Lq Chr **13**, 1001-1011.

*p***Perindopril** & **[1,2]mtbs** (urine also):- anion-exch. col.; formic acid, step grad., giving 3 fractions: **RIA** on each. Doucet, L. et al. ('90) J Ph Sc **79**, 741-745.

Pindolol.- (1): LE, ⇥dry; C-18; aq. MeCN/TEA pH 3.5; fluor; sens. 2 ng/ml. Chmieloweic, D. et al. ('91) J. Chr. Sci. **29**, 37-39. (2) (rat, incl. tissues): LE, ⇥dry; **chiral** deriv; C-18; MeCN/phos. pH 3.4; 258 nm; sens. (each enantiomer; 0.1 ml sample) <50 ng/ml. Hasegawa, R. et al. ('89) J Chr **494**, 381-388.

....

Consult # the introduction; #p. 148 for Journal-title & other | * means scrutiny
ABBREVIATIONS. *They include:-* | warranted

p = prodrug | auto = automated system | MeCN = acetonitrile
mtb(s) = metabolite(s): **invstg.** | deriv = derivatization | HPLC cols.: e.g. C-18;
  if investigated; Phase I, | LE = liq.-liq. extraction | nm value = UV detection
  [1], or **[1]** if incl. *N*-des (oxid. = | ⇥dry = drying-down | sens. = sensitivity
  oxidative metabolism); Phase | pptn. = precipitation | *SAMPLE usually*
  II, [2] (hydrol. = hydrolyzed) | swtch = switching | *human plasma/serum*

Propranolol.- (1), + oxid. [1]**mtb** (active): LE; C-8; MeCN/phos. pH 2.9; fluor; sens. 10 or (**mtb**) 5 ng/ml. Fu, C.J. et al. ('89) Analyst **114**, 1219-1223. (2): LE, →dry; C-18 cols., swtch; MeCN/phos. pH 3.5 & (1st col.; ion-pair) octanesulphonate; 215 nm; sens. 0.06 ng/ml.   Yamashita, K. et al. ('90) JChr **527**, 196-200. (3) & (4): rat/mouse, incl. liver/kidney/heart; swtch.- (3), + **mtb**
* as in (1): sucrose/phos. homogenate (pH 3 to stop **mtb** autoxidation) direct onto protein-coated cation-exch. pre-col.; C-18; fluor; sens. ~0.2 mg/0.1 ml homogenate. (4); **chiral** col. (ovomucoid); phos./EtOH pH 4.6; fluor.- Tamai, G. et al.: (3) ('90) JChr **528**, 542-546; (4) ('90) Biomed. Chr. **4**, 157-160. #*Other chiral refs.* [*see also* ('91) Xenobiotica **21**, 453-461 (blood)].- (5) & (6): LE, →dry; **chiral** deriv; (5) C-18 & aq. MeOH, (6) RP-diol & DCM/TEA; fluor; sens. 1 ng/ml.- (5) Prakash, C. et al. ('89) JPhSc **78**, 771-775.   (6) (peak
* problems discussed): Karlsson, A. et al. ('89) JChr **494**, 157-171.

Ramipril:- **chiral** deriv, distinguishing 2 **chiral** moieties; appropriate cols.; 210 nm. Ito, M. et al. ('90) JLqChr **13**, 991-1000.

Sesmodil fumarate & [1]**mtb:-** SPE & LE, →dry; C-18; MeCN/Bu$_2$NH/phos. acid pH 3; EC; sens. 0.4 ng/ml. Morishima, K. et al. ('90) JChr **527**, 381-388.

Sotolol:- pptn. (PCA); C-18; MeCN/phos. pH 4.6; fluor; sens. 50 ng/ml. Boutagy, J. et al. ('91) JChr **565**, 523-528.

Terbutaline: see Bambuterol in Group #**2**.

Timolol (urine too):- LE; C-18; MeCN/TEA, pH 3; 295 nm; sens. (plasma) 0.5 ng/ml.   Kubota, K. et al. ('90) JChr **533**, 255-263.

Verapamil:- pptn. (MeCN); C-8; MeCN/phos. pH 7.1; 220 nm; 30 ng/ml. Rustum, A.M. ('90) JChr **528**, 480-486.

Yohimbine & oxid. [1]**mtbs, invstg.** (incl. hydrol. urine):- LE, →dry; NP col.; acet./MeOH pH 5; EC; sens. 2 ng/ml.   MS, GC-MS & NMR also used. Le Verge, R. et al. ('92) JChr **574**, 283-292.

*p***Zofenopril** & [1]**mtb, invstg.** (urine also):- TLC; UV & radio. Singhvi, S.M. et al. ('90) JPhSc **79**, 970-973.

#4. CARDIOVASCULAR AGENTS, besides those in #3
Previous coverage: Vol.18 (compendium: p. 219)

Adenosine agonist PD117519 - a dihydro-inden-adenosine (dog; urine also):- SPE; C-8 col., 35°; MeCN/ammon. sulphate/TBA-HSO$_4$/acetic acid, grad.; 280 nm; sens. 25 ng/ml. Reynolds, D.L. et al. ('91) JPhBmAn **9**, 345-349.

Amidepin:- LE; C-18; MeOH/phos./TEA pH 3; fluor; sens. 10 ng/ml. Stehlik, P. et al. ('91) JChr **565**, 477-481.

Bromocriptine (rat, incl. brain):- pptn. (MeCN); C-8 col.; MeCN/isoPrOH /amm. carb. pH 9.0; 310 nm; sens. 12 ng/ml.   Phelan, D.G. et al. ('90) JChr **533**, 264-270.

Carvedilol (incl. monkey):- LE, →dry; **chiral** deriv; C-18,50°; MeOH/EtOH/phos. pH 7; fluor; sens. 1.5 ng/ml. Fujimaki, M. et al. ('90) JPhSc **79**, 568-572.
* **Cibenzoline** & [1]**mtb:-** LE; CN col.; 'PIC' reagents/MeCN/butylamine; 214 nm; sens. 10 ng/ml. Kühlkamp, V. et al. ('90) JChr **528**, 267-273.

Cicletanine (auto, swtch):- C-18 cols. (back-flush); MeCN/phos. pH 2.5; fluor; sens. 50 ng/ml. Antoniewicz, S.M. et al. ('92) JChr **572**, 93-98.

Cromakalim (racemic) & diastereoisomeric [2]**mtbs** (urine): cation-exch. pre-col.; C-18, 35°; MeCN/acet. pH 3.2, grad.; fluor; sens. 0.3 ng/ml.   Kudon, S. & Nakamura, H. ('90) JChr **515**, 597-602.

Debrisoquine & **chiral** oxid. [1]**mtbs** (urine; MODEL for '**hydroxylation status**' assessment).- (1): deriv (pyrimidines formed); LE, →dry; cap. GC-MS; sens. 0.2 μg/ml. Daumas, L. et al. ('91) JChr **570**, 89-97. (2): LE, →dry; CN col.; MeCN/phos. pH 4.8; 208 nm; sens. <0.2 μg/ml. Goubier, C. et al. ('91) Chr'ia **32**, 523-526.   (3): LE, →dry; **chiral** (Pirkle) col.; non-aq. eluent; fluor; sens. 0.9 (drug) or 4 (**mtb** enantiomer) ng/ml. Meese, C. et al. ('87) JChr **423**, 344-350.

Diprafenone:- + SDS & alkali, then SPE; NP; MeOH/hexane/DCM/NH$_3$; 250 nm; sens. 5 ng on-col. Austin, D. et al. ('90) J Chr **527**, 182-188.

* **Dipyridamole:**- ultrafilter (cf. protein binding); LE; C-18; MeOH/acet. pH 5; 280 nm; sens. 10 ng/ml. Barberi, M. et al. ('91) J Chr **565**, 511-515.

Disopyramide.- (1): ultrafilter; LE, ⇥dry; **chiral** col. (AGP); phos./NaCl /PrOH pH 7.0; 254 nm; sens. <0.2 µg/ml for each enantiomer [but racemate had to be assayed too: silica col.; DCM/MeOH/PCA]. Disfavoured option: C-18 followed by **chiral** (AGP) col.   Enquist, M. et al. ('89) J Chr **494**, * 143-156.   (2) (urine also): LE, ⇥dry; **chiral** col. (cellulose carbamate); hexane/PrOH/DEA; 260 nm; sens. 40 µg/ml. Takahashi, H. et al. ('90) J Chr **529**, 347-358.

Etilefrine:- SPE, LE, ⇥dry; C-18; MeOH/phos. pH 4.0; sens. 1.1 ng/ml. Kojima, K. et al. ('90) J Chr **525**, 210-217.

Falipamil & [1]**mtb** (incl urine; auto, swtch):- C-18; MeOH/cyclohexylamine; + acid, then fluor; sens. <25 ng/ml. Roth, W. et al. ('90) J Ph Sc **79**, 415-419.

Fendiline:- LE, ⇥dry; cap. GC-NPD; sens. 1 ng/ml. Lohmann, A. et al. ('91) J Chr **564**, 289-295.

Flecainide:- LE; **chiral** deriv; NP col.; hexane/Et acet./TEA; fluor; sens. 2.5 ng/ml.   Turgeon, J. et al. ('90) J Ph Sc **79**, 91-95; cf. 257-260 → Alessi-Severini, S. et al.:- *similar* (urine too), but C-18; 280 nm; sens. 50 ng/ml.

Flecainide & [1]**mtbs:**- LE; **chiral** deriv (OR **chiral** col.); deriv; GC-MS(NCI); sens. 0.4 ng/ml. Fischer, C. et al. ('90) Biomed. Environ. MS **19**, 256-266.

Glyceryl trinitrate & [1]**mtbs** (dinitrates; tissues also, but vague):- *(if homogenate) LE, ⇥dry; SPE; cap. GC-ECD; sens. <0.5 ng/ml. Adsorption problems discussed. Torfgard, K. et al. ('90) J Chr **534**, 196-201.

Hydrochlorothiazide:- LE, ⇥dry; C-18; MeCN/acetic acid/heptanesulphonate /TEA; 272 nm; sens. 1 ng/ml. Azumaya, C.T. ('90) J Chr **532**, 168-174.

Isoxsuprine (horse):- SPE, cation-exch.; C-18; phos./MeOH/MeCN/octane sulphonate pH 3; EC; sens. 1 ng/ml. Hashem, A. et al. ('91) J Chr **563**, 216-223.

Itazigrel:- pptn. (MeOH); C-8 col.; aq. MeOH; fluor; sens. 0.5 ng/ml. Bombardt, P.A. et al. ('90) J Chr **527**, 233-237.

Mepamil [& **Metazosin**]:- LE, ⇥dry: C-18; MeOH/phos. [MeOH/TEA/phos. acid]; fluor; sens. 0.5 [5] ng/ml.   Stehlík, P. et al. ('90) J Chr **529**, 251-257 [**Met......**: 245-250].

Mexiletine:- LE; CN col.; MeCN/acet./TEA pH 6.0; 210 nm; sens. 4 µg/ml. Vasbender, E. & Annesley, T. (1991) Biomed. Chr. 5, 19-22. *Method adaptable for assay of* **tocainide** and (280 nm) **N-acetylprocainamide** or **procainamide**.

Mexiletine:- LE; deriv; C-18, 60°; EtOH/2-PrOH/TRIZMA pH 9; fluor; sens. (100 µl sample) 5 ng/ml. Shibata, N. et al. ('91) J Chr **566**, 187-194.

*p***Molsidomine** & active [1]**mtb** (**Linsidomine**):- deriv; LE, ⇥dry; C-18; MeCN/phos.; 312 nm;   sens. 0.5 ng/ml. Dutot, C. et al. ('90) J Chr **528**, 435-446.

Naftopidil & oxid. [1]**mtbs:**- LE; C-18; phos./MeCN/MeOH pH 1.8; fluor; sens. <20 ng/ml. Niebch, G. et al. ('90) J Chr **534**, 247-252.

Nicorandil (nitrate):- SPE; C-18; MeOH/phos. pH 8.4; 254 nm; sens. 0.1 µg/ml. Gomita, Y. et al.('90) J Chr **528**, 509-516.

Nitrates, organic mono-, e.g. **Isosorbide 5-mononitrate** [cf. **Glyceryl......**] (rat):- LE, ⇥dry; cap. GC-ECD; sens. <0.5 ng/ml. Tzeng, T-B. et al. ('91) J Chr **570**, 109-120.

Pentoxifylline.- (1), + oxid. [1]**mtb:**- pptn. (MeCN), ⇥dry; then SPE, ⇥dry. CN col.; aq. MeCN; 280 nm; sens. 10 ng/ml. Musch, G. et al. ('89) J Chr * **495**, 215-226.   (2) (rabbit):- pptn. (MeOH); 'shielded hydrophobic' col.: aq. MeCN; 280 nm; sens. 11 ng/ml. Lockemeyer, M.R. & Smith, C.V. ('90) J Chr **532**, 162-167.   (3), + [1]**mtb:**- LE, ⇥dry; C-18; MeCN/phos.; 273 nm; sens. 5 ng/ml. Ostrovska, V. et al. ('90) Cesk. Farm. **39**, 158-160.

Pirmenol (dog):- LE, →dry; **chiral** (Chiracel) col.; hexane/isoPrOH/DEA; 262 nm; sens. 20 ng enantiomer/ml.   Janiczek, N. et al. ('91) J Chr **571**, 179-187.

* **Propafenone:-** LE (unusual solvent); **chiral** deriv; NP; hexane/isopropanol /isobutanol; 220 nm; sens. 6 ng/ml. Mehvar, R. et al. ('90) J Chr **527**, 79-89.

Stobadine:- SPE, →dry; cap. GC-NPD; sens. 5 ng/ml.   Marko, V. ('88) J Chr **433**, 269-275.

Tocainide:- LE, →dry; **chiral** deriv; NP col.; hexane/CHCl$_3$/MeOH; fluor; sens. 0.25 µg/ml. Carr, R.A. et al. ('91) J Chr **566**, 155-162.

Vincamine:- LE; C-18; MeCN/phos./TEA pH 3; 273 nm; sens. 0.3 ng/ml. Bo, L.D. et al. ('91) J Chr **572**, 158-162.

Xipamide (urine too)- LE; C-8;   MeCN/MeOH/AcOH; 231 nm; sens. 10 ng/ml. Bodenan, S. et al. ('90) J Chr **533**, 275-281.

#5. NEUROACTIVE AGENTS incl. (ANTI)CHOLINERGICS
   [See #1 & #7 for analgesics, #7 for anaesthetics]
                                        Previous coverage: Vol. 16

Adosupine (a tricyclic) & **¹mtbs** (rat, incl. brain):- LE, →dry; C-18, 38°; MeCN/MeOH/phos./nonylamine pH 4.5; 240 nm; sens. 50 ng/ml.   D'Aranno,V. et al. ('92) J Chr **574**, 319-325.

Alentanol hydrobromide:- SPE; C-18; MeCN/TEA pH 2.5; 230 nm; sens. 0.1 ng/ml. Schwende, F.J. et al. ('91) J Chr **565**, 488-496.

Alprazolam (ther. monitoring):- LE, →dry; GC-ECD (GC-MS for comparison); sens. <4 ng/ml. Greenblatt, D.J. et al. ('90) J Chr **534**, 202-207.

Alprazolam & oxid. active **¹mtbs** incl. 4-OH (unstable!):- LE, →dry; deriv; C-18; MeOH/phos./hexyltriethylammon. phos. pH 7.4; 221 nm; sens. <1 ng/ml. Schmith, V.D. et al. ('91) J Chr **568**, 253-260.

Amineptine & **¹mtb** (ther. monitoring):- SPE & LE, →dry; C-18; MeCN/phos.; 210 nm; sens. 2 ng/ml. Rop, P. et al. ('90) J Chr **532**, 351-361.

Amitriptyline & other tricyclic antidepressants (ther. monitoring; auto):-
* 2 pre-cols. (Carbopack & cation-exch.); NP (CN) col.; MeCN/MeOH/phos. pH 7.8 (recycled); 215 nm; sens. <5 ng/ml. Carfagnini, G. et al. ('90) J Chr **530**, 359-366.

Aniracetam & **¹mtb:-** LE (only for **mtb**), →dry; C-18; (parent drug: swtch); MeOH/phos./TEA pH 3.0; 255 nm; sens. 5 or (**mtb**) 50 ng/ml.   Guenzi, A. & Zanetti, M. ('90) J Chr **530**, 397-406.

* ***Anticonvulsants;* methylxanthines** (ther. monitoring):- ISRP pre-col.; C-18 (2 cols.; swtch, **well explained**); phos./MeCN, grad.; 230 & 275 nm. Haginaka, J. et al. ('90) J Chr **529**, 455-461.

Apomorphine:- LE; C-18; MeCN/phos./octanesulphonate pH 3; EC; sens. 0.5 ng/ml. Essink, A.W.G. et al. ('91) J Chr **570**, 419-424.

p**Aprophen** & **¹mtbs, invstg.** (one oxid./active; urine):- LE, →dry; NP; MS. Brown, N.D. et al. ('91) J Chr **563**, 466-471.

Bemegride (brain also):- LE, →dry; C-18; MeCN/phos. pH 4.8; 200 nm; sens. 1 µg/ml or /g. Soto-Otero, R. et al. ('91) J Ph Bm An **9**, 177-182.

Benperidol & **¹mtb:-** RP- & cation-exch. SPE, →dry; CN col.; acet./MeCN pH 4.7; EC; sens. <0.5 ng/ml. Suss, S. et al. ('91) J Chr **565**, 363-373.

....

Consult ₰the introduction; ₰p.148 *for Journal-title & other* | *means scrutiny
ABBREVIATIONS. They include:- | warranted

p = prodrug	auto = automated system	MeCN = acetonitrile
mtb(s) = metabolite(s): **invstg.**	deriv = derivatization	HPLC cols.: e.g. C-18;
if investigated; Phase I,	LE = liq.-liq. extraction	nm value = UV detection
¹, or ꟾ if incl. *N*-des (oxid. =	→dry = drying-down	sens. = sensitivity
oxidative metabolism); Phase	pptn. = precipitation	*SAMPLE usually*
11, ² (hydrol. = hydrolyzed)	swtch = switching	*human plasma/serum*

Bromocriptine (rat, plasma/brain):- LE; C-8; ammon. carb./MeCN/iPrOH pH 9; 310 nm; sens. 12 ng/ml. Phelan, D.G. et al. ('90) J Chr **533**, 264-270.

Buprenorphine (incl. rabbit; pharmacokinetics):- LE, ⇥dry; NP col.; MeCN/acet. pH 3.75; fluor; sens. 1 ng/ml. Ho, S-T. et al. ('91) J Chr **570**, 339-350.

Bupropion (ther. monitoring):- pptn. (MeCN), then assay quickly (instability); C-18; aq. MeCN pH 3.15; 250 nm; sens. 5 ng/ml. Al-Khamis, K. ('89) J Lq Chr **12**, 645-655.

Buspirone:- SPE; C-18 × 2: #1 heart-cut onto #2; MeCN/phos./TEA/[in #2] lauryl sulphate pH 2.5 (% compositions different for #1 & #2); 235 nm; sens. 0.2 ng/ml. Kristiansson, F. ('91) J Chr **566**, 250-256.

* **Carbamazepine** (& see **Phenobarbital**) (ther. monitoring):- protein-excluding col.; aq. acet.; 285 nm. Gish, D.J. et al. ('88) J Chr **433**, 264-268.

Carbamazepine (I) & oxid. **mtbs**; and 2 **Triazolecarboxamides** (CGP 33101),
* as MODELS for 'fast-LC' (with 3 μm packings):- SPE (**or** LE, ⇥dry); C-18; MeCN/phos./(where needed) MeOH; 230 nm; sens. 50/25 ng/ml. Rouan, M.C. et al. ('92) J Chr **572**, 59-64; also 65-68:- (1), auto, as in Vol. 20 art. from same lab. (Lecaillon, J.B. et al.).

m-**Chlorophenylpiperazine** (rat; brain also):- LE, ⇥dry; deriv; cap. GC-MS; sens. 4 ng/ml. Andriollo, O. et al. ('90) J Chr **533**, 215-233.

* **Chlorpromazine** & ¹**mtbs** incl. oxid., **invstg.:-** SPE [cf. traditional LE]; C-8; MeOH/MeNH₂/phos. pH 6.75; 254 nm; sens. <10 ng/ml. Smith, C.S. et al. ('87) J Chr **423**, 207-216.

Citalopram & ¹**mtbs:-** SPE; LE (wash); C-18; MeCN/phos.; 239 nm; sens. 0.8 ng/ml. Rop, P.P. et al. ('90) J Chr **527**, 226-232.

Clebopride & ¹**mtb:-** LE, ⇥dry; deriv; GC-MS(neg.-ion CI); sens. 0.1 ng/ml. Robinson, P.R. et al. ('91) J Chr **564**, 147-161.

Clonazapam:- LE, ⇥dry; C-18; MeCN/MeOH/phos.; 242 nm; sens. 3 ng/ml. Boukhabza, A. et al. ('90) J Chr **529**, 210-216.

Codeine & ¹,²**mtbs**, pattern **invstg.**, viz. O-glucuronide, nor-codeine, & morphine equivalents (in 'codeine-converters'; incl. urine, usually hydrol. by appropriate enzyme):- SPE; C-8 col.; MeOH/phos./heptanesulphonate; EC. Verwey Van Wissen, C.P.W.G.M. et al. ('91) J Chr **570**, 309-320. See also
[Morphine.

Denbufylline & ¹**mtbs** (some active; incl. oxid.):- SPE, ⇥dry; C-18; acet./MeCN, grad.; 280 nm; sens. <5 ng/ml. Deeks, N.J. et al. ('90) Anal. Proc. **27**, 179-180.

Denopamine (incl. dog):- LE, ⇥dry; C-18; MeCN/phos.; EC; sens. 2 ng/ml. Tagawa, K. et al. ('90) J Chr **529**, 500-506.

Diazepam & other **benzodiazepines** as source of N-desMe ¹**mtb**, **invstg.** (rat microsomes), by HPLC: Reilly, P.E.B. et al. ('90) Mol. Pharmacol. **37**, 767-774 - cited in Biochem. Pharmacol. ('92) **43**, 1377-1380.

Eseroline: see Physostigmine.

Etoperidone & active ¹**mtbs** incl. oxid.:- LE; C-8 col.; MeOH/MeCN/acet./pentyl-amine/heptanesulphonate pH 3.7; 254 nm; sens. 2 ng/ml. Holland, M.L. et al. ('91) J Chr **567**, 433-440.

Felbamate (rat, rabbit, dog):- LE, ⇥dry; C-18; aq. MeOH; 210 nm; sens. 0.4 μg/ml. Adusumali, L.E. et al. ('91) Drug Metab. Disp. **19**, 1116-1125.

Fenethylline & ¹**mtbs**, incl. oxid., **invstg.** (urine):- GC-MS. Von Rücker, G. et al. ('88) DrRs **38**, 497-501 [in German].

★ **Fluoxetine** & active [1]**mtb**:- SPE (LE inefficient), ⇢dry; deriv, ⇢dry; cap. GC-ECD; sens. <20 ng/ml. Dixit, V. et al. ('91) J Chr **563**, 379-384.

Fluparoxan:- LE, ⇢dry; deriv; GC-ECD; sens. 10 ng/ml. Gristwood, W.E. ('90) J Chr **527**, 436-440.

Flurazepam & [1]**mtbs**:- LE; C-8 col.; MeCN/MeOH/phos. pH 4.1; 230 nm; sens. 0.2 ng/ml. Selinger, K. et al. ('89) J Chr **494**, 247-256.

Fluvoxamine (rat also):- LE, ⇢dry; NP col.; MeOH/MeCN/DEA; 254 nm; sens. 0.5 ng/ml. Van Der Meersch-Mougeot, V. et al. ('91) J Chr **567**, 441-449.

Haloperidol & [1]**mtb**:- SPE, ⇢dry; CN col.; MeCN/PrOH/phos. pH 6.8; EC; sens. <10 ng/ml. Eddington, N.D. et al. ('88) J PhSc **77**, 541-543.

Hexobarbital (rat):- LE; chiral (AGP) col.; phos./PrOH pH 5.4; 210 nm; sens. 0.05 μg enantiomer/ml. Vermuelen, A.M. et al. ('91) J Chr **567**, 472-479.

Idebenone & [1]**mtb** (rat; brain also):- LE, ⇢dry; C-18; MeOH/ClO$_4^-$; EC; sens. (mini-samples) 0.2 ng/ml (or /g).   Wakabayashi, H. et al. ('92) J Chr **572**, 154-157.

Isocarboxazid (Marplan):- LE; C-18; octanesulphonate/MeOH; 230 nm; sens. 100 ng/ml. Powell, M.L. et al. ('90) J Chr **529**, 237-244.

★ **Lamotrigine** & quaternary ammonium glucuronide, [2]**mtb, invstg.** (guinea-pig):- SPE, ⇢dry; C-8 col.; MeCN/phos./SDS pH 2.2, grad.;   277 nm; sens. 0.1 μg/ml. Remmel, R.P. et al. ('91) Drug Metab. Disp. **19**, 630-636 /J Chr **571**, 217-230.[≠]

★ **Levodopa** & oxid. [1]**mtbs**.- (1), continuous monitoring: **microdialysis**; C-18; MeCN/THF/citric acid/EDTA/octanesulphonate pH 2; EC; sens. 5 μg/ml. With similar HPLC: (2), also **Carbidopa**: pptn. (PCA); EC; sens. <0.5 ng/ml. (1) Deléu, D. et al. ('91) J PhBmAn **9**, 159-165; (2) Wikberg, T., as for(1), 167-176.

Medifoxamine [1]**mtbs**, some oxid. (urine):- LE, ⇢dry; C-18; MeOH/phos. pH 8; 266 nm; sens. 5-30 ng/ml. Saleh, S. et al. ('90) J Chr **528**, 531-536.

Metamphetamines [cf. **Seleg....**] (rat urine):- SPE; deriv, ⇢dry; **chiral** col. (cellulose-based), 48°; hexane/PrOH; 220 nm; sens. unclear. Nagai, T. et al. ('90) J Chr **525**, 203-209.

Methylxanthines (hepatocyte incubates; cf. above, *Anti.......*:- pptn. (PCA); LE, ⇢dry; deriv; cap. GC-MS. Benchekroun, Y. et al. ('90) J Chr **532**, 262-275.

Metrifonate & active [1]**mtb, Dichlorvos**:- LE, ⇢dry; C-18; MeOH/THF/octane-sulphonate (different for drug and **mtb**); 210 nm; sens. 1 μg/ml. Unni, L.K. et al. ('92) J Chr **572**, 99-103.

★ **Midazolam** (total & unbound; ther. monitoring):- extraction; GC-NPD (GC-MS validation); sens. 0.02 [cf. HPLC, 0.1] μg/ml.   De Vries, J.X. et al. ('90) Biomed. Chr. **4**, 28-33.

Midazolam & oxid. [1]**mtbs**:- pptn. (MeCN); SPE, ⇢dry; C-18; MeOH/MeCN/THF/phos. pH 5.6; 254 nm; sens. 50 ng/ml. Sautou, V. et al. ('91) J Chr **571**, 298-304; cf. ('90) **528**, 526-530.

Milacetamide & within-brain active [1]**mtb, Glycinamide** (CSF also):- deriv; pptn. (MeCN); C-18; MeCN/borate pH 6.5 or 7; fluor; sens. 2 μg/ml. Semba, J. et al. ('91) J Chr **565**, 357-362.

Morphine [& see **Apo.....** & **Codeine**] & [2]**mtbs** (glucuronides).- (1): SPE, ⇢dry; C-18; MeCN/phos. acid; 214 nm; sens. <3 ng/ml. Breda, M. et al. ('91) Int. J. Clin. Pharmacol. **11**, 93-97.   (2), also **Codeine** (urine & CSF also): SPE, ⇢dry; C-8 col.; MeCN/phos. pH 4.7; sophisticated scanning, incl. 230/255/280 nm; sens. 0.5 ng/ml. Chari, G. et al. ('91) J Chr **571**, 263-270.

───────────

[≠]see also J Chr **554**, 181-189.

* **Nefazodone** & [1]**mtbs** incl. oxid. (auto, robotics):- LE, ⇾dry; phenyl col.; MeCN/MeOH/phos.; 254 nm; sens. 10 ng/ml.   Franc, J.E. et al. ('91) J Chr **570**, 129-138.

Nitecapone (urine also):- LE; C-18; MeOH/THF/phos./citrate pH 2.0; EC; sens. 5 ng/ml. Wikberg, T. et al. ('91) J PhBmAn **9**, 59-64.

Oxcarbamazepine & oxid. active [1]**mtb** (ther. monitoring):- LE, ⇾dry; C-8; aq. MeCN; 215 nm; sens. 0.1 µg/ml. Elyas, A.A. et al. ('90) J Chr **528**, 473-479.

Phenobarbital, Carbamazepine, Phenytoin (ther. monitoring):- ISRP pre-col. & col.; phos./MeCN/THF pH 7; 254 nm; sens. unclear.   Rainbow, S.J. et al. ('90) J Chr **527**, 389-396.

Phenytoin oxid. [1]**mtb**, 5-(p-hydroxyphenyl)-5-phenylhydantoin enantiomers (rat hepatocyte incubates):- SPE, ⇾dry; **chiral** [cellulose tris(4-Me-benzoate)]; aq. EtOH; 228 nm. Eto, S. et al. ('91) J Chr **568**, 157-163.

Physostigmine & [1]**mtb**, **Eseroline**.(1) LE, ⇾dry; silica col.; MeOH/HClO₄; fluor; sens. 0.1 ng/ml. Wu, Y. et al. ('90) J Lq Chr **13**, 575-590. (2)

Piribedil & [1]**mtbs**:- LE; C-18; phos./MeCN, grad.; 240 nm; sens. <5 ng/ml. Sarati, S. et al. ('91) J Chr **563**, 323-332.

Piritramide (ther. monitoring):- **LE**, ⇾dry; GC-NPD; sens. 1 ng/ml. Michaelis, H.C. et al. ('91) J Chr **571**, 257-262.

Progabide & active [1]**mtb**:- LE, ⇾dry; C-18; MeOH/MeCN/NaCl/phos. pH 5; 340 nm; sens. 20 ng/ml. Decourt, J.P. et al. ('90) J Chr **527**, 214-219.

Protriptyline: see **Tripelennamine** in Group #1.

Selegiline (**L-Deprenyl**) & [1]**mtbs** - cf. Metamphet.... (dog; urine also):- LE, ⇾dry; cap. GC-ECD; sens. 1 ng/ml. Salonen, J.S. ('90) J Chr **527**, 163-168.

Sertaline:- LE, ⇾dry; deriv, ⇾dry; cap. GC-ECD; sens. 1 ng/ml. Tremaine, L.M. & Joerg, E.A. ('89) J Chr **496**, 423-429.

Tacrine & [1]**mtbs**, incl. oxid.-   (1): LE, ⇾dry; C-18; MeCN/phos. pH 2.7; 240 nm; sens. 0.3 ng/ml: Ekman, L. et al. ('89) J Chr **494**, 397-402.   (2) (**mtb invstg.**; incl. urine & rat incubates): C-18; MeOH/phos./TEA pH 8.5; fluor: Truman, C.A. et al. ('91) Biochem. Pharmacol. **42**, 956-959.

Temazepam (urine also, hydrol.):- LE, ⇾dry; C-8; MeCN/acetic acid; 228 nm; sens. <5 ng/ml.   Kunsman, G.W. et al. ('91) J Chr **568**, 427-436.

Thioridazine & [1]**mtbs** incl. oxid. (monitoring; auto):- SPE; C-18; MeOH/phos. /TEA pH 4.1; 254 nm. Svensson, C. et al. ('90) J Chr **529**, 229-236.

Tiagabine (incl. dog):- SPE, ⇾dry; C-18; MeCN/MeOH/phos./octanesulphonate; EC; sens. 2 ng/ml. Gustavson, L.E. et al.('92) J Chr **574**, 313-318.

Trimipramine & [1]**mtbs** (incl. oxid., **invstg.**).- (1): LE, ⇾dry; deriv; cap. GC-MS. Maurer, H. ('89) DrRs **39**, 101-103. (2) (rat urine, hydrol.):-
* LE; deriv; GC-MS(EI; fragmentation discussed).   Coutts, R.T. et al. ('90) Biomed. Environ. MS **19**, 793-806.   (3):- LE, ⇾dry; CN col.; phos./MeOH/MeCN pH 6.5; EC; sens. 1 ng/ml. Gulaid, A.A. et al. ('91) J Chr **566**, 228-233.

Valproic acid.- (1): pptn. & LE; deriv; NP or RP col.; 254 nm. Accord with EMIT values. Gentil de Illiano, B. et al. ('90) J. High Resol. Chr. **12**, 540-543.   (2) + [1]**mtbs** incl. oxid./active, **invstg.** (urine also): **LE** (discussed), ⇾dry; deriv; GC-MS(NCI); sens. 1-8 ng/ml. Kassahun, K. et al.('90) J Chr **527**, 327-341.

Vanoxerine:- SPE, ⇾dry; C-8; aq. MeOH/PCA, grad.; fluor; sens. 1 ng/ml. Ingwersen, S.H. ('91) J Chr **571**, 305-311.

Velnacrine (urine also):- LE, ⇾dry;   RP ('SAS') col.; MeOH/formate pH 2.7; 325 nm; sens. 4 ng/ml. Hsu, R. et al. ('89) J Chr **494**, 347-353.

Vercuronium & [1]**mtbs** (dog/mouse/rat also):- SPE, ⇾dry; CN col.; MeCN/phos. pH 5.55; EC; sens. <0.4 ng/ml (less good for **mtbs**). Ducharme, J. et al. ('92) J Chr **572**, 79-86.

* **Vinpocetine**:- LE [transesterification if MeOH, not iPrOH], ⇾dry; cap. GC-MS; sens. 0.1 ng/ml. 'Anti-adsorptive' vessels. Lohmann, A. & Dingler, E. ('90) J Chr **529**, 442-448.

#6. ANTI-INFECTIVE/ANTI-CANCER AGENTS & immunosuppressives
Previous coverage: Vols. 20 (compendium: p. 235) & (anti-cancer) 14

REVIEW.- **Antimalarials** assay: Bergqvist, Y. & Churchill, F.C. ('88) JChr **434**, 1-20.

Acetretin, *cis*- & all-*trans*-, auto:- pptn. (EtOH); LE; NP col.; hexane/Me salicylate/acetic acid; 360 nm; sens. 3-4 ng/ml.   Meyer, E. et al. ('91) JChr **570**, 149-156.

Acyclovir:- pptn. (TCA); C-18; MeOH/octanesulphonate; 254 nm; sens. 0.1 µg/ml. Molokhia, A.M. et al. ('90) JLqChr **13**, 981-989.

†**Adriamycin & Adriamycinol**:- SPE, →dry; C-18; MeCN/MeOH/formate pH 4.0; fluor; sens. 50 pg/ml.   Leca, F.R. et al. ('89) Chr'ia **28**, 375-378.

†**Adriamycin** (incl. biopsies):- SPE; phenyl col.; formate/MeCN, grad.; fluor; sens. 2 ng/mg.   Cox, S.K. et al. ('91) JChr **564**, 322-329.

Albenazole & oxid. active ¹**mtb**.- (1): SPE, →dry; C-18; MeOH/phos. pH 5.7; 295 nm; sens. 20 ng/ml.   Hurtado, M. et al. ('89) JChr **494**, 403-407. (2): LE, →dry; RP (phenyl) col.; MeOH/MeCN pH 3.1; 254 nm; sens. <20 ng/ml. Hoaksey, P.E. et al. ('91) JChr **566**, 244-249.

Amifloxacin, Enoxacin & Norfloxacin (fluoroquinolones):- pptn.; C-18; MeCN/phos./lauryl sulphate/TBA Br⁻ pH 2; fluor; sens. respectively 10, 100 & 20 ng/ml.   Nangia, A. et al. ('90) JPhSc **79**, 988-991. Cf. Griggs, D. & Wise, R. ('89) J. Antimicrob. Ch'ther. 24, 437-445:- C-18; variants of pptn., eluent, etc., to suit the analyte.

* **Amikacin**:- SPE (CM Sephadex); C-18; pentanesulphonate pH 3.3 (no organic solvent); deriv (OPA); fluor; sens. 25 ng/ml.   Wichert, B. et al. ('91) JPhBmAn **9**, 251-254.

Amopyroquin & (species differences) ¹**mtbs** (urine also):- LE; C-18; MeCN/phos. pH 3; 340 nm; sens. 7 ng/ml.   Verdier, F. et al. ('89) AnMcAgCh **33**, 316-321 (& JChr **421**, 192-197).

Amphotericin B (tissues incl. liver, kid.):-MeOH; C-18; MeOH/MeCN; 405 nm;
* sens. 0.1 µg/g. Bioassays also; values < HPLC values.   Collette, N. et al. ('89) AnMcAgCh **33**, 362-368; cf. JChr **419**, 401-407 (serum assay).

Ampicillin & **Sulbactam**, a β-lactamase inhibitor (incl. urine):- ultra-filter; C-18; MeOH/phos./TBA-Br; post-col. deriv; 270 nm (or 230 nm); sens. ~1 µg/ml. Haginaka, J. et al. ('90) JChr **532**, 87-94.

Anthracyclines: see **Dauno**...... (& **Doxo**......).

Arabinosyl-5-azacytosine & ¹**mtb** (mouse):- SPE (phenylboronic, to remove endogenous ribosides); C-8, then C-18; phos./MeCN pH 6.8; 240 nm; sens. 50 ng/ml. Heideman, R.L. et al. ('89) JLqChr **12**, 1613-1633.

Artemisinin:- LE, →dry; deriv; C-18; UV; sens. 2.5 ng/ml. Titulaer, H.A. et al. ('90) J. Pharm. Pharmacol. **42**, 810-813.

* **Aspoxycillin**:- 'micellar electrokinetic chromatography' in capillaries; phos./borate/SDS; 210 nm; sens. 1.3 µg/ml. Nishi, H. et al. ('90) JChr **515**, 245-255.

Azidothymidine: see Zidovudine.

Azithromycin (incl. tissues/animal samples):- LE, →dry; phenyl col.; phos./MeCN pH 11; EC; sens. 0.01 µg/ml (tissues: 0.1 µg/g). Shepard, R.M. et al. ('91) JChr **565**, 321-337.   [†See also p. 163, **Doxorubicin** (synonym).

Consult # the introduction; #p. 148 for Journal-title & other | *means scrutiny
ABBREVIATIONS. They include:-* | warranted

p = prodrug	auto = automated system	MeCN = acetonitrile
mtb(s) = metabolite(s): **invstg.**	deriv = derivatization	HPLC cols.: e.g. C-18;
if investigated; Phase I,	LE = liq.-liq. extraction	nm value = UV detection
¹, or **²** if incl. *N*-des (oxid. =	→dry = drying-down	sens. = sensitivity
oxidative metabolism); Phase	pptn. = precipitation	*SAMPLE usually*
II, ² (hydrol. = hydrolyzed)	swtch = switching	*human plasma/serum*

Benzylpenicillin (lymph also):- pptn. (MeCN); LE & concentrate (optional); deriv, → mercuric mercaptide; C-18; 325 nm; sens. through 'digital subtraction': 1 ng/ml. Wiese, B. et al. ('89) JPhBmAn **7**, 107-118.

Busulphan.- (1): deriv, →dry; SPE; C-18; MeCN/THF; 278 nm; sens. 0.4 ng/ml. (2): SPE; CN col.; aq. MeOH; photochem. 'deriv'; 226 nm; sens. 20 ng/ml. ('90) JChr **532**:-(1): MacKichan, J.J. & Bechtel, T.P., 424-428; (2): Blanz, J. et al., 429-437.

Camptothecin (incl. urine):- LE (pH invstg.), →dry; C-18; aq. MeOH; fluor; sens. 1 ng/ml. Loh, J-P. & Ahmed, A.E. ('90) JChr **530**, 367-376.

Cefadroxil:- pptn. (MeCN); C-8 col.; MeOH/phos. pH 2.6; 230 nm; sens. 0.5 µg/ml. Nahata, M.C. & Jackson, D.S. ('90) JLqChr **13**, 1651-1656.

Cefamandole & its nafate ester (urine also; swtch):- esterase inhibitor added to blood at start; direct onto pre-col.; C-8; MeOH/TBABr⁻; 270 nm; sens. 0.5 µg/ml. Lee, H.S. et al. ('90) JChr **528**, 425-433.

Cefibuten (urine also; auto, swtch):- direct onto pre-col.; C-18; MeCN /MeOH/phos./TBA Br⁻ pH 5.0; 256 nm; sens. 1 µg/ml. Matsuura, A. et al. ('89) JChr **494**, 231-245. Cf. Wise, R. et al. ('90) AnMcAgCh **34**, 1053-1055:- similar assay (*cis* & *trans*), also microbiological (incl. skin & blister exudate); sens. (HPLC) 0.05 µg/ml.

Cefoxitin/Cefuroxime/Cephalexin/Cephaloridine (rat; swtch):- direct onto pre-col.; C-18; MeCN/acet. pH 3.5; 254 nm; sens. 0.5 µg/ml. Lee, Y.J. & Lee, H.S. ('90) Chr'ia **30**, 80-84. | **Cefoperazone/Cefepime/Ceftazidine:**

Cefaclor, Cephalexin: see **Loracarbef.** | Kalman, D. ('92) AnMcAgCh **36**, 453-457.

Cefpirome.- (1) (paediatric monitoring): C-18; MeOH/acet.; 240 nm; sens. <0.5 µg/ml. Nahata, M.C. ('91) JLqChr **14**, 193-200. (2) (incl. human milk; auto): LE, →dry; C-18; aq. TEA pH 5.1; 240 nm; sens. 0.6 µg/ml. Kearns, G.L. et al. ('92) JChr **574**, 356-360.

*p***Cefpodoxime** proxetil (incl. urine):- pptn. (MeCN); C-18; acet./MeCN pH 5; 254 nm; sens. 0.02 µg/ml, or (urine) 0.3 µg/ml. Molina, F. et al. ('91) JChr **563**, 205-219. Cf. Tremblay, D. et al. ('90) AnMcAgCh **26** (Suppl. E), 21-28.

Ceftretam(e):- SPE; RP (phenyl); MeOH/phos. pH 5.2; 225 nm; sens. 60 ng/ml. Hicks, C.M. & Powell, M. ('89) JChr **497**, 349-354.

⋆ **Cephalosporin DQ-2556** (urine also):- pptn. (TCA) or (easier/costly) ultrafilter; C-18; MeCN/acetic acid; 306 nm; sens. 0.1 or (urine) 2 µg/ml. Matsubayashi, K. et al. ('90) JChr **515**, 547-554.

Cephalosporins with side-chain α-NH₂ (urine also):- C-18; MeCN/phos.,
⋆ optimized composition; 262 nm, or fluor after post-col. deriv. Blanchin, M.D. et al. ('88) JLqChr **11**, 2993-3010.

Chloramphenicol (horse, dog & cat):- pptn. (EtOH & MeCN), concentrative ultrafilter; C-18, 50°; MeOH/acet.; 278 nm; sens. <0.5 µg/ml. Tyczkowska, K. et al. ('88) J. Chromatog. Sci. **26**, 533-536.

⋆ **Chloramphenicol**, paediatric monitoring:- pptn. (MeOH); microbore C-18, 50°; acet./THF/MeCN pH 6.0; 254 nm; sens. <3 µg/ml. Wong, S.H-Y. et al. ('88) JLqChr **11**, 1143-1158 [cf. Davidson, D.F., 1139-1142: similar, but 278 nm].

Chloroquine & ¹mtbs (incl. blood & urine):- LE, →dry; C-18; DEA/MeCN/heptane-sulphonate pH 3.4; 343 nm; sens. 12 ng/ml. Houzé, P. et al. ('92) JChr **574**, 305-312.

Cilastatin (imipem protector) & ¹mtb (rat; incl. urine & bile):- SPE, →dry; C-18; MeCN/phos. pH 4.0; 210 nm; sens. 1 µg/ml. Chen, I-W. et al. ('90) JChr **534**, 119-126.

Ciprofloxacin:- LE; C-18; acet./MeCN/TBA Br⁻ pH 3; 277 nm; sens. 0.1 µg/ml. Pou-Clave, L. et al. ('91) JChr **563**, 211-215. Cf. assay in rat brain & CSF:- Katagiri, Y. et al. ('90) Chem. Pharm. Bull. **38**, 2884-2886.

Cisplatin (unstable! urine also).- (1): direct onto anion-exch. col.; MeCN/NaCl; post-col. deriv, then 290 nm; sens. 80 ng/ml.   Kinoshita, M. et al. ('90) J Chr **529**, 462-467.   (2): ultrafilter (plasma); C-18; citrate/ hexadecylammon. hydrox. pH 6.5; EC.   Treskes, M. et al. ('90) J Lq Chr **13**, 1321-1338.   (3)(plasma; swtch): cation-exch. (pre-col. too) at 40°; phos.; 210 nm; sens. 30 ng/ml. Kizu, R. et al. ('89) Biomed. Chr. **3**, 14-19.

Clarithromycin & oxid. [1]mtb (urine also):- LE, ⇥dry; C-8; MeCN/MeOH/phos. pH 6.8; EC; sens. 10 ng/ml.   Chu, S-Y. et al. ('91) J Chr **571**, 199-208.

* **Clavulanic acid** (penicillin protector; saliva):- deriv; SPE, ⇥dry; micro-bore C-18; MeCN/phos./octanesulphonate/TE-ammon.-Br⁻ pH 2.5; 311 nm; sens. <8 ng/ml. Low, A.S. et al. ('89) AnMcAgCh 25 Suppl. **B**, 83-86.

Cyclosporin (A) (**Ciclosporin**) (blood) [Ther. monitoring REVIEW: Furlanut, M.
* et al. ('89) J Lq Chr **12**, 1759-1789]:- pptn. (Zn^{2+} & solvents); SPE (polymer-based RP col., preferable to silica-based RP col.);   C-18; MeCN/TFA; 210 nm; sens. 10 ng/ml.   Kabra, P.M. et al. ('89) J Lq Chr **12**, 1819-1834.

Cyclosporin(e) (plasma, ther. monitoring):- LE, ⇥dry; C-18; MeCN; 214 nm;
* sens. 10 ng/ml.   Easier alternative (compared): specific RIA.   Vernillet, L. et al. ('89) Clin. Chem. **35**, 608-611.

Cyclosporin A (whole blood, auto):- pptn. (MeCN); C-18; EtOH/hexane; 210 nm; sens. 12 ng/ml. Lachno, D.R. et al. ('90) J Chr **525**, 123-132.

Daunorubicin & **mtbs**, incl. oxid. (heart also):- SPE, ⇥dry; C-18; MeCN/phos. pH 4; fluor; sens. 0.1 ng/ml.   De Jong, J. et al. ('90) J Chr **529**, 359-369.

Daunorubicin & †**Doxorubicin**.- (1), swtch:- direct onto pre-col.; CN col.; MeCN/phos. acid; 254 nm; sens. 10 ng/ml.   Mikan, A. et al. ('90)
* Biomed. Chr. **4**, 154-156.   (2) (**Epirubicin** also):- LE; CE; LIF; sens. ~0.2 ng/ml.   Reinhoud, N.J. et al. ('92) J Chr **574**, 327-334.   See also *N*-L-**Leucyl......**, & (**Dox...** + [1]**mtb**) Mross, K. et al. ('90) J Chr **530**, 192-199: sens. 1 ng/ml

Deoxy(ribo)nucleosides: azido-/fluoro-**deoxythymidine**, **dideoxy-thymidine**/ -**inosine** (incubates too):- pptn. [plasma; $(NH_4)_2SO_4$] or freeze-thawing (pelleted hepatocytes; MeOH); phenyl col.; for AZT, acet./MeOH, pH 6.55; 267 nm; sens. 50 ng/ml. Frijus-Plessen, N. et al. ('90) J Chr **534**, 101-107.

* **Deoxyspergualin** (dog):- SPE (CM-Sephadex, then C-18); C-18; MeCN/phos. /pentanesulphonate pH 3; 205 nm; sens. ~50 ng/ml.   Nakanuma, R. et al. ('90) J Chr **527**, 208-213.

Enoxacin:- see Amifloxacin.   [† See also p. 161, **Adriamycin** (synonym).

Etoposide & **Teniposide** (auto, swtch):- C-18 pre-col., phos. pH 7; CN col.; MeOH/phos. pH 7; 254 nm or EC; sens. 0.1 or (EC) 0.02 µg/ml. Van Opstal, M.A.J. et al. ('89) J Chr **495**, 139-151.

Fleroxacin & [1,2]**mtbs** (urine & dialysis fluid also):- for serum, pptn. (TCA), etc.: method of Dell, D. et al. ('88) J Lq Chr **11**, 1299-1312 -
* but C-18 pre-col. (swtch) to obviate late-running peaks.   Stuck, A. et al. ('89) AnMcAgCh **33**, 373-381.

Fluconazole.- (1): pptn. (MeCN); C-18; aq. MeCN; 210 nm; sens. 0.5 µg/ml. Hosotsubo, K.K. et al. ('90) J Chr **529**, 223-228.   (2) & (3) (CSF also):
* LE, ⇥dry; (2): C-18/CN '**mixed-phase**' col., MeCN/phos. pH 3, 210 nm, sens. <0.2 µg/ml; (3): GC-ECD, sens. 0.1 µg/ml.- ('92) AnMcAgCh 36: (2) Wallace, J.E. et al., 603-606; (3) Rege, A.M. et al., 647-650.

....

Consult #the introduction; #p. 148 for Journal-title & other | *** means scrutiny**
ABBREVIATIONS. They include:- | **warranted**

p = prodrug	auto = automated system	MeCN = acetonitrile
mtb(s) = metabolite(s): **invstg.**	deriv = derivatization	HPLC cols.: e.g. C-18;
if investigated; Phase I,	LE = liq.-liq. extraction	nm value = UV detection
[1], or [1] if incl. *N*-des (oxid. =	⇥dry = drying-down	sens. = sensitivity
oxidative metabolism); Phase	pptn. = precipitation	*SAMPLE usually*
II, [2] (hydrol. = hydrolyzed)	swtch = switching	*human plasma/serum*

Fluorouracil.- (1): pptn. (EtOH); C-8 col.; aq. MeOH; 276 nm; sens. 0.15 µg/ml. Jäger, W. et al. ('90) J Chr **532**, 411-417.   (2): LE; deriv; GC-MS(NCI); sens. ~0.4 ng/ml. Bates, C.D. et al. ('91) J Ph Bm An **9**, 19-21. (3) **(Floxuridine** also): pptn. (PCA, removed from supernatant by KOH); Zorbax Rx col.; phos./MeCN, pH 2.5, grad.; 270 nm; sens. <5 ng/ml. Smith-Rogers, J.A. et al. ('91) J Chr **566**, 147-154 (peritoneal fluid also assayed). (4) (rat, incl. urine): LE, →dry; deriv, →dry; open-tubular GC-MS(EI); sens. 1 ng/ml. Kubo, M. et al. ('91) J Chr **564**, 137-145.   (5) (rabbit also; swtch): LE, →dry; deriv, & LE to wash; RP (CN, then C-18) cols.; MeOH/acet.;   fluor; sens. 0.5 ng/ml. Kindberg, C.G. et al. ('89) J Chr **473**, 431-444.

Fotemustine (a nitrosourea phosphonate; rat, incl. tissues):- LE, →dry; * EC - polarography with no chromatography; sens. 0.3 µg/ml. Berry, D.J. et al. ('91) Xenobiotica **21**, 1211-1216.

Furazolidone (urine & CSF also):- SPE, →dry; C-18; MeOH/acetic acid; 365 nm; sens. 50 ng/ml. Valdez-Salazar, A. et al. ('89) J. Antimicrob.Ch. **23**, 589-595.

Ganciclovir (ther. monitoring):- pptn. (PCA); C-18; phos. pH 3.   Boulieu, R. * et al. ('91) J Chr, (1): **567**, 480-484; & (2) [interference cured]: **571**, 331-333. *   **4-Hydroxyanisole** (ther. monitoring):- direct onto ISRP col.; phos./MeOH pH 6.8; 280 nm; sens. 2 µg/ml. Dawson, C.M. et al. ('90) J Chr **534**, 267-270.

Ifosfamide & oxid. [1]mtbs - chloroacetaldehyde splits off. (1) (urine):- LE, →dry; C-18 (phenyl col. tried); MeCN, grad.; 190 nm; sens. <0.5 µg/ml. Goren, M.P. ('91) J Chr **570**, 351-359.   (2) (plasma & urine; paediatric monitoring):- LE, →dry; cap. GC-NPD; sens. (50 µl sample) 2 ng/ml. Kaijser, G.P. et al. ('91) J Chr **571**, 121-131.

Isoniazid & [1]mtb (rabbit, incl. CSF):- LE; deriv; C-18; MeCN/TEA/heptane-sulphonate pH 6; 320 nm; sens. 0.2 or (mtb) 10 µg/ml. Walubo, A. et al. ('91) J Chr **567**, 261-266.

Isoniazid & [1]mtb (urine; relevant to assessing **acetylation status**):- pptn. (by mobile phase); C-18; phos./propanesulphonate pH 7.0; deriv, then fluor; sens. <50 ng/ml. Kubo, H. et al. ('90) Chr'ia **30**, 69-72.

Ivermectin (cattle).- (1): SPE (C-18 & silica); C-18; MeCN/MeOH; 245 nm; sens. <10 ng/ml.   Oehler, D.D. et al. ('89) J. Assoc. Off. Anal. Chem. **72**, 59. (2) (incl. blood & muscle; method fails with liver): SPE, →dry; C-18; MeCN/MeOH; 254 nm; sens. 2 ng/g muscle. Dickinson, C.M. ('90) J Chr **528**, 250-257.

p**N**-L-**Leucyldoxorubicin** & [1]mtbs incl. **Doxorubicin** (auto):- **pre-dilution** with MeCN/phos. acid; C-8 pre-col.; C-18; MeCN/phos. pH 3.5; fluor; sens. ~0.3 ng/ml. de Jong, J. et al. ('92) J Chr **574**, 273-281.

Loracarbef & oxid. [1]mtb; **Cefaclor, Cephalexin**:- SPE, →dry; C-18; MeOH/THF * /TEA/heptanesulphonate; 265 nm; sens. 0.5 µg/ml. Good accord with bioassay values. Kovach, P.M. et al. ('91) J Chr **567**, 129-139.

Mebendazole & [1]mtbs:- SPE; RP col. ('SAS'); phos./MeCN pH 3; EC; sens. ~0.2 ng/ml. Betto, P. et al. ('91) J Chr **563**, 115-123.

Merbarone:- pptn.; C-18; MeOH/acet./SDS/Mg^{2+}; 306 nm; sens. (on 50 µl) 50 ng/ml. Malpeis, L. & Supko, J.G. ('90) J Lq Chr **13**, 1301-1319.

Meropenem & (by RIA) [1]mtb, **invstg.** (urine also):- direct onto C-18 col.; MeOH/(for urine) phos. pH 7.4; 296 nm. Burman, I.A. et al. ('91) J. Antimicrob. Chemother. **27**, 219-224; cf. ('89) **24A**, 265-277 & 311-320.

Methotrexate.- (1): SPE, →dry; C-18; MeCN/phos. pH 3.9; 313 nm; sens. 9 ng/ml. Fluor.-polarization **immunoassay** equally satisfactory. Cosolo, W. et al.('89) J Chr **494**, 201-208. (2) (paediatric monitoring): pptn. (acetone) & (if concn. low) LE too; C-18; THF/phos. pH 4.85; 313 nm; sens. unclear. Assadullahi, T.P. et al. ('91) J Chr **565**, 349-356.

Methylmercaptopurine riboside (MMPR; incl. urine):- SPE, →dry; C-18; phos./MeOH pH 5.3; 289 nm; phos./MeOH   pH 5.3;   289 nm;   sens.   0.1 µg/ml. Tinsley, P. et al. ('91) J Chr **564**, 303-309.

Metronidazole.- (1): pptn. (MeCN); C-18; MeOH/phos.; 325 nm; sens. 0.1 µg/ml. Paton, D.M. & Webster, D.R. (1988) Int. J. Clin. Pharm. Res. **8**, 227-229. (2) (rat; incl. prostate): LE, →dry; C-18; MeOH/ammon. carbonate; 254 nm; sens. 0.1 µg/ml. Tu, Y-H. et al. ('90) Int. J. Pharm. **61**, 119-125.

Mezlocillin (incl. urine; rat):- pptn. (serum; MeCN, LE to remove); C-18; MeCN/phos. pH 7.0; 214 nm; sens. 0.1 µg/ml.   Jungblath, G. et al. ('89): AnMcAgCh **33**, 839-843, & J Chr **494**, 376-380.

Minocycline & **[1]mtbs** (one oxid.), **invstg.** (urine also):- LE, →dry; * C-18; MeCN/formic acid/DEA, grad.; 352 nm; sens. <0.1 µg/ml. Bioassays also. Bocker, R.H. et al. ('91) J Chr **568**, 363-374.

Mitomycin C (guinea pig; swtch):- ultrafilter; C-18; MeCN/Na perchlorate; EC; sens. <0.2 µg/ml. Ohkubo, T. et al. ('90) J Chr **527**, 441-446.

Mitoxanthrone & oxid. **[1]mtbs:-** LE, →dry; C-18; MeOH/phos. pH 3.0; 242 nm; sens. 1 ng/ml. Hu, O. et al. ('90) J Chr **532**, 337-350.

Mizoribine:- pptn. (PCA); C-18; MeOH/phos./octanesulphonate pH 3; 275 nm; sens. 0.25 µg/ml. Erdmann, G.R. et al. ('89) J Chr **494**, 354-360.

Navelbine (urine also):- LE; C-18; MeCN/MeOH/phos./heptanesulphonate; EC; sens. 1 ng/ml. Nicot, G. et al. ('90) J Chr **528**, 258-266.

* **Neomycin** (cattle; urine also):- for plasma, pptn. (TCA; MeCN caused problems); C-8 col., 33°; acetic acid/pentanesulphonate; deriv (OPA) & fluor; sens. <0.3 µg/ml. Shaikh, B. et al. ('91) J Chr **571**, 189-198.

Norfloxacin: see Amifloxacin.

(S-)Ofloxacin & **[1]mtbs** (incl. urine):- SPE (**better than pptn. or LE**); C-18; THF/phos./acet. pH 2; fluor; sens. 1 ng/ml.   Okazaki, O. et al. ('91) J Chr **563**, 313-322.

Oltipraz (urine also):- LE, →dry; C-18; MeOH/ammon. acet.; 313 nm; sens. 2 ng/ml. Bennett, J.L. et al. ('92) J Chr **572**, 148-149.

Ornidazole & **[1]mtbs** (urine & CSF also):- LE, →dry; C-18; MeOH/phos. pH 6; 313 nm; sens. 0.5 µg/ml. Heizmann, P. et al. ('90) J Chr **534**, 233-240.

* **Oxytetracycline** (blood & plasma, cattle & salmon; auto, swtch):- dialysis; trace-enrichment col.; polymer-RP col.; MeCN/phos. acid/heptanesulphonate; 350 nm; sens. 50 ng/ml. Agasøster, T.J.B. et al. ('91) J Chr **570**, 99-107.

Pyrazinimide:- LE, →dry; HPTLC; benzene/EtOH; 269 nm (scanner); sens. * 1 µg/ml (= HPLC; good accord, but HPTLC fast and simple). Guermouche, M-H. ('91) J. Planar Chromatog. **4**, 166-167.

Rifabutin [a spiropiperidylrifamycin] & **[1]mtbs** (incl. oxid.), **invstg.:-** LE; C-18; MeCN/phos.; 275 nm & radio. Battaglia, R. et al. ('90) J. Antimicrob. Chemother. **26**, 813-822.

* **Rimantadine** & a carbamyl-O-glucuronide **[2]mtb** (**invstg.**):- SPE, →dry; TLC; deriv; GC-MS(NCI). Brown, S.Y. et al. ('90) Drug Metab. Disp. **18**, 546-547.

* **Rufloxacin.-** (1) (urine also): LE, →dry; anion-exch. col.; MeCN/phos. pH 7; 296 nm; sens. 0.1 µg/ml. Carlucci, G. et al. ('91) J Chr **564**, 346-351. (2), + **[1]mtbs** (monkey, urine): C-18; MeCN/phos./DEA, grad., pH 2.1; 246 nm; sens. 0.5 µg/ml. Vree, T.B. et al. ('92) J Chr **572**, 168-172.

Consult # the introduction; #p.148 for Journal-title & other | *means scrutiny
ABBREVIATIONS. They include:- | warranted

p = prodrug
mtb(s) = metabolite(s): **invstg.**
  if investigated; Phase I,
  [1], or **[1]** if incl. N-des (oxid. =
  oxidative metabolism); Phase
  II, **[2]** (hydrol. = hydrolyzed)

auto = automated system
deriv = derivatization
LE = liq.-liq. extraction
→dry = drying-down
pptn. = precipitation
swtch = switching

MeCN = acetonitrile
HPLC cols.: e.g. C-18;
  nm value = UV detection
  sens. = sensitivity
SAMPLE usually
human plasma/serum

Spiramycin (eye also):- SPE; C-18; MeCN/phos.; 231 nm; sens. 50 ng/ml. Carlhant, D. et al. ('89) Biomed. Chr. **3**, 1-4.

Sulbactam: see Ampicillin, above.

Sulfadoxine, Pyramethamine & Mefloquine & oxid. [1]**mtbs** (co-assay, as in ther. monitoring):- pptn. (MeCN & Zn^{2+}); LE (+ TBA-H sulphate), >dry; C-18; MeCN/phos. pH 3.5; 229 nm; sens.: analyte differences (rec. 2% for sulfadoxine). Bergqvist, Y. et al. ('91) J Chr **570**, 169-177.

Sulfalene (incl. blood & red cells):- LE, >dry; NP; DCM/MeOH/PCA; 254 nm; sens. <0.5 ng/ml. Dua, V.K. et al. ('91) J Chr **563**, 333-340.

* **Sulphamidine** & [1]**mtbs** incl. oxid. [synthesis described], **invstg.** (incl. hydrol. urine, & incubates, hydrol. if needed):- LE, >dry; C-18; MeOH/phos. pH 6.7; 265 nm; sens. 0.05 µg/ml.   Van'T Klooster, G.A.E. et al. ('91) J Chr **571**, 157-168.

Suramin:- pptn. (solvent); C-18; MeOH/phos. pH 7.5; 313 nm; sens. 0.1 µg/ml. Tjaden, V.R. et al. ('90) J Chr **525**, 141-149.

Tamoxifen acidic [1+2]**mtbs, invstg.** (rat, urine/liver/uterus/faeces):- LE (MeOH), >dry; TLC; radio. Ruenitz, P.C. et al. ('90) Drug Metab. Disp. **18**, 645-648.

Tamoxifen (auto):- filter & SPE (ISRP) on-line; C-18; MeCN/THF/phos. pH 7.8; 260 nm; sens. 6 ng/ml. Matlin, S.A. et al. (1991) J Lq Chr **13**, 2261-2268.

Tazobactam (β-lactamase inhibitor) & **Piperacillin** (incl. urine & bile):- LE; C-18 in 'ion-suppression' mode; 220 nm; sens. 1 or (urine) 50 µg/ml. Ocampo, A.P. et al. ('89) J Chr **496**, 167-179.

* **Teicoplanin.**- (1) (with microbiol. assay for comparison): SPE (anion-exch.); C-8; MeOH/acet./heptanesulphonate pH 4.0; 240 nm; sens. 0.3 ng/ml. Georgopoulos, A. et al. ('89) J Chr **494**, 340-346. (2): SPE; microbore C-18; MeCN/phos. pH 6.0; 210 nm; sens. 50 ng/ml.   Taylor, R. et al. ('91) J Chr **563**, 451-457.

Temafloxacin (urine also):- ultrafilter, or LE, >dry; C-18; MeCN/phos./SDS /N-acetylhydroxamic acid; fluor; sens. 10 or (if LE) 1 ng/ml. Granneman, G.R. et al. ('91) J Chr **568**, 197-206, also ('92) An McAg Ch **36**, 378-386.

* **Teniposide** (swtch, auto):- add SDS.   (1) Micellar HPLC: C-18, PrOH/phos. /SDS pH 7.0, EC; **or** (2) - simple, fast, inexpensive - C-18 col. for surfactant-mediated clean-up, then as in (1) but phenyl col. and no SDS; sens. 10 ng/ml. Van der Horst, F.A.L. et al. ('91) J Chr **567**, 161-167. See also Etoposide.

Terbinafine & [1]**mtbs** (incl. milk & hydrol. urine; auto):- different plasma/urine methods depending on which **mtb** assayed:- pptn. (MeOH) or LE; C-8 or C-18, grad., EC or 283 nm; **or**, for naphthoic acid (urine): C-18, then deriv & cap. GC-FID; sens. 0.1 or (urine) 0.3 µg/ml. Schatz, F. et al. ('89) Dr Rs **39**, 527-532.

Tobramycin:- deriv; C-8 col.; MeCN/phos. pH 3.5; 340 nm.   Dash, A.K. et al. ('91) J Ph Bm An **9**, 237-245.

Toremifene & [1]**mtbs**:- pptn. (MeCN); C-18; MeCN/acet./TEA pH 6.4; 277 nm; sens. 200 ng/ml. Webster, L.K. et al. ('91) J Chr **565**, 482-487.

Trimethopterin (urine also; ther. monitoring):- SPE; C-18; MeCN/acet.; 254 nm; sens. 0.05 µg/ml. Svirbely, J.E. et al. ('88) J Lq Chr **11**, 1075-1085.

Trospectomycin, complementing colleagues' work [Vol. 20, & Simmonds, R.J. et al. ('91) J Lq Chr **13**, 1125-1142]:- radiolabel disposition. Nichols, D.J. et al. ('91) Xenobiotica **21**, 827-837.

Xanthenone-4-acetic acid (mouse):- pptn.; LE; C-18 cartridge; MeCN/acetic acid; fluor; sens. 0.05 µg/ml. Prone to photodecomposition. Kestell, P. et al. ('91) J Chr **564**, 315-321.

Zidovudine (AZT) & [2]**mtb** (glucuronide).-   (1): pptn. (TCA), then 1 h at 60°; C-18; MeCN/phos. pH 2.7; 267 nm; sens. 10 ng/ml. Kamali, F. & Rawlins, M.D. ('90) J Chr **530**, 474-479.   (2) (monkey; urine also): C-18; aq. MeCN; 267 nm. Qian, M. et al. ('91) J Ph Bm An **9**, 275-279.

#7. UNCLASSIFIED AGENTS, e.g. anti-diabetics, diuretics, anaesthetics, relaxants, anti-ulcer drugs, hormone-related agents

Vol. 16 featured bioactive peptides (also chiral studies)

N-Acetylcysteine. - DTT (dithiothreitol) incubation (reduces disulphides); pptn. (acetone); LE, ⇥dry; deriv; GC-MS(EI). DeCaro, L. et al. ('89) Dr Rs **39**, 382-386.

Alcuronium (incl. urine).- SPE, ISRP pre-col.; C-18;   MeOH/acetic acid; 280 nm; sens. 5 ng/ml. DeBros, F. et al. ('90) J Chr **529**, 449-454.

Amiloride.- (1) (analogues too; rat): LE, ⇥dry; C-18, 35°; MeCN/PCA pH 2.2; 361 nm; sens. <10 ng/ml. Meng, Q. et al. ('90) J Chr **529**, 201-209. (2) (urine also): pptn. (PCA); C-18; MeOH/PCA; fluor; sens. 0.5 ng/ml. Xu, D-K. et al. ('91) J Chr **567**, 451-458.

ε-Aminocaproic acid (e.g. 'Amicar'; incl. urine):- pptn. (Zn²⁺/MeOH
* [cf. Clin. Chem. **26**, 963]); deriv (OPA); C-18; MeCN/Cu²⁺-proline complex; fluor; **sens.** 50 ng/ml. Lam, S. ('90) Biomed. Chr. **4**, 175-177.

Bisphosphonates, e.g. APD = 3-amino-1-hydroxy- (urine also; auto):- co-pptn. with Ca²⁺ phos.; endogenous phosphates removed by anion-exch. col.; dil. nitric acid, pre-purged of metal ions; post-col. (flow-injection) oxid. deriv to release phos., + Mo Blue; 820 nm; sens. 25 or (urine) 250 ng/ml. Hoggarth, C.R. et al. ('90) Anal. Proc. **27**, 18-19.

Bumetanide (urine also; auto):- pptn. (MeCN); SPE, ⇥dry; C-18; MeOH/acetic acid; fluor; sens. 5 ng/ml. Wells, T.G. et al. ('91) J Chr **571**, 235-242.

Bupivacaine:- LE; phenyl col.; MeCN/phos.; 210 nm; sens. 1 ng/ml. Michaelis, H.C. et al. ('90) J Chr **527**, 201-207.

Calcipotriol [vit. D kinship] & ¹**mtbs**, shown to be inactive (rat, incl. cell studies):- pptn. (MeCN); SPE, ⇥dry; C-18; aq. MeOH, grad.; 264 nm. Kissmeyer, A-M. & Binderup, L. ('91) Biochem. Pharmacol. **41**, 1601-1606.

S-Carboxymethyl-L-cysteine.- (1): pptn.; deriv; C-18 (swtch); fluor; sens. 0.1 µg/ml. Guinebault, P. et al. ('88) Chr'ia **26**, 377-382. (2) (swtch): pptn. (MeOH); pre-col., heart-cut; C-18; MeCN/THF/phos. pH 7.0; deriv (OPA); fluor; sens. ~0.5 µg/ml. de Schutter, J.A. et al. ('88) J Chr **428**, 301-310. (3), S-oxide ¹**mtbs** (urine): deriv; C-18; EC. Woolfson, A.D. et al. ('87) Analyst **112**, 421-425.

Cimetidine (tissues):- LE, ⇥dry; C-8 col.; MeCN/phos. pH 4.9; 228 nm; sens. ~10 ng/g. Imamura, T. et al. ('90) J Chr **534**, 253-259.

Crotamiton, *trans* isomer (incl. urine):- LE, ⇥dry; C-18; MeCN/phos.; 220 nm; sens. 9 or (urine) 60 ng/ml. Sioufi, A. et al.('89) J Chr **494**, 361-367.

(Des)ferrioxamine (Deferoxamine):- SPE; C-18; MeOH/acet. pH 4.7; EC; sens. ~5 µg/ml. Glennon, I.D. & Senior, A.T. ('90) J Chr **527**, 481-489.

Dextromethorphan & Dextrorphan (urine):- SPE; phenyl col.; MeCN/MeOH/phos. /octanesulphonate; pH 2.5; fluor. Wenk, M. et al. ('91) J Ph Bm An **9**, 341-344.

Enoxolone: see Glycyrrhetinic acid.

Ethacrinic acid:- LE, ⇥dry; C-18; MeOH/MeCN/TEA/phos.; 280 nm; sens. 0.1 µg/ml. LaCreta, F.P. et al. ('91) J Chr **571**, 271-276.

pEtophyllin|clofibrate & active ¹**mtbs** (split|as shown):- LE; C-18; MeCN /heptane sulphonate; 275 nm; sens. 0.5 µg/ml.   Ostrovska, V. et al. ('89) J Lq Chr **12**, 2793-2799.

....

Consult # the introduction; #p.148 for Journal-title & other |* means scrutiny
ABBREVIATIONS. *They include:-* | warranted

p = prodrug	auto = automated system	MeCN = acetonitrile
mtb(s) = metabolite(s): **invstg.**	deriv = derivatization	HPLC cols.: e.g. C-18;
if investigated; Phase I,	LE = liq.-liq. extraction	nm value = UV detection
¹, or ▪ if incl. *N*-des (oxid. =	⇥dry = drying-down	sens. = sensitivity
oxidative metabolism); Phase	pptn. = precipitation	*SAMPLE usually*
II, ² (hydrol. = hydrolyzed)	swtch = switching	*human plasma/serum*

Fellavine (plant-derived) & nameless [1]**mtbs** (rat & human; incl. urine), swtch (back-flush transfer from RP pre-col. that replaces sample loop):- C-18; MeCN/acetic acid; 280 nm; sens. 0.05 µg/ml.   Tjukavkina, N.A. et al. ('91) J Chr 571, 312-317.

Fenfluramine & [1]**mtbs** incl. oxid., **invstg.** (urine, hydrol.):- XAD-2 resin & other SPE (garbled!); clean-up on C-18, grad.; deriv; cap. GC-MS(EI/CI). Brownsill, R. et al. ('91) J Chr 562, 267-277.

Finasteride:- CN-SPE; C-8; MeOH/MeCN; 210 nm; sens. 1 ng/ml.   Constantzer, M.L. et al. ('91) J Chr 566, 127-134.

Furosemide (Frusemide; urine also):- LE, freeze & →dry; C-18; MeCN/phos. acid, grad.; fluor; sens. 10 ng/ml. Saugy, M. et al. ('91) J Chr 564, 567-578.

Gemfibrozil & [1,2]**mtbs** (urine also; auto):- col. (C-18/phenyl)/eluent/fluor setting depends on matrix/analytes; sens. 100 ng/ml. Takagawa, H. et al. ('91) Biom. Chr 5,68-73.

Glimepiride & oxid. [1]**mtb** (urine also):- LE, →dry; thermolytic deriv; C-18; MeCN/PCA, grad.; 350 nm; sens. 5 ng/ml.   Lehr, K.H. et al. ('90) J Chr 526, 497-505.

Glipizide, Glyburide: see *Sulphonyl...*

Glycyrrhenetic acids, 18α & (**Enoxolone**) **18β.-**   (1) (rat): pptn. (MeCN); C-18; MeOH/ammonia/PCA; 254 nm; sens. 0.1 µg/ml.   Tsai, T-H. et al. ('91)
* J Chr 567, 405-414. (2), β isomer:+urea (disrupts binding); SPE, ion-pairing; C-18; MeOH/acetic acid; 248 nm; sens. 10 ng/ml.   Brown-Thomas, J.M. et al. ('91) J Chr 568, 232-238.

Ketamine & **Norketamine:**- LE, →dry; **chiral** (AGP) col.; iPrOH/phos. pH 7; 215 nm; sens. 40 ng enantiomer/ml. Geisslinger, G. et al. ('91) J Chr 568, 165-176.

Lansoprazole & [1]**mtbs** incl. oxid. (urine also):- LE, →dry; C-18, 40°; MeCN/EtOH/octylamine/phos. pH 7; 285 or 303 nm; sens. (plasma) 5 ng/ml. Aoki, I. et al. ('91) J Chr 571, 283-290.

Levodropropizine:- SPE; cap. GC-MS(EI); sens. 5 ng/ml.   Zaratin, P. et al. ('88) Dr Rs 38, 1156-1158.

Lonapalene & oxid. [1]**mtb** (rat urine, hydrol.):- phenyl- & silica-SPE; C-18; MeCN/THF/citric acid/TBA-acet.; 280 nm; sens. <0.1 µg/ml. Kiang, C-M. et al. ('91) J Chr 565, 339-347.

Methazolamide (whole blood):- LE, pH 3; C-18; acet./MeCN; 290 nm; sens. <1 ng/ml. Way, S.L. et al. ('87) JPhSc 76, S13.

Metyrapone & oxid. [1]**mtbs** ('tool' in drug-metabolism & ACTH studies; liver fractions besides biofluids):- LE; C-18; MeCN/phos. pH 7.4; 261 nm; sens. ~1 ng/ml. Usansky, J.I. et al. ('91) J Chr 563, 283-298.

Nitecapone: in Group #5.

Nizatidine & [1]**mtbs** (one oxid.):- LE, →dry; NP col.; MeCN/MeOH; 320 nm; sens. 6-18 ng/ml. Tracqui, A. et al. ('90) J Chr 529, 369-376.

* **Nonoxy-9** (urine/vaginal fluid too):- pellic'r amino col.; THF/MeCN ('NP' eluent); fluor; sens. (urine) 0.2 µg/ml. Beck, G.J. et al. ('91) JPhSc 79, 1029-1031.

Onapristone & [1]**mtb:**- seemingly direct onto guard-col.; C-18; MeCN/phos. pH 7.2; 315 nm; sens. 5 µg/ml. Zurth, Chr. et al. ('90) J Chr 532, 115-123.

Pethidine & [1]**mtbs** (urine, hydrol.):- LE, →dry; GC-NPD; sens. 0.01 µg/ml. Chan, K. et al. ('91) J Chr 565, 247-254.

Phenazopyridine & oxid. [1]**mtbs** (incl. animals; urine, hydrol.):- pptn. (MeOH), →dry; C-18; MeOH/MeCN/amm. formate grad.; 248 & 430 nm; sens. ~3 µg/ml. Thomas, B. et al. ('90) JPhSc 79, 321-325.

Phenprocoumon & [1,2]**mtbs** (urine, hydrol. where needed):- SPE (NR_4^+); Hyperchrome col.; MeCN/MeOH/THF/acetic acid pH 2.7; 312 nm; sens. <25 ng/ml. Edelbroek, P.M. et al. ('90) J Chr 530, 347-358.

Phenylpropanolamine (urine also):- LE, →dry; 2-stage C-18, swtch; MeCN/phos./butanesulphonate pH 3.5; 205 nm; sens. 0.4 or (urine) 8 ng/ml. Yamashita, K. et al. ('90) J Chr 527, 103-114.

Prednis(ol)one & [1]**mtbs** (perfusion media):- SPE; C-18; MeOH; 242 nm; Cannel, G.R. et al. ('91) J Chr 563, 341-347.

* **Prilocaine:**- SPE, **preceded by** LE (advantageous!); CN col.; MeCN/phos.
acid; EC; sens. 5 ng/ml. Whelpton, R. et al. ('90) JChr **526**, 215-222.

Propofol (whole blood):- pptn. (MeCN); SPE; RP (phenyl) col.; MeOH/phos.
pH 6.8; EC; sens. 20 ng/ml. Nazzi, G. & Schinella, M. ('90) JChr **528**, 537-541.

Propylthiouracil:- LE, ⇥dry; C-18; MeOH/phos. pH 7.4; 214 nm; 40 ng/ml.
Cannel, G.R. et al. ('91) JChr **564**, 310-314.

Retinoids: see Acitretin in Group #6, & Rissler, K. et al. ('91) JChr 565, 375-382.

Ritalinic acid, active [1]**mtb** from *p*Methylphenidate:- LE; chiral invstg.;
cap. GC-ECD; sens. 5 ng/ml. Srinvas, N.R. et al. ('90) JChr **530**, 327-336.

Simvastatin:- SPE; 'deriv' (lactone → acid); GC-MS; sens. 0.1 ng/ml.
Takano, T. et al. ('90) Biomed. Environ. MS **19**, 577-581.

Sparteine & oxid. [1]**mtbs** (urine):- direct onto CN col.; MeOH/MeCN/phos.
pH 2.5; EC; sens. 4 ng/ml. Moncrieff, J. ('90) JChr **529**, 194-200.

Stanozol & oxid. [1]**mtb** (urine):- SPE; hydrol.; deriv; cap. GC-MS; sens.
1 ng/ml.   Choo, H.P. et al. ('90) J. Anal. Toxicol. **14**, 109-112.

Succinylcholine:- pptn. (TCA); C-8; MeCN/MeOH/phos. pH 5.0; fluor; sens.
100 ng/ml. Lagerwerf, A.J. et al. ('91) JChr **570**, 390-395.

Sulphonylureas, e.g. therapeutic monit'g.- **Chlorpropamide, Glibenclamide,
Glipizide, Tolazamide, Tolbutamide:**- LE, ⇥dry; deriv; C-18; phos. acid/MeCN;
360 nm; sens. 40 ng/ml. **Starkey, B.J.** et al. ('89) JLqChr. **12**, 1889-1896.
Tolbutamide & [1]**mtbs** incl. oxid. (rat also; incl. urine):- LE, ⇥dry;
C-18; MeCN/phos./TBA-phos. pH 7; 254 nm; sens. 100-200 ng/ml. St.-Hilaire, S.
& Belanger, P.M. ('89) JPhSc **78**, 863-866. **Glyburide:**- SPE; C-18; MeCN
/Me₄N⁺ClO₄⁻ pH 3.5; fluor; sens. 10 ng/ml. Gupta, R.N. ('89) JLqChr **12**, 1741-1758.

Telenzepine:- SPE, ⇥dry; deriv; GC-MS; sens. 2 ng/ml.   Sturm, E. &
Junker, A.J. ('88) JChr **430**, 43-51.

Theophylline (saliva also; ther. monitoring; caffeine &c. don't interfere):-
pptn. (PCA); C-18; MeCN/MeOH/phos. pH 4.7; 247 nm; sens. 25 ng/ml. Moncrieff, J.
('91) JChr **568**, 177-185.

Thiazolidine-type anti-diabetic (CP 68,722) & [1]**mtbs** incl. oxid., **invstg.**
(rat; bile, after hydrol., & incubates):- (bile) pptn. (MeOH); C-8 col.;
MeCN/acet./(where needed) THF pH 4.7; 230 nm & radio. Fouda, H.G. et al.
('91) Xenobiotica **21**, 925-934.

Tor(a)semide & [1]**mtbs** incl. oxid. (urine also):- SPE × 2 (⇥dry), #1: C-2,
& #2: propylsulphonyl; C-18; MeCN/phos. pH 4.5, grad.; 290 nm; sens. 10
(urine, 20) ng/ml. March, C. et al. ('90) JPhSc **79**, 453-457. Previous
* study (auto; SPE comparisons):- Karnes et al. ('89) JLqChr 12, 1809-1818.

17β-Trenbolone & oxid. [1]**mtbs** (urine; relevant to residues in beef):-
C-18; MeOH/MeCN, grad.; 340 nm & radio; **mtb** verified by GC-MS. Spranger, B.
et al. ('91) JChr **564**, 485-492.

* **Warfarin** (swtch):- SPE, ⇥dry; ISRP col., & drug peak put onto chiral
(AGP) col.; iPrOH/phos. pH 7; fluor; sens. 10 ng enantiomer/ml. McAleer, S.D.
& Chrystyn, H. ('90) J.Pharm. Pharmacol. 42 (Suppl.), 21P.

* **Zeranol** & [1]**mtb**, β**-Zearalanol** (calf urine, hydrol.):- affinity-SPE (Ab),
⇥dry; deriv; GC-MS(NCI); sens. unclear (& **mtb** quantitation would need
special Ab). Bagnati, R. et al. ('91) JChr **564**, 493-502.

Zopiclone (ther. monitoring):- LE, ⇥dry; cap. GC-NPD; sens. 2 ng/ml.
Kennel, S. et al. ('90) JChr **527**, 169-173.

COMMENTS ON THE COMPENDIUM
supplementing the guidance that appears initially

#An indication of analyte **structure** is obtainable from its assigned category in the ANALYTE INDEX - which, conversely, should be consulted for **other text material** on the analyte.

#A **'state-of-the-art'** implication for other analytes (cf. Sect. #D) is evident in some entries, e.g. 'dialysis' for Oxytetracycline (Group 6), 'deriv' for S-Carboxymethylcysteine (Group 7), 'polarography' for Fotemustine (Group 6).- Cf. the alerting mark, *.

#Where a particular drug seems bioanalytically *déja vu*, having featured repeatedly in the past decade (cf. ANALYTE INDEX), justification of yet another paper (not prestige-seeking?) may be:
- devising instrumental processing ('auto', 'swtch');
- simplification of sample-preparation, e.g. through SPE (replacing LE) or even direct injection;
- in the latter and other contexts, a newer column packing may have been used, e.g. ISRP or polymer-RP packings [for these and classical packings too, there is deplorable use of uninformative trade names only];
- 'chiral' distinction (as investigated for Terfenadine, Group 2) as an assay innovation which will survive the anticipated marketing trend away from racemic drugs in favour of the active form: possible formation of the other enantiomer (as a metabolite?) *in vivo* will still have to be investigated;
- improvement in **sensitivity**, which nowadays may have to cope with plasma levels below 1 ng/ml for drugs given in low dosage or topically, or with tiny plasma or biopsy samples:-
    # values given on a molar rather than a weight basis are a minor irritation; an on-column basis or lack of any sensitivity indication is a more serious offence;
    # sensitivity demands are reflected in detector choice - increasingly MS or (not for GC) EC or fluorimetry; but the first choice is still UV [whilst 254 nm, instrumentally advantageous, is a dubious fashion now resurrected, many authors now take the interference-prone 200-230 nm region in their stride].

#Certain drugs serve to establish **metabolic capabilities** (e.g. for hydroxylation) in individuals or in incubates [hepatocyte example, using diazepam:- Blankson, E.A. et al. ('91) Biochem. Pharmacol. **42**, 1241-1245], as in the entries for Antipyrine (Group 1) and Debrisoquine (Group 4).

EARLIER ANALYTICAL ('A') BOOKS IN THE SERIES (ed. E. Reid.....)

Group headings above indicate 'therapeutic specializations'; see also the list facing title p. of present book. In summary:
#Sole source of V.5 is now *via* 'GAA' (72 The Chase, Guildford, U.K.). #V. 7 & (O/P) 10: Horwood. #V. 12, 14, 16 & 18: Plenum. V.20: as for present vol. *Use as a reference set is aided by their indexing.* [Vols. not listed are in the Biochemistry ('B') subseries.]

Section #C

ANTI-ASTHMATICS AND KINDRED AGENTS

Art. #B-7 is also pertinent, and Sect. 2 (p. 151) – also Sects. 1 & 3 – in #ABC

For diagram of **eicosanoid** area, see p. 225

#C-1

ASTHMA AND ANTI-ASTHMATIC DRUGS: A SURVEY

I.F. Skidmore*

Biochemical Pharmacology Division,
Glaxo Group Research Ltd., Ware, Herts. SG12 0DG, U.K.

This article serves as background for the bioanalytical articles that follow. Asthma is characterized by reversible obstruction of the airways caused by factors of varying importance - bronchoconstriction, inflammation and hypersecretion of mucus. The immediate life-threatening symptoms respond to drugs, now surveyed in respect of type of action. Administration is now commonly by inhalation, posing problems for the analyst and provoking questions about the value of kinetic studies for inhaled drugs. Problems in relating systemic exposure to therapeutic effect are particularly difficult with the inhaled steroids where events downstream from receptor occupancy may control duration of action. Future therapy may be centered on the pathobiological basis of the disease, and this will provide a fresh set of problems for the bioanalyst.

This article has three main purposes:- (1) to place the current drugs used in the treatment of asthma into the context of what we understand about the disease; (2) by looking at the pathobiology of asthma, to suggest where new drugs may arise; finally (3), this Section being primarily concerned with drug bioanalysis, to discuss what benefits may arise from being able to measure these drugs in biological fluids.

Asthma is a very complex and remarkably common disease. It is estimated that between 5% and 10% of the population of developed countries suffer from asthma to some degree. Furthermore, because the disease has been, and to some extent still is, under-diagnosed, it is still seriously under-treated. Asthma is also a very variable disease, ranging from mild intermittent episodes which have very little effect on the life-style of the subject to, in some cases, attacks of such severity, *status asthmaticus*, that life is threatened.

------

*Current address: Vice President - Development, Glaxo Research Institute, Glaxo Inc., 5 Moore Drive, Research Triangle Park, NC 27709, U.S.A.

Editor's note.- Dr Skidmore was co-organizer of Forum sessions on which Sect. #C is based. [Cross-refs. appear in Table 1 to accompanying arts., wherein the structural formula for mentioned drugs may appear.]

Asthma is characterized by variability in the flow of air in the bronchi, associated with hyper-responsiveness of the airways to irritant stimuli such as cold air, dusts, irritant gases and pharmacological agents such as histamine, PG's*, adenosine and methacholine⊕. Frequently, but not invariably, there is an immunological component to the disease and the patient demonstrates allergy to one or more common airborne antigens such as pollens [cf. art. #ncC-6], mould spores and animal dander which is manifested by the symptoms of asthma.

The physical origin of the primary symptom of asthma, a reduction in the flow of air, is a narrowing of the airways. There are three major causes of this:- bronchoconstriction caused by contraction of the circular smooth muscle of these airways, inflammatory swelling caused by oedema and cellular infiltration (i.e. the leakage of fluid out of the capillaries and into the surrounding tissues and the active movement of inflammatory cells from the blood into the tissues) and hypersecretion of mucus leading in extreme cases to the blocking of the airways.

Until relatively recently, physiological measurements were the only way of assessing the pathological status of a patient unless a severe asthmatic attack had actually led to death and could be investigated post mortem. As a consequence, a view of asthma developed which considered that the primary cause of airway obstruction in the mild disease was broncho-constriction, and inflammation became important only as the condition became more severe. With the advent of invasive investigative techniques such as bronchoscopy with biopsy or cellular lavage it has become clear that even in mild asthma the airways are inflamed, and now it is considered not only that inflammation is present throughout all stages of asthma but also that the severity of the inflammation is directly related to the severity of the disease and is a cause of the hyper-responsiveness of the airways that is characteristic of the disease (Fig. 1).

Fig. 1. Features
of asthma.

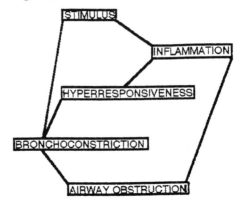

*Abbreviations.- NSAID, non-steroidal anti-inflammatory drug; PAF, platelet-activating factor; PG, prostaglandin.
⊕$CH_3COOCH(CH_3)CH_2\overset{+}{N}(CH_3)_3$; for some other drugs cited, later arts. may show structure.

The modern view of asthma is, then, that inflammation and bronchoconstriction, linked by the dependence of hyper-responsiveness on the inflammatory state, are the critical components in asthma in all its stages. It is, therefore, not surprising that the two most important therapeutic classes used in the treatment of the disease are targetted at these two components. However, both the bronchodilators and the anti-inflammatory steroids predate our current views of asthma by at least 15 years, their development and use being based on a pragmatic response to the belief that both bronchoconstriction and inflammation were important in the disease but in ignorance of their interrelationship.

BRONCHODILATORS

The bronchodilators are physiological antagonists, that is they have an action in the body which is opposite to that which we wish to treat; in this case they relax smooth muscle, opposing the contraction of bronchoconstriction. Almost without exception they are based on the natural catecholamine neurohormone adrenaline and are potent and selective agonists at the β-adrenergic receptor present in the plasma membrane of the smooth muscle cells. Activation of this receptor by these drugs leads, *via* a series of complex biochemical reactions, to the relaxation of smooth muscle fibres. Predominant among these drugs are the five bronchodilators listed below (Table 1), all but the last being given mainly either as metered dose aerosols or as dry powders directly into the lung to avoid the systemic side-effects associated with oral administration of these drugs.

Table 1. Some drugs used in the treatment of asthma.

Class	Drug (& ref. elsewhere in book)
Bronchodilators	Salbutamol (#D-6)
	Terbutaline (#C-3)
	Fenoterol
	Salmeterol (#A-2, #ncC-4, #ncC-4)
	Clenbuterol (#ncC-2)
Anti-bronchoconstrictors	Ipatropium bromide
Xanthines (mechanism unclear)	Theophylline
Prophylactic/anti-inflammatory	Disodium cromoglycate (Cromolyn sodium)
	Nedocromil (#C-2) [(#C-2)
	Ketotifen (p. 249)
Steroid anti-inflammatory	Beclomethasone dipropionate (Vol. 18)
	Budesonide (#C-3)
	Fluticasone propionate (#C-5)
In future? -	
Anti-eicosanoid agents	(#C-7, #C-8, #ncC-1)

The other class of drugs working on bronchoconstriction are the anticholinergics such as ipratropium bromide. These strictly speaking are not bronchodilators but antibronchoconstrictors, that is they are pharmacological antagonists blocking the action of the bronchodilator acetylcholine at its receptor. In some asthmatics acetylcholine released from vagal nerve endings in the lung causes bronchoconstriction, and in these patients anticholinergics are effective. Obviously in patients without a vagal component to their asthma they are inactive.

STEROIDS

Anti-inflammatory steroids but not NSAID's such as aspirin and indomethacin are highly effective in asthma. The fact that the former, exemplified by beclomethasone diproprionate and budenoside (Table 1), but not the latter are useful in asthma is clear evidence that the inflammatory process in the asthmatic lung is complex, an indication borne out by the results of bronchoscopy, bronchiolar lavage and biopsy which show not only oedema but also the influx of a variety of inflammatory cell types into the tissues. The effectiveness of steroids can also be shown using these techniques; regular treatment of an asthmatic with inhaled steroids leads to a resolution of the pulmonary inflammation associated with a dramatic improvement in lung function and a reduction or even elimination of attacks of bronchoconstriction.

MAST CELL STABILIZERS

There is one other class of anti-asthmatic drugs which must be considered - the 'mast cell stabilizers', specifically disodium cromoglycate (cromolyn sodium), nedocromil and ketotifen. There is no doubt that in certain classes of patients these are effective drugs, but not all subjects respond. There is uncertainty about the mechanism of action of these drugs. Almost certainly they do not work by stabilizing mast cells. They may be anti-inflammatory but are by no means as effective as the steroids in patients where inflammation predominates. They may also block sensory fibres in the lung, thus preventing axon reflex-dependent constriction and neurogenic inflammation. The search for how these drugs work gave a major stimulus to the investigations of the pathobiology of asthma from which the next generation of anti-asthma drugs may come.

Unusually for a new type of drug, the activity of disodium cromoglycate was first identified in man; indeed it was some time before meaningful activity could be found in any pharmacological test in experimental animals. Eventually it was found that the drug blocked the release of histamine from rat mast cells when they had been sensitized to specific antigen with a newly discovered class of antibody which bound to these cells.   Because histamine was known to be a powerful bronchoconstrictor   and   inflammatory   agent   it   was,   quite reasonably, assumed that this was the basis of the therapeutic activity   of   cromoglycate,   and throughout   the   pharmaceutical industry   programmes   to   make   mast cell   stabilizers   were   set up.   They   were   very   successful,   producing   compounds   many hundreds   of   time   more   potent   than   cromoglycate   in   the   rat mast   cell.   Unfortunately   this   enhanced   potency   was   not   seen in the asthmatic, and it now appears that mast-cell stabilization was   not   the   mechanism   by   which   this   drug   worked   and   that data in the rat were not predictive for man.

CLARIFICATION OF THE PATHOBIOLOGY OF ASTHMA

By   the   time   the   difference   between   rat   and   man   was realized,   both   academic   and   industry-based   groups   had   become intensely   interested   in   the   pathobiology   of   asthma.   What were the cellular and subcellular mechanisms involved in broncho-constriction   and   inflammation?   What   substances   released   from which   cells   caused   smooth   muscle   to   contract   in   response to   stimuli   as   diverse   as   cold   air   and   grass   pollen?   Why were   eosinophils   and   macrophages   abundant   in   lung washes from asthmatics?   -   Were   they   cause   or   effect?   How   did   all   these inflammatory   cells   get   there?   It   is   beyond   the   scope   of this   article   to   do more   than   summarize   the   results   of   15   years' research   and   to   indicate   what   pharmaceutical   targets   this research has identified.

There   are   two   major   theories   of   the   pathobiology   of asthma,   the   Cell-Mediator   hypothesis   and   the   Neurogenic hypothesis.   They   are   not   mutually   exclusive;   components   of both are probably involved in the overall response.

The   **Cell-Mediator hypothesis**   takes   the   view   that   cells in   the   lung   respond   directly   to   stimuli   by   releasing chemical mediators   which   cause   bronchoconstriction   and   inflammation directly.   The   inflammation   itself   involves   other   cell   types which   then   release   more   mediators,   perpetuating   the   response. This   raises   obvious   questions.-   What   cells   are   stimulated? What   do   they   release?   What   cells   are   summoned   up?   What do they release?

While   the   most   obvious   cell   type   to   be   considered   is the   mast   cell,   in   view   of   both   its   sensitivity   to   antigen and   of   the   range   of   potential   mediators   that   it   releases, the   lung macrophage,   the   neutrophil   and   the   eosinophil   are also   established   sources   of   mediators   and   certainly   play   a role   in   the   later   stages   of   the   cellular   response.   The

Table 2. Postulated mediators of asthma and their activities.

Agent	Constriction	Oedema	Chemotaxis
Histamine	+	+	
Leukotriene C/D	+	+	
Leukotriene B			+
Prostaglandins (PG's)	+	+	
Thromboxane	+		
Neuropeptides	+	+	
PAF	+	+	+
Peptide chemotaxins			+

list of mediators released is extensive, and their properties
are consistent with the induction of both bronchoconstriction
and inflammation (Table 2).

The **Neurogenic hypothesis** also involves cells responding
to an initial stimulus that releases mediators which cause
pulmonary inflammation that leads to activation of sensory
fibres in the lung: this stimulates an axon reflex which
in turn induces both further inflammation and bronchoconstriction.   Questions then arise.- What are the initial target
cells?   What do they release?   What are the transmitters
involved in the axon reflex?   Again the same cell types
are candidates for initiating the response, and the same
group of mediators is probably involved but the transmitters
involved in the axon reflex are in all probability the neurokinin
group of peptides.

Drug targets suggested by these hypotheses

When we consider the potential therapeutic targets offered
by both of these hypotheses, it is obvious that there are
several important considerations.   Inflammation is a complex
process, involving many mediators and several cell types.
Targetting an individual mediator will be of value only if
this mediator plays a key controlling role in the response.
For example, although histamine is an important component
of mast cells and is released in quantity in response to
antigenic stimuli, highly potent H_1 antagonists are not effective
even in the acute bronchoconstriction of asthma (although
quite effective in hay fever).   It can therefore be assumed
that histamine does not play a controlling role in bronchospasm.
Similarly, although PG's with pro-inflammatory activity are
generated by the cells involved in the asthmatic response
and are known to be important in inflammation, PG sythetase
inhibitors either are ineffective or exacerbate asthma.
Leukotriene antagonists are also without useful activity, and
no therapeutically effective lipoxygenase inhibitors have yet
been developed.

A potentially more promising target is PAF. This product of phospholipid metabolism is produced by many inflammatory cells and has a range of biological properties which suggest that it could be a key mediator in asthma (Table 2).   Potent antagonists of PAF have been developed*, but so far clinical results have been disappointing.   Neuropeptide antagonists offer an attractive target for the pharmacologist and the medicinal chemist, and developments in this area are awaited with interest. However, none of these agents will be effective unless their target mediator plays a central role in asthma. The adrenergic bronchodilators of course do not have this problem as they are not mediator-specific; they relax smooth muscle that has been contracted by any agent.   In the same way the anti-inflammatory steroids will counteract inflammation of whatever origin.   The mechanism of the anti-inflammatory action of these drugs is still not clear. They evidently prevent the synthesis of PG's and leukotrienes by interfering with the availability of arachidonic acid required for their synthesis; but we know that inhibitors of PG synthetase or lipoxygenase and antagonists of leukotrienes are ineffective. It seems much more likely that these drugs have fundamental inhibitory actions on a variety of cell types and on those molecules, cytokine peptides or interleukins, which they use to communicate with and activate each other. This inhibitory activity is likely to be operating at or close to the level of gene expression and its control.

ROLE OF BIOANALYSIS

Despite the concerns shared by many investigators that targetting individual mediators is not the way forward in the development of new drugs for asthma, we can expect antagonists of many of the proposed mediators of asthmatic bronchoconstriction and inflammation to enter pharmaceutical development over the next few years, providing a series of challenges for the analyst.   Before we get to this stage there are already many challenges before us.   Most drugs for the treatment of asthma are now given by the inhaled route, to maximize

*Editor's interpolation.- For the biochemical pharmacology of PAF, see art. #B-4 (D.P. Tuffin & P.J. Wade) in Vol. 17, this series [Cells, Membranes and Disease, including Renal, ed. E. Reid et al.; Plenum, 1987].   The eicosanoid area (diagram: p. 225 of present vol.) features in Vol. 18 [ed./publ. as for Vol. 17; 1988] together with bioanalytically oriented arts. on various anti-inflammatories including steroids (beclomethasone dipropionate: #B-2, W.N. Jenner & D.J. Kirkham). For interleukins see art. #A-3 (L.A.J. O'Neill et al.) in Vol. 21 (Cell Signalling, ed. E. Reid et al.; Roy. Soc. Chem., 1991).   The 'Compendium' (#ABC) that precedes the present art. gives, in Sect. 2 (also 1 & 3), apposite bioanalytical refs.

effective delivery of the drug to the target organ - the
lung - and to reduce the potential for systemic side-effects
that would be present if the drugs were given orally. Indeed
it is accepted and expected that for orally administered
drugs there is a regulatory requirement to investigate the
pharmacokinetics and biopharmaceutics thoroughly, to establish
the basic parameters and the effects that formulation may
have on them, to establish dose linearity (or demonstrate
non-linearity), to establish bioavailability and bioequivalence,
and to demonstrate dosage-form proportionality. However, it
has generally been accepted that such information is not
required for inhaled drugs.

     There are three reasons for this.-
(1) When the drug is administered into the lung, drug reaching
the systemic circulation represents drug which has either
left the target organ or has never reached it because it
has been swallowed, the proportions may vary.
(2) It is very difficult if not impossible to relate systemic
concentrations to therapeutic effects when the drug has been
given by inhalation. This is recognized by the requirement
by the authorities for clinical rather than bioequivalence
studies to be done when introducing new formulations or generic
versions.
(3) The inhaled doses given have been too low to permit
accurate measurement of systemic concentrations at therapeutic
dose levels.

     However, as manifested in some articles that follow,
improvements in analytical techniques and instrumentation and
the introduction of new technologies now make it possible
for the analyst to measure systemic concentrations after inhaled
doses. The authorities are aware of this, and are starting
to ask for the information. As I have indicated above,
there is little that can be done with such information.
We should under such circumstances discuss with the authorities
the value of conducting such studies.

'Background' references

This list of references is not and is not meant to be comprehensive. It focuses not on original papers, but rather on
books useful to any reader who wishes to learn about asthma
and the drugs used to treat it.

Asthma, Basic Mechanisms and Clinical Management. Barnes, P.J.,
     Rodger, I.W. & Thomson, N.C., eds. (1988): Academic
     Press, London/New York [ISBN 0-12 079825-4].

Handbook of Experimental Pharmacology, Vol. 98. The Pharmacology of Asthma. Page, C.P. & Barnes, P.J., eds. (1991): Springer Verlag, Berlin. [ISBN 3-540-52839-3.]
New Drugs for Asthma. (1989): IBC Technical Services, London. [ISBN 1 85271 0551.]
Asthma. Clinical Pharmacology and Therapeutic Progress. Kar, A.B., ed. (1986): Blackwell, Oxford. [ISBN 0 632 0246 52.]
The Pharmacology of Asthma. Morley, J. & *Rainsford, K.D. (1982): Birkhauser Verlag, Basel. [ISBN 3-7646 1503-2.]
The Development of Anti-asthmatic Drugs. Buckle, D.R. & Smith, H., eds. (1984): Butterworths, London. [ISBN 0-408 11576-9.]

---

*This author has an art. on interleukin/leucocyte interactions in Vol. 18 *(see footnote on p. 179)*. - *Ed.*

#C-2

EXPERIENCES WITH THE ANALYSIS OF ANTI-ASTHMATIC DRUGS OVER 25 YEARS

J.J. Gardner, D. Wilkinson and N. Entwistle

Department of Physical Chemistry, R & D Laboratories,
Fisons Pharmaceuticals, Loughborough LE11 0RH, U.K.

The discovery of potent drugs for the treatment of asthma coupled with an increased regulatory requirement has placed heavy demands on bioanalytical techniques. To meet these demands, analytical methods employing separation science, spectroscopy and immunoassay techniques have been developed, tuned and enhanced. How this has been achieved is exemplified by a survey of the needs, problems and successes in the in-house development of current and new-generation anti-asthmatics.

The development of bioanalytical methods for cromolyn sodium* started with a colorimetric urine assay method in the late 1960's. This approach provided a detection limit of ~500 ng/ml. Improved methodology was obviously needed, particularly to cope with plasma as the matrix and to increase the sensitivity of the method. Hand-in-hand with the discovery of new compounds, several stages of method improvement and development were pursued, including the use of quantitative TLC, of ion-exchange HPLC and, later, of RIA. The development and application of these techniques has been exploited to solve many critical problems in drug development programmes. In particular, methods to support the development of nedocromil sodium and tipredane have gained from these experiences. LOQ's†<1 ng/ml are now routinely obtained and required. The latest generation of drugs, notably tipredane, continues to extend the analytical techniques needed to solve problems associated with low doses, complex metabolism and species differences.

The bioanalysis of anti-asthma drugs over the past 25 years has seen many advances. The different methods used in this period reflect the continuing changes in technology and analytical instrumentation. The rate and progress of these changes has been driven by the scientific needs of clinical and safety studies together with growing expectations of regulatory

---

*(Note by Ed.) DRUG NAMES are generic versions, e.g. cromolyn sodium (authors' synonym: cromoglycate disodium); '(di)sodium' may be analytically irrelevant. Proprietary names appear thus, [ ]. †**Abbreviations.**- Ab, antibody; LOQ, limit of quantification; RIA, radioimmunoassay; RP, reversed-phase; SPE, solid-phase extraction.

authorities. The development of modern methods for the analysis
of drugs in biological matrices started at Fisons in the
late 1960's with the inhalation agent cromolyn sodium.   The
different methods used for it exemplify the changing needs
for bioanalysis.   These methodologies have subsequently been
successfully used in the support of other anti-asthma drugs.
The present survey includes the methodology used for further
inhalation agents, particularly nedocromil sodium, and finally the
on-going development of methods for tipredane, an inhalation
steroid.   The successive approaches were as follows:

- cromolyn sodium (~1966
to 1982):   colorimetry,
TLC-fluorimetry,   HPLC,
RIA;
- oral chromones (1970's):
TLC, GC-MS, HPLC;
- nedocromil sodium and
other inhalation drugs
(1980's): RIA, HPLC;
- tipredane   (1990's):
RIA, HPLC.

**SODIUM CROMOGLYCATE
(CROMOLYN SODIUM)**

PROXICROMIL

TIPREDANE

NEDOCROMIL SODIUM
(Tilade)

SODIUM CROMOGLYCATE (CROMOLYN SODIUM)

Cromolyn sodium [Intal] was introduced in 1968 as capsules
containing 20 mg for administration to the lungs *via* a spinhaler
device.   Subsequently the compound has also been administered
via pressurized aerosol devices delivering 1 or 5 mg.   It
is a di-Na salt of a bis-chromone-2-carboxylic acid. A notable
physico-chemical characteristic, besides strong UV absorbance
and weak native fluorescence, is the very strong acidity
of the carboxyl groups (each pK_a <1.8), as utilized to provide
selectivity in several of the methods described.

The first method [1] developed for the compound in urine
used an anion-exchange resin procedure to selectively concentrate
the highly acidic chromone from the weaker endogenous acids
excreted in the urine.   The analyte was then base-hydrolyzed
and derivatized with a diazotized amine to produce a yellow
product.-
 1. Mix urine (10 ml) with formic acid to 30% (v/v) and
apply mixture slowly to the resin.
 2. Elute with formic acid (98% w/w), evaporate to dryness
and dissolve in $NaHCO_3$ solution.

3. Hydrolyze the analyte to bis-acetophenone by heating.
4. Couple with diazotized *p*-nitroaniline; measure 490 nm absorbance.

The ion-exchange resin work-up was not robust. A very slow and controlled rate of sample application was necessary for reproducible recoveries. The method was used for monitoring clinical studies although its detection limit was only 0.5 µg/ml.

A method for cromolyn sodium in plasma was also developed in the late 1960's [1]. It involved a complex multi-stage sample work-up, including a TLC separation, with final detection using the compound's native fluorescence.-

1. Deproteinize plasma (2 ml) with tungstic acid.
2. Adjust supernatant to pH 4, extract with ethyl acetate (5 ml × 2) and discard extracts.
3. Acidify aqueous phase to pH <1, extract as above (but × 3) and combine extracts.
4. Back-extract the analyte into dil. aq. NH_4OH and take extract to dryness.
5. Apply extract to silica-gel TLC plate for chromatographic purification.
6. Scrape appropriate spots from plate and elute the drug.
7. Measure fluorescence: ex, 350 nm; em, 450 nm.

To correct for variable blank responses in different subjects the response for each subject's pre-dose blank sample was deducted from that subject's post-dose responses. The method measured down to 0.1 µg/ml; its use necessitated administering three times the normal therapeutic dose, yet the plasma curves were poorly defined.

Although the methodology for plasma and urine satisfied the regulatory requirements of the late 1960's the sensitivity limits, particularly for plasma, were not ideal and for pharmaco-kinetic needs were inadequate. It was not till nearly 10 years later that the development of improved methodology was attempted. The impetus for this effort was the introduction of the low-dose aerosol devices and the need to compare concentrations obtained after the use of different devices. The scientific and regulatory need to generate plasma pharmacokinetic data was a further factor.

The first improved method for urine became possible due to the advent of HPLC. This method [2] used a strong anion-exchange column (Partisil SAX), which was found to be more selective than an RP column. An extraction and back-extraction procedure gave selective recovery of the analyte. Elution of the HPLC column with low-pH buffer, and 325 nm measurement gave selectivity and specificity.-

1. Urine acidified with conc. HCl (1 ml), excess solid NaCl added, and diethyl ether extraction performed (10 ml × 2).
2. Extracts combined and back-extracted into 1 M glycine-HCl buffer (pH 3.5, 1 ml).
3. Extract (~1 ml urine) analyzed on Partisil SAX column eluted with pH 2.3 phosphate buffer (0.9 M) at 3.2 ml/min.
4. Detection at 325 nm (absorption max. at high wavelength).

Although the ion-exchange HPLC system was not highly efficient the method was specific. It was standardized over the concentration range 50 ng/ml to 20 µg/ml. Peak height data were best fitted to two standard curves for the range, up to and then beyond 1 µg/ml. The 50 ng/ml detection limit represented a 10-fold improvement compared to the earlier colorimetric method. The HPLC method was more specific, robust and precise.

The next, and more important, new method was RIA for plasma analysis [3]. It allowed a plasma concentration of 0.5 ng/ml to be detected with 0.1 ml samples. Compared to the earlier plasma method (which used a 2 ml sample and had a detection limit of 100 ng/ml) the sensitivity had been improved 200-fold.

The development of the RIA was a lengthy process and involved much synthetic work by our chemists. A conjugate was synthesized in which the compound was covalently linked to bovine serum albumin as a protein carrier. This conjugate was used to raise an anti-cromolyn antiserum, in sheep. The second requirement for the RIA was a radio-labelled form of cromolyn possessing high specific activity, a radio-ligand. The radio-label selected was ^{125}I, and a heterogeneous radioligand was synthesized using a mono-tyramide derivative of cromolyn. The third requirement was a separation procedure for antiserum-bound radioligand and free non-bound radioligand. For this separation a second antiserum, an anti-sheep IgG antiserum raised in a donkey, was used. Initially the RIA employed a 2-stage incubation procedure. The immune competition reaction occurred in the first incubation, and the free and bound radioligand were separated in the second incubation, using the double Ab and polyethylene glycol (PEG) as an accelerator.-
1. Mix plasma (0.1 ml) + radioligand in buffer (0.5 ml) + anti-cromolyn antiserum in buffer (0.5 ml).
2. Incubate at ambient temperature for 2.5 h.
3. Add PEG solution (0.25 ml) + donkey anti-sheep IgG antiserum (double Ab) solution (0.25 ml).
4. Incubate at ambient temperature for a further 40 min.
5. Centrifuge, decant supernatant and count pellet.

The method had a detection limit of 0.5 ng/ml, allowing plasma pharmacokinetics to be satisfactorily investigated in volunteers and, more importantly, in patients. However, with the resulting deluge of samples the many manual additions in the 2-stage procedure and the need to follow a timed protocol became major disadvantages.

Method simplification was found to be feasible. The double Ab was added with the radioligand and the first anti-cromoglycate Ab, the glycol accelerator was omitted, and a single overnight incubation of non-critical duration was used. This permitted much larger analysis batches to be processed, and improved sample throughput. The introduction of automation of the preparation of the assay tubes using a Micromedic Pipetting Station further improved the throughput. These changes also improved the precision of the method. The plasma RIA was also applied to the analysis of urine samples. Each urine was analyzed using many dilutions in order to obtain a result falling within the limited range of the method.

Consequently, by the early 1980's bioanalysis methods were available with sufficient sensitivity, good specificity and precision and adequate throughput for the determination of cromolyn sodium in both plasma and urine.

PROXICROMIL

During the 1970's much effort was directed to the development of oral anti-asthmatic chromones. The compounds studied were lipophilic mono-carboxylic acids with structures similar to that of proxicromil (a drug not taken to the market stage). The methods initially developed used solvent extraction followed by TLC separation and *in situ* detection on the TLC plate; the measurements were performed with a Flying Spot Densitometer. The instrument measured compounds which showed native fluorescence and compounds such as proxicromil which absorbed high-wavelength UV or visible light. Whereas with compounds of the former type ~50 ng/ml was detectable, with the latter type the sensitivity limits were considerably higher, ~500 ng/ml, and the standard curves showed considerable curvature. This curvature and the limited number of samples that could be applied to a TLC plate were major disadvantages.

When it was decided to progress proxicromil with a low therapeutic dose, only 6 mg, it was clear that TLC-based methodology was inadequate. The first alternative sensitive method developed was a GC-MS method which had a 20 ng/ml detection limit. The disadvantages of this method included a very slow analysis rate, the frequent need to service the mass spectrometer and a long chromatographic analysis time. Soon after the GC-MS method had been validated an ion-pair RP-HPLC method was developed. This method also had a detection limit of 20 ng/ml but was simpler and used generally more reliable instrumentation than the GC-MS method.

NEDOCROMIL SODIUM

The most important of the new generation of inhalation drugs was nedocromil sodium [Tilade; has been available in the U.K. for >4 years]. The compound has anti-inflammatory properties in reversible obstructive airways diseases including asthma. The development of a plasma bioanalysis method for the compound was the first priority. Using the approaches which had been previously used for cromolyn sodium, a plasma RIA was rapidly developed. For plasma the assay had a detection limit of 0.25 ng/ml and could measure up to 8 ng/ml. The RIA was also applied to the analysis of diluted urine samples [4].

For urine an HPLC method was also developed, with a 50 ng/ml detection limit, involving solvent extraction and chromatography on a C-18 RP column. Comparison of the results obtained by the two independent urine analysis methods (Fig. 1) served to confirm the validity of the urine RIA method, as manifested by near-unity values for slope (0.981) and the correlation coefficient (0.99).

The validation of urine methods often has some uncertainty due to the extreme variability of the urine matrix in contrast with plasma; hence accord between two independent methods gives important reassurance concerning both methods. Subsequently a more sensitive urine HPLC method was developed using direct injection onto a column switching system. This employed an RP column to concentrate from the urine the analyte which was then transferred to an anion-exchange column. The system provided efficient and selective chromatography for nedocromil. The range of the method was 20 to 10,000 ng/ml, with precision values of <8% over this extremely wide range. Results obtained by this second urine method and the urine RIA similarly served to cross-validate both methods.

Later in the development of nedocromil sodium the human plasma RIA was applied to the analysis of rat and dog plasma. This reflects the increasing bioanalytical requirement for methods to support toxicology studies.

TIPREDANE

Tipredane is a novel anti-inflammatory steroid with metabolically labile thio-ether substituents at the 17 position. The compound was developed by Squibb in the U.S.A. for dermal use and licensed to Fisons for inhalation and intra-nasal

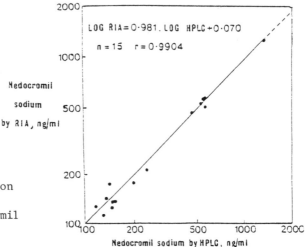

$$LOG\ RIA = 0{\cdot}981.\ LOG\ HPLC + 0{\cdot}070$$
$$n = 15 \quad r = 0{\cdot}9904$$

Nedocromil sodium by RIA, ng/ml

Nedocromil sodium by HPLC, ng/ml

Fig. 1. Comparison of RIA and HPLC assay for nedocromil sodium in urine.

use. It was designed to produce minimal systematic steroidal side-effects; it is rapidly metabolized in the liver to inactive products.

As a consequence of rapid metabolic clearance, plasma concentrations of tipredane were expected to be very low, and it was decided to provide bioanalytical support to toxicological studies by the determination of metabolites in urine. The metabolism in man and primates appeared simple: two major metabolites, the 6-hydroxy sulphoxide and sulphone, are excreted in the urine. In other species the metabolism is more complex with many products, none of which clearly predominates in the urine. The metabolic pattern is different in the rat, mouse and dog with, moreover, evidence of sex-dependent differences in the rat. A separate method measuring a different metabolite was therefore required for each of the animal species used in toxicology. HPLC methods are the more appropriate where complex metabolism occurs. HPLC methods have already been developed for monkey and rat urine. Work continues to provide methods for mouse urine and a more sensitive method for rat urine. Support to toxicology is also being given by the measurement of tipredane in the lungs immediately after dosing. Support to the clinical areas is being given by the development of RIA procedures for tipredane in plasma and an HPLC method for the sulphoxide and sulphone in urine.

Several methods investigated for urine analysis have encountered specificity problems with urine samples from some subjects. When extraction with a mixture of dichloromethane and isopropanol was used to isolate the metabolites from urine there was a large interference with the measurement of the sulphone. Changes to the extraction system to improve the specificity resulted in poor recoveries. The use of SPE with a Varian AASP system was therefore investigated.

After much optimization using a range of supports and wash systems, improved specificity was obtained using C-2 cartridges and, to reduce the urine interferences, an alkaline wash which had to be followed by a buffer rinse to obtain a reasonable recovery.-

1. Filtered urine (0.5 ml) is diluted with aqueous 0.5% ammonium acetate/5 mM EDTA (0.5 ml).

2. Apply to pre-conditioned C-2 AASP cartridge.

3. Apply successive washes (1 ml): 1.0% NaOH; 50 mM NaH_2PO_4; water.

4. Elute onto a Resolve C-18 column with methanol/acetonitrile /water (26:9:65 by vol.).

5. After HPLC with gradient elution, measure at 240 nm.

Developing a tipredane RIA involved antiserum production in sheep and an ^{125}I radioligand.   Synthesis of the bovine serum albumin conjugate (to raise the antiserum) and the iodohistamine radioligand used the same precursor, a 3-O-carboxymethyl oxime derivative.   The hapten's free carboxyl was linked to the protein and iodohistamine by a mixed anhydride reaction.   In a plasma-free system the RIA had a detection limit of 50 pg/tube; but the presence of plasma caused large interferences, reduced by clean-up of the plasma by acetonitrile precipitation.   Thereby a detection limit of 0.4 ng/ml was obtained, worsened to 0.8 ng/ml by inter-individual variations in the blank-plasma responses.

The method has been used to analyze plasma samples from subjects given tipredane by various routes. The method could only just measure the maximum plasma concentrations found, and improved sensitivity is being sought, centered on a more selective pre-RIA clean-up to reduce the matrix interferences.

CONCLUSIONS

During the last 25 years the technologies and techniques available to the bioanalyst have improved immeasurably. The development of the many different and sensitive methods required in support of compounds such as tipredane shows that the job of the bioanalyst remains as challenging today as in the early days with cromolyn sodium.

Acknowledgements.- J.R. Preston, P.R. Baker, W.J.S. Lockley, D.J. Wilkinson, C.M. Gilbert, M.S.J. Bayliss and M. Jones made valuable contributions to methods described.

References

1. Moss, G.F., Jones, K.M., Ritchie, J.T. & Cox, J.S.G. (1971) *Toxicol. Appl. Pharmacol.* 20, 147–156.
2. Gardner, J.J. (1984) *J. Chromatog.* 305, 228-232.
3. Brown, K., Gardner, J.J., Lockley, W.J.S., Preston, J.R. & Wilkinson, D.J. (1983) *Ann. Clin. Biochem.* 20, 31-36.
4. Gardner, J.J., Preston, J.R., Gilbert, C.M., Wilkinson, D.J., Lockley, W.J.S. & Brown, K. (1988) *J. Pharm. Biomed. Anal. 6*, 285-297.

#C-3

HPLC-MS AND COUPLED COLUMN HPLC FOR BIOANALYSIS OF ANTI-ASTHMATIC DRUGS

[1]Claes Lindberg[‡], [1]Jan Paulson, [1]Ann Blomqvist,
[1]Lars-Erik Edholm and [2]Agneta Walhagen

[1]Bioanalytical Chemistry, Astra Draco AB,
Box 34, S-221 00 Lund, Sweden

[2]Technical Analytical Chemistry, University of Lund,
S-221 00 Lund, Sweden

We have found CC-HPLC* and HPLC-MS very useful for exploratory as well as for routine work in drug development. Both techniques afford high selectivity, permitting simple clean-up procedures to be used or even direct injection of biological samples. Drug enantiomers in such samples have been directly separated and determined by CC-HPLC with chiral columns; compression of peaks may increase the relatively low efficiency often found for columns packed with immobilized proteins, in this case α_1-AGP.

Quantitative methods based on TSP* HPLC-MS have been developed for a number of anti-asthmatic drugs. Terbutaline together with its prodrug bambuterol can be measured after separation on a gradient HPLC system. The sensitivity of TSP HPLC-MS for these compounds is comparable to what can be achieved by GC-MS. The enantiomers of terbutaline have also been determined, by TSP HPLC-MS using a β-cyclodextrin column. Derivatization can be performed to increase sensitivity as shown for budesonide assay by TSP HPLC-MS.

The bioanalytical laboratory, involved as a service function in drug development, is faced with a wide variety of analytical problems. Initial pharmacological and pharmacokinetic studies require rapid development of methods, applicable to several, chemically similar compounds and to samples from different animal species. Often only a limited number of samples are analyzed. In this stage of drug development high flexibility of the bioanalytical methodology is required. Later, when a compound with interesting pharmacological properties is found to be safe and is taken into clinical research, clinical

[‡]to whom any correspondence should be addressed.
Abbreviations.- CC (*prefixing* HPLC), coupled column; MS, mass spectrometry; TSP, thermospray; SPE, solid-phase extraction; α_1-AGP, α_1-acid glycoprotein; TEA, triethylamine; s.i.m., selected ion monitoring.

studies will generate thousands of samples. Robust, automated methods with short analysis time and high sample throughput are needed in this phase. We have found CC–HPLC and TSP HPLC–MS very useful for exploratory as well as routine work.

CC–HPLC

Bioanalytical methods require high selectivity so as to discriminate the analytes from matrix components. When drug enantiomers are analyzed in biological matrices, selectivity has two meanings: selectivity towards matrix components and selectivity between the enantiomers. We have used CC–HPLC to increase selectivity towards matrix components and to overcome some of the problems which may arise when drug enantiomers are analyzed in biological matrices [1].

Two basic configurations of coupled columns can be used (Fig. 1). If close-to-baseline separation of the enantiomers is obtained on the chiral system, **configuration A** should be used. The analyte is first separated from endogenous components on a non–stereoselective system. The analyte fraction is trapped in a loop or on a small column, and the mixture of enantiomers is then transferred to the chiral column where they are separated. Quantification can be performed either by using external calibration curves, one for each enantiomer, or by calculating the area ratio between the enantiomers. In the latter case, the amount of each enantiomer can be calculated if the total amount is known. This approach was utilized for determination of the enantiomers of terbutaline, a β_2–receptor agonist, in plasma and intestinal juice (Fig. 2) [2]. The chiral separation of the terbutaline enantiomers was obtained on a chemically bonded β–cyclodextrin column. By using CC–HPLC with pre-separation on a reversed phase column with neat aqueous eluent, sufficient selectivity and sensitivity could be obtained, allowing quantitative analysis down to 0.4 nM (90 pg/ml) using 2 ml of plasma.

Where the drug enantiomers are baseline-separated on the chiral system and the efficiency is low, **configuration B** (Fig. 1) is preferred. After chiral-column separation the enantiomers are trapped on two separate trapping columns. Each enantiomer is subsequently transferred to the non–chiral column where the fraction is compressed and endogenous compounds are separated from the analyte. In order not to introduce extra band broadening by the transfer process the mobile phase used in the chiral system must be a stronger eluting solvent on the chiral system than on the trapping column. Further, a mobile phase with higher eluotropic strength must be chosen for eluting the enantiomer from the trapping column and for final separation on the non–chiral column. The same calibration curve can be used for the two enantiomers, because they act as identical compounds in the non–chiral system.

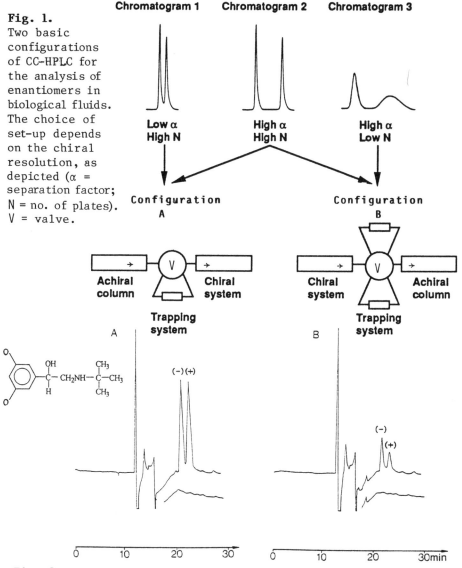

Fig. 1. Two basic configurations of CC-HPLC for the analysis of enantiomers in biological fluids. The choice of set-up depends on the chiral resolution, as depicted (α = separation factor; N = no. of plates). V = valve.

Fig. 2. CC-HPLC chromatograms of intestinal juice (**A**) and plasma (**B**) after administration of racemic terbutaline. Lower traces represent blank samples. The calculated concentrations for the enantiomers were: **A**: (−) 20.1 and (+) 20.1 µM; **B**: (−) 21.4 and (+) 13.1 nM.   Chiral column: β–cyclodextrin; eluent: aq. ammonium acetate, ± methanol (to 10% v/v); electrochemical detection. *From ref. [2], courtesy of publisher (John Wiley & Sons).*

GENERAL INSTRUMENTATION (noted by Ed. in cited refs.):   Switching valves were from Valco, generally #CV-6-UHPa-N60 (#C6W for the Fig. 6 experiments). Where a mass spectrometer was used, it was a Finnigan instrument (#4500 MS; #TSP1 thermospray interface), with selected ion monitoring (s.i.m.).

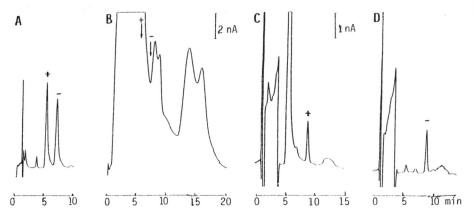

Fig. 3. Chromatograms obtained after separation of racemic
terbutaline on a CC-HPLC system according to configuration **B**
(Fig. 1), with electrochemical detection. (**A**) Standard solution
(50 pmol racemate) injected directly into the α_1-AGP column
(column 1). (**B**) Blank plasma after separation on the α_1-AGP
column, demonstrating insufficient selectivity obtained by a
single-column approach. *Arrows* indicate positions of the drug
isomers. (**C**) Spiked plasma, containing racemic terbutaline
(50 nM), after separation by CC-HPLC: (+)-terbutaline, the
first-eluted enantiomer, injected (21 pmol). (**D**) As for (**C**),
but (-)-terbutaline, the second-eluted enantiomer, injected.
Taken by permission from ref. [3], which gives details, e.g. of:
initial SPE; eluents, C-18 trapping col. $\xrightarrow[\text{pH 7.5 phos.}]{}$ AGP col. (self-made).

    The potential of the CC **B** configuration was demonstrated
by analyzing terbutaline enantiomers in plasma on an α_1-AGP
phase (Fig. 3) [3]. This kind of phase often affords good
separation factors for the enantiomers but rather poor efficiency.
Mobile phases with a high aqueous content and a relatively
low proportion of organic modifier are often used for protein
columns, which facilitates trapping of the enantiomers. The
example shown in Fig. 3 clearly demonstrates the selectivity
and efficiency achieved by using CC-HPLC.

HPLC-MS

    MS can be used as a highly selective yet general detection
technique in HPLC. The high selectivity permits simple extraction procedures to be used, or even direct injection of
biological samples. Less stringent requirements for sample
clean-up procedures facilitate method development and also
reduce the risk of artefact formation during sample pre-treatment,
making HPLC-MS an almost ideal reference method for the validation of other bioanalytical methods.

We have used TSP HPLC-MS to measure terbutaline enantiomers in plasma [1].   Due to the high selectivity achieved by s.i.m. the eluate obtained after SPE could be directly analyzed on a β-cyclodextrin column connected to the mass spectrometer. Fig. 4 shows mass chromatograms obtained from blank plasma and from plasma following a terbutaline dose.   Endogenous compounds, eluting close to the solvent front, caused considerable disturbances on the baseline, impairing sensitivity.   Yet the terbutaline enantiomers could be accurately quantified. The concentrations of (+)- and (-)-terbutaline, determined from the chromatogram shown in Fig. 4, were 15 and 25 nM respectively, in accord with the total concentration of terbutaline (40.3 nM) measured for the same plasma sample by a non-chiral GC-MS method [4].   The accuracy and precision of quantitative measurements using MS are improved by the use of ^{2}H-labelled analogues as internal standards.

In spite of the high technical complexity of HPLC-MS the technique can be made robust and reliable and useful for routine analysis.   We have designed an automated TSP HPLC-MS system [5] which is currently in routine use for quantitative analysis of some anti-asthmatic drugs.   Terbutaline can be measured together with its prodrug bambuterol and an intermediate metabolite (D2439) after separation on a gradient HPLC system, as shown in Fig. 5 for spiked plasma.   The sensitivity of TSP HPLC-MS for these compounds is comparable to what can be achieved by GC-MS.   The reliability of the HPLC-MS system was demonstrated for this type of gradient analysis by replicate injections of standard solutions, containing all 3 compounds together with their ^{2}H-labelled internal standards, during a 19-h run [5].   The relative S.D. of the peak-area measurement of the internal standards was in the range 6.5-8.2%, and the calibration curves constructed from the standards showed excellent linearity (r= 0.997-0.999).

Another example is provided by budesonide, which is clinically used in treatment of asthma and rhinitis.   TSP HPLC-MS following SPE can be used to measure the drug in plasma down to 0.1 nM (0.04 ng/ml)[θ]   with adequate precision (C.V. 10-18%).   A chromatogram is shown in Fig. 6 for plasma spiked with budesonide (to 0.5 nM).   The simple and rapid derivatization procedure ([6]; see legend) increased sensitivity ~10-fold, making the method applicable to clinical samples.

_____

[θ] C. Lindberg, J. Paulson & A. Blomqvist, to be published

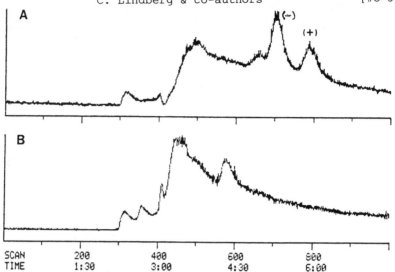

Fig. 4. Mass chromatograms (m/z 226) obtained with TSP HPLC-MS for terbutaline enantiomers using a β-cyclodextrin column (200 × 4.5 mm; eluent: 0.1 M ammonium acetate, pH 5). **A,** plasma sample, post-dose (racemate given); **B,** blank plasma. See text for comment. SPE was done initially, and the extract dried down. ['CC option', to remove plasma interferences:- first run on an RP (phenyl) column with pH 4.6 ammonium acetate.]   *From ref. [1], by permission.*

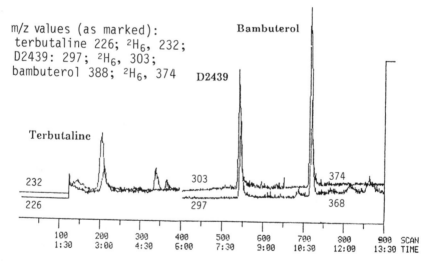

Fig. 5. Mass chromatogram obtained by thermospray CC–HPLC of a plasma sample spiked with terbutaline, D2439 and bambuterol (each 4 nM). ^{2}H-Labelled analogues were used as internal standards *(upper traces).* SPE was done initially. C-18 column; gradient from 6.3% up to 40% methanol (v/v) in 0.1 M ammonium acetate buffer pH 5.

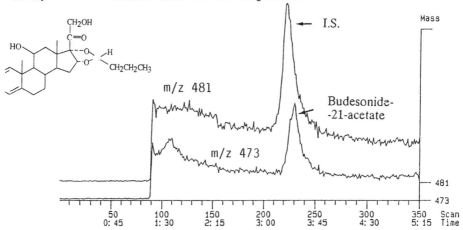

Fig. 6. Mass chromatogram obtained by automated thermospray HPLC-MS for budesonide-21-acetate in a plasma sample containing budesonide, after conversion to the 21-acetate by derivatization with acetic anhydride/TEA in acetonitrile. $^2H_3$-budesonide was added as internal standard. SPE (C-18) was done initially. Column: Supelcosil LC-8-DB (33 × 4.6 mm); eluent: 64% methanol in 0.1 M ammonium acetate pH 5.

From ref. [5], by permission.

References

1. Edholm, L-E., Lindberg, C., Paulson, J. & Walhagen, A. (1988) *J. Chromatog. 424*, 61-72.
2. Walhagen, A., Edholm, L-E., Kennedy, B.M. & Xiao, L.C. (1989) *Chirality 1*, 20-26.
3. Walhagen, A. & Edholm, L-E. (1989) *J. Chromatog. 473*, 371-379.
4. Jacobsson, S-E., Jönsson, S., Lindberg, C. & Svensson, L-Å. (1980) *Biomed. Mass Spectrom. 7*, 265-268.
5. Lindberg, C., Paulson, J. & Blomqvist, A. (1991) *J. Chromatog. 554*, 215-226.
6. Paulson, J. & Lindberg, C. (1991) *J. Chromatog. 554*, 149-154.

#C-4

DEVELOPMENT OF AN HPLC METHOD FOR ASSAYING BRL 38227, A K+ CHANNEL ACTIVATOR, IN HUMAN PLASMA AND URINE

**A. Beerahee, J.N. Bullman, E. Doyle, C.M. Kaye,
E.A. Mead, A.C. Taylor and N.A. Undre**

Drug Metabolism and Pharmacokinetics Department,
SmithKline Beecham Pharmaceuticals Research Division,
The Frythe, Welwyn, Herts. AL6 9AR, U.K.

Require-   *A rapid and specific method for determining BRL 38227*
ment   *in human plasma and urine with LOQ* 0.1 ng/ml.*

End-step   *Isocratic RP-HPLC (C-18) and fluorimetric detection.*

Sample   *Alkalinization so that acidic and polar endogenous*
handling   *components remain in the aqueous phase when analytes*
           are selectively extracted with toluene containing iso-
           amyl alcohol.

Comments   *Past sample-preparation experience with the compound*
           helped quick achievement of an optimized analytical
           method which has been successfully implemented inter-
           nationally. In its final definitive form the method
           is simple, highly sensitive, rapid, robust and precise.

(-)-*Trans*-6-cyano-3,4-dihydro-2,2-di-
methyl-4-(2-oxo-1-pyrrolidinyl)-2H-1-
benzopyran-3-ol, BRL 38227, is a novel,
orally active K+-channel activator,
under development for the treatment
of hypertension. Such activators are
potentially useful in the treatment
of asthma and other disorders also.

Structure of BRL 38227 and
(relevant part) BRL 38226

BRL 38227 is the pharmacologically active form of cromakalim
- a racemate containing BRL 38227 and BRL 38226. To support
pre-clinical and clinical development, both stereospecific and
non-stereospecific assays utilizing GC-MS [1, 2] and HPLC [3, 4]
have been used to determine BRL 38227 in biofluids. However,
GC-MS requires extensive equipment maintenance, and all these
methods include tedious sample preparation or long analysis
time, resulting in low sample throughput. A more rapid
HPLC procedure has therefore been developed which also has
the advantage of reliably high sensitivity.

---

*Abbreviations.- i.s., internal standard; LOQ, limit of quanti-
fication; SPE, solid-phase extraction.

STRATEGY FOR METHOD DEVELOPMENT

At the outset the main requirements for this assay were high sensitivity, robustness and especially rapidity. The compound is neutral and has strong native fluorescence [4]. At the time (mid-1980's), because of the limitations of column phase technology (both SPE and HPLC packings) and of HPLC fluorimeter instrumentation it was difficult to achieve, with biofluid samples, clean extracts, good chromatographic separation and a sufficiently low LOQ (<1 ng/ml). Hence the main assays developed to support the initial development programme of BRL 38227 had low throughput, being based on GC-MS following solvent extraction and derivatization [1, 2]; optimal assay conditions with GC-MS need strict equipment maintenance.

Chromatography and detection.- Firstly an evaluation was made of the most sensitive fluorescence spectrophotometers available: the Perkin-Elmer LC 240 appeared to have the best sensitivity and precision (reproducibility). The next step was to study the chromatography of the compound and devise some simple preliminary HPLC conditions. Of the C-18 (ODS) bonded silica phase columns examined (using a gradient; acetonitrile/0.05 M phosphate pH 7), the Apex column (Jones Chromatogy.) was chosen as giving suitable peak shape, symmetry and retention time. A data-base search for i.s. candidates quickly indicated fluorescent structural analogues of BRL 38227, and from screening using the unrefined HPLC conditions two were selected based on a capacity factor (k') of <10. BRL 36603, as used here, has strong native fluorescence.

Sample preparation.- Although preliminary experiments with SPE seemed promising, development time was rate-limiting in the optimization of recovery and selectivity. Effort was therefore concentrated on solvent extraction, in parallel with exploration of SPE as a longer-term option relevant to possible automation. Guided by previous experience with solvent extraction in the preceding assays for BRL 38227 [1, 2], and with reinforcing information on the inactive and polar metabolites of the compound, an optimized sample preparation procedure was arrived at, as described below.

FINAL ANALYTICAL METHOD

Materials.- All analytes and the i.s. were dissolved in ultra-pure water from an Elgastat Spectrum System (Elga Products, High Wycombe). All organic solvents used were of high purity HPLC grade (Rathburn Chemicals, Walkerburn, Scotland).

Sample treatment

1. To each plasma or urine sample (1 ml) was added the i.s. (aqueous solution), 0.5 M NaOH (0.5 ml), and 5% (v/v) isoamyl in toluene (3.5 ml).

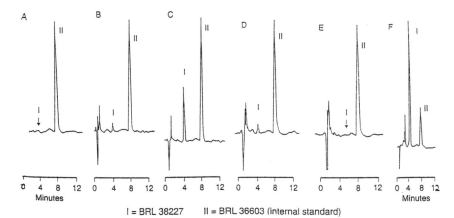

I = BRL 38227   II = BRL 36603 (internal standard)

Fig. 1. Typical chromatograms for plasma (**A-C**) and urine (**D-F**). All samples spiked with i.s. (to 5.0 ng/ml). Controls: no other spike (**A, D**), or spiked with BRL 38227, to 0.1 ng/ml (**B, E**). Samples from a volunteer after an oral dose of 0.75 mg BRL 38227: plasma taken at 48 h (**C**) and 12-24 h urine (**F**).

2. The vials were capped, mechanically shaken (reciprocal action) for 15 min, and centrifuged (300 **g**, 10 min).

3. The upper organic layers were transferred to fresh vials and evaporated to dryness under a stream of O_2-free N_2 at 55°.

4. The residues were dissolved in 150-200 µl water, capped, vortex-mixed for 30 sec, and transferred into tapered vials for injection.

Chromatography

Aliquots (50-100 µl) were injected onto a 15 cm Apex ODS 1 (5 µm) HPLC column protected by a 15 mm 'New guard' RP18 (7 µm) guard column and run at ambient temperature. The analytes were eluted with 70:30 (by vol.) 0.01M pH 6.0 phosphate buffer/acetonitrile at 1.5 ml/min flow-rate. The fluorimetry wavelengths were 254 (ex.) and 306 (em.) nm. The retention times were 4.5 min for BRL 38227 and 8.0 min for the i.s. (Fig. 1).

Peak height ratios were determined using the Waters 860 Chromatography Data System. For calibration curves an unweighted regression line was fitted by the Waters 860 and used to calculate the concentration of BRL 38227 in each sample.

RESULTS

Linearity.- In both plasma and urine, linear calibration curves were obtained (Fig. 2) covering the normal therapeutic range, 0.1-20.0 ng BRL 38227/ml, whilst linear up to at least 50 ng/ml.

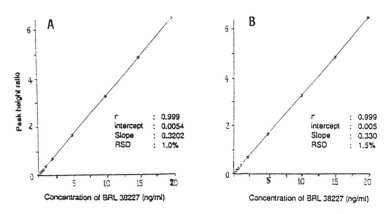

Fig. 2. Typical calibration graphs for BRL 38227: **A**, human plasma; **B**, human urine.

Recovery and reproducibility.- Extraction efficiencies were high (>80%) for BRL 38227 and the i.s. in both plasma and urine, the peak heights of the extracted analyte being compared with those obtained by direct injection of aqueous standards.

Accuracy and reproducibility.- The within- and between-day variation of assay accuracy was tested by analyzing replicate samples (plasma and urine containing BRL 38227) at four different concentrations (LOQ, and low, medium and high *vs.* ranges found in plasma and urine samples). The results (Table 1) demonstrated that the assay was reproducible, with within- and between-day C.V. values <10%, and accurate, with <15% deviations from set values at all concentrations in each biofluid.

Inter-assay and inter-laboratory comparison.- Further valid-ation of the method was obtained by comparing results with those from a specific GC-MS assay [2] of authentic plasma and urine samples from healthy volunteers dosed orally with BRL 38227. A correlation coefficient of 0.96 for BRL 38227 was obtained (Fig. 3) for both plasma (n = 40) and urine (n = 25) with slope and intercept not differing significantly from unity and zero respectively. To implement the assay inter-nationally, an inter-laboratory comparison was performed by analyzing spiked sets of plasma samples with BRL 38227. Results in Table 2 show that there was good agreement among the replicates in different laboratories, with deviations <10%.

Sample-preparation 'rationale'.- Selective extraction, with few interfering peaks, was obtained through use of toluene, supplemented with isoamyl alcohol as a well-known stratagem for maximizing recovery and minimizing the analyte's tendency to adsorb onto glassware (e.g. onto vials). SPE extracts were less satisfactory.

Nominal Sample Conc. (ng/ml)	Biofluid	Within-day C.V.(%)	Between-day C.V.(%)	Accuracy (%)
0.10	Plasma	10.2	13.7	2.7
	Urine	8.0	10.3	3.0
0.40	Plasma	3.3	4.3	13.5
	Urine	2.5	2.8	5.3
10.0	Plasma	3.2	4.0	11.8
	Urine	2.0	2.2	3.0
20.0	Plasma	0.7	1.0	6.0
	Urine	1.0	1.1	11.0

Table 1.   Assay of BRL 38227: mean precision and accuracy values.

Spiked Conc.* (ng/ml)	Measured conc.(ng/ml; 3 laboratories)		
	Mean	Range	C.V., %
0.34	0.38	0.33 - 0.48	13.9
0.86	0.85	0.75 - 1.01	12.3
2.37	2.25	1.96 - 2.44	7.5
2.70	2.91	2.60 - 3.16	10.4
5.41	5.33	5.00 - 5.64	4.8
6.48	6.67	5.89 - 6.41	9.8
11.8	12.0	11.1 - 14.6	12.3
13.5	14.8	13.4 - 16.1	7.8
20.3	20.6	18.9 - 25.6	11.6

* Duplicate samples at each conc. assayed in each laboratory.

Table 2.   Assay of BRL 38227: inter-laboratory reproducibility. Control human plasma was prepared in bulk and spiked with the indicated concentrations, then split into 1.0 ml replicates and stored at -20°. Participating analytical laboratories (Germany, Japan and U.K.) were sent the samples which were assayed by the described method.

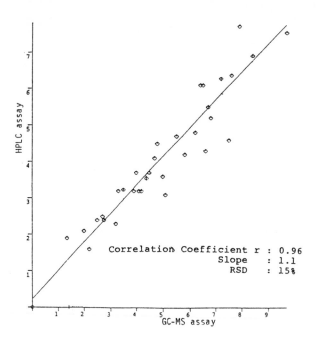

Fig. 3.
Correlation of plasma concentrations (ng/ml) of BRL 38227 determined by HPLC and specific GC-MS assays in samples obtained from healthy volunteers given various doses of the drug.

Correlation Coefficient r : 0.96
                   Slope   : 1.1
                   RSD   : 15%

CONCLUDING COMMENTS

The simple HPLC conditions contributed to a long column life, which exceeds 1800 injections provided that the guard column is changed after 100-200 injections or at the earliest sign of deterioration. The method has now also been validated and applied to rat and dog plasma. The selective solvent extraction in conjunction with a short HPLC run time resulted in a rapid method, the sample throughput being 70 samples per working day per analyst even without automation.

References

1. Gill, T.S., Allen, G.D. & Davies, B.E. (1988) *Br. J. Clin. Pharmacol. 26*, 227P.
2. Taylor, A.C., Gill, T.S., Allen, G.D. & Nash, M. (1992) *Anal. Proc. 29*, in press.
3. Kudoh, S. & Nakamura, H. (1989) *Anal. Sci. 5*, 39-42.
4. Kudoh, S. & Nakamura, H. (1990) *J. Chromatog. 515*, 597-602.

#C-5

ENHANCING THE SENSITIVITY OF RIA, THROUGH SOLID-PHASE EXTRACTION, FOR DETERMINING pg/ml CONCENTRATIONS OF FLUTICASONE PROPIONATE IN PLASMA

K.D. Jenkins, E.V.B. Shenoy, A.J. Pateman and M.J. Daniel

Drug Metabolism Department, Glaxo Group Research Ltd.,
Greenford Road, Greenford, Middx. UB6 0HE, U.K.

For FP*, a novel glucocorticoid, an RIA was developed to meet the need for measuring low concentrations in pharmaco-kinetic studies. Applied to unprocessed plasma, the RIA could measure down to 500 pg/ml in a 100 μl sample, and showed no cross-reactivity with the principal metabolite. However, improved sensitivity was needed, and was achieved by use of SPE prior to the RIA, whereby measurements could be made down to 50 pg/ml, allowing assays on human plasma after drug administration by inhalation.

FOR COMPARISON:

Fluticasone propionate (FP)   Pregnane   Androstane

FP, a trifluorinated steroid, has potent topical anti-inflamma-tory properties. It is a propionate ester with an androstane-based structure, akin to pregnane but with characteristic C-17 substituents (S at C-21). The molecule was designed to maximize topical anti-inflammatory activity and minimize the unwanted systemic activity (suppression of the HPA axis) associated with glucocorticoids. This separation of activities is possible because FP is rapidly metabolized to a 17β-carboxylic acid which has negligble glucocorticoid and anti-inflammatory activity. The excellent topical activity of FP combined with the low side-effects make it an ideal compound to treat asthma and seasonal and perennial rhinitis. In the McKenzie vasoconstrictor test it had twice the activity of betamethasone dipropionate.

---

*Abbreviations.- Ab, antibody; FP, fluticasone propionate; NSB, non-specific binding; QC, quality control; RIA, radioimmunoassay; SPE, solid-phase extraction; TBS, Tricine-buffered saline.

CHOICE OF ANALYTICAL METHOD

Once the compound had been selected for development, an assay for plasma-level determinations was required. The options were evaluated against criteria such as the need for high sample throughput and good sensitivity. The decision to concentrate on RIA was made after considering several factors.-

1. The suitability of steroid molecules for raising high-affinity selective Ab's and the experience of many other groups in developing RIA methods for both natural and synthetic steroids.

2. The lack of a suitable chromophore or fluorophore for the development of a sensitive assay method using optical detection techniques.

3. The unrewarding experience in most laboratories where analytical methods for steroids were developed using MS as the means of detection.

With the above points in mind, RIA offered the most promising route to a selective and sensitive assay method capable of handling the numbers of samples that both toxicological and human studies required.

RIA DEVELOPMENT

RIA is based on the reactions of a limited, fixed concentration of specific Ab with varying amounts of antigen (drug). Complexes arising from the reaction are measured by adding a fixed amount of radiolabelled antigen and separating the bound complexes from the free drug. Measurement of either bound or free radioactivity by liquid scintillation counting will enable drug concentrations to be estimated by comparison with standards. Drug molecules are not usually immunogenic and therefore have to be conjugated with a hapten. The site of conjugation of FP to the hapten was chosen carefully so that the likely sites of metabolism would be at the opposite side of the drug molecule. This position was the 3-keto moiety, whereby Ab's produced would be able to distinguish between FP and metabolites. Changes near the conjugation site

Fluticasone propionate 3-carboxymethoxime

would be less distinguishable and conducive to cross-reactivity.

Conjugation.- The FP molecule was firstly combined with 3-carboxymethoxime. This molecule was then combined with thyroglobulin to form FP-carboxymethoxime thyroglobulin. This molecule enabled an immunogenic response to FP to be induced.

Ab production.- For immunization, the components of the initial injections were: FP-thyroglobulin conjugate (4 mg), 0.9% NaCl (1 ml) and 1% Tween 80 (1 ml). The components were mixed in a homogenizer, and 1 ml of Freund's complete adjuvant was added. The mixture was administered to New Zealand white rabbits, 1 ml i.m. to each of the hind limbs and 1 ml in 0.2 ml aliquots to 5 sites on the back (intradermally).

Booster injections were given in Freund's incomplete adjuvant. After the boosters, Ab titre tests were performed on small serum samples removed from the rabbits. The procedure comprised testing various dilutions of serum to ascertain whether they could bind 50% of a known small amount of radio-labelled FP. When titres of ~1:15,000 were achieved, serum was collected from the rabbits and stored frozen.

Cross-reactivity.- It is important to test Ab's for cross-reactivity with other molecules, lest there be false-positive results due to molecules of similar structure. The FP Ab's were tested with synthetic and naturally occurring steroids. Cross-reactivity was <0.2% for each of the following:
#hydrocortisone (cortisol), corticosterone, progesterone, cholesterol, testosterone, betamethasone, oestriol and GR 36264 (17-COOH analogue; the only metabolite identified in man);
#cross-reactivity was 13.6% for GR 40775 (11-keto analogue), and 45.0% for GR 98145 (6-hydroxy analogue; absence from systemic circulation confirmed by GC-MS).

METHODOLOGY

As FP is not very water-soluble, stock solutions were prepared in ethanol. Dilutions were performed in Tricine buffer, in order to achieve a calibration range of 125 to 2,500 pg/ml; ethanol concentrations were kept to 5% (v/v) in all tubes. Due to ethanol being present in the standards, it was also added to samples and QC's in the buffer in order to minimize any denaturing effects that it may have on the Ab's.

Reagents and buffers were added in 0.1 ml amounts to the different tubes - calibration tubes ('calib.'), sample tubes ('test') and NSB tubes, as follows:-
  Calibration standard (5% ethanol): to calib. only;
  Blank matrix: to calib. and NSB only;
  Test samples: to test only;
  Tricine buffer (5% ethanol: to test & NSB only;
  Tricine buffer (no ethanol): to NSB only;
  Ab (1 in 14,500): to calib. and test only.

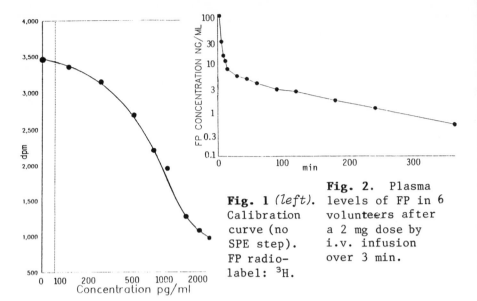

Fig. 1 *(left).* Calibration curve (no SPE step). FP radio-label: ³H.

Fig. 2. Plasma levels of FP in 6 volunteers after a 2 mg dose by i.v. infusion over 3 min.

The tubes, after mixing the contents, were incubated at 4° for 30 min, and 0.1 ml of radiolabel was then added. After mixing, the tubes were incubated for a further 30 min at 4°; then 1 ml of dextran-coated charcoal was added. After mixing and centrifugation at 2000 rpm for 7 min, the supernatant was decanted into scintillation vials containing 10 ml of scintillant and counted.

This method is a disequilibrium method. It does not function well if taken to equilibrium, as the calibration line becomes flatter, thereby reducing sensitivity. There is no sample preparation, resulting in high sample throughput.

Fig. 1 shows a typical calibration curve using the above method. It has a usable range of 500-2500 pg/ml, with 5% NSB.

Pharmacokinetic applicability

The time curve after i.v. dosage is illustrated in Fig. 2 for 6 healthy volunteers. As is shown for the 24-h period studied, FP has a high clearance and a short half-life. Plasma levels fell rapidly, and by 6 h had decreased below the limit of quantification.

THE NEED FOR IMPROVED SENSITIVITY

Clinical studies have established that the likely inhaled dose of FP will be ~200 µg. In general, inhaled formulations of this type result in as little as 20% of the dose reaching

Table 1. Accuracy, precision and 10-month performance of the assay employing SPE. Concentration (C) values are pg/ml; ± values are S.D.'s for the tabulated means (spiked plasma samples).

Nominal C	Within-batch		Between-batch		QC's over 10 mths.	
	C, found	% C.V.	C, found	% C.V.	C, found	% C.V.
25	29±6	21	15±12	80*		
50	55±12	22	46±6	13	51±12	24
75	82±15	18	57±10	18		
100	110±10	9	88±9	10	96±14	15
125	140±20	14	100±10	10		
150	170±10	6	140±20	14	150±30	20
200	220±30	14	180±30	17		
250	240±30	13	220±50	23		

*Calibration line used only to quantify results for 50 pg/ml upwards.

the lungs; the remainder enters the GI tract. FP is poorly absorbed by the GI epithelium and has a negligible oral bioavailability as a consequence of virtually complete first-pass metabolism to the 17β carboxylic acid which possesses <1% of the glucocorticoid activity of FP. This is highly advantageous in inhalation therapy, but for the bioanalyst it means notably low plasma levels. Any systemic levels seen would be due almost entirely to topical absorption in the lung. Administration by this route coupled with high hepatic clearance is likely to result in very low plasma levels. Accordingly, an initial SPE step was introduced with a view to improving assay sensitivity.

C-18 EXTRACTION ASSAY

FP, being lipophilic, is ideal for SPE with C-18 cartridges. It was possible to put a 1 ml sample through a standard 200 mg Bond Elut column to extract FP from the plasma. Retained FP could be successfully recovered from the column by elution with methanol, with 80% recovery. The extraction system was as follows.-
1. Conditioning: 1 ml methanol, 1 ml TBS buffer.
2. Loading: 1 ml standard/sample/QC.
3. Wash: 1 ml TBS buffer, 1 ml dist. water.
4. Elution: 1 ml ethanol (collecting into assay tubes).
5. Evaporation: take to dryness (N_2 stream, 50°).
Following extraction the residue obtained was reconstituted using 100 μl of drug-free plasma. The sample was then treated as described above for the direct RIA.

Spiked plasma standards were extracted as above, and a 50-250 pg/ml calibration range established (typical curve: Fig. 3). Table 1 shows accuracy and precision values, within-batch

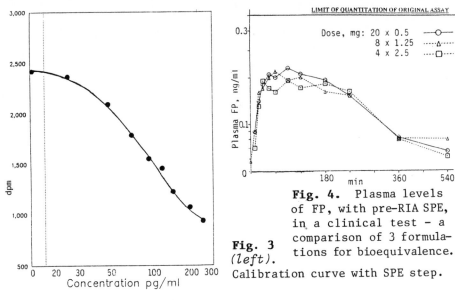

Fig. 3 *(left).* Calibration curve with SPE step.

Fig. 4. Plasma levels of FP, with pre-RIA SPE, in a clinical test - a comparison of 3 formulations for bioequivalence.

and (6 assays) between-batch, and QC values obtained over a 10-month period. The NSB remained at 5%.

Application to a bioequivalence study

Fig. 4 shows a plot of the results from a clinical bioequivalence test. It can be seen that the C_{max} is ~250 pg/ml and that finite amounts are detectable up to ~6 h. The horizontal line at 500 pg/ml shows the lower limit of quantification of the direct method. The data show clearly the equivalence of the three formulations tested, and also that only with the increased sensitivity offered by the extraction method could the information revealed have been obtained, the systemic concentrations being below the quantification limit achievable with the earlier calibration curves (e.g. Fig. 1).

CONCLUDING COMMENTS

The SPE step in the assay increased sensitivity with no significant change in NSB. This extra step increases the assay time, but the information gained is worth the decrease in sample throughput. Indeed, without the increased sensitivity there would have been little point in assaying most of the samples generated during the development of FP for inhalation therapy (feasible since 50 pg/ml was measurable).

#C-6

DETERMINATION OF BAY u 3405, A NOVEL THROMBOXANE ANTAGONIST, IN PLASMA AND URINE BY HPLC AND GC

Wolfgang Ritter

Institute of Clinical Pharmacology,
Bayer AG, 5600 Wuppertal 1, Germany

BAY u 3405

	HPLC assay	GC assay
Require-ment	*An assay for BAY u 3405 in plasma and urine, sensitive to a few ng/ml, suitable for pharmacokinetic investigations in humans. The dosage may be only 6 µg/kg.*	
End-step	*RP-HPLC* with acetonitrile /phosphate buffer pH 5.5; detection at 230 nm.*	*Capillary GC-ECD with temperature-programming.*
Sample handling	*Extraction (+ i.s.) with diethyl ether/hexane; dried down.*	*Extraction (+ i.s.) with DCM; dried down and derivatized with diazomethane.*
Comments	*The two assays had good accuracy and precision, and correlated well. For GC the LOD was 1 ng/ml (2 ng/ml for urine). The HPLC assay, slightly less sensitive, has performed well in other laboratories.*	

TXA$_2$* is a major metabolite of the arachidonate cascade$^\oplus$ and one of the most powerful naturally occurring substances. It is produced by platelets and in cells of various tissues, and has platelet-activating and smooth muscle-contracting effects. TXA$_2$ receptors have recently been identified in bronchial, tracheal and lung-parenchymal smooth muscle, and are regarded as playing a role in the pathogenesis of bronchial asthma.

BAY u 3405, (+)-(3*R*)-3-(4-fluorophenylsulphonamido)-1,2,-3,4-tetrahydrocarbazole-9-propanoic acid [1, 2], is a novel selective and competitive TXA$_2$ receptor antagonist with pronounced efficacy in various pharmacological model systems [2-6]; thus,

*Abbreviations.- DCM, dichloromethane; ECD, electron-capture detection; i.s., internal standard; LOD, limit of detection; RP, reversed-phase; TX, thromboxane:- p.   has a diagram$^\oplus$ also showing PG = prostaglandin. *Assay of* TX's: *see* Vol. 18 (ed. E. Reid *et al.*; Plenum, 1988). 'Cascade' details: p. 130 in Vol. 21 (Royal Society of Chemistry).

it can reverse PGD_2-induced bronchioconstriction in the guinea-pig [6].   BAY u 3405, the (+)-enantiomer, is more potent *in vitro* by one or two orders of magnitude than the (-)-enantiomer [1].

For the pharmacokinetic investigation during clinical development, a specific assay procedure for determination in human plasma and urine was required. Since rising-dose studies in healthy volunteers started with an oral dose of 0.5 mg/subject, an LOD of 5 ng/ml or better seemed requisite. Two chromatographic procedures, HPLC and GC, were developed in parallel. Because the assay methods were to be employed not only within the company but also in clinical laboratories, much time   and effort was invested to obtain assay procedures which could easily be performed by different operators without compromising specificity and reliability.   The descriptions now given are followed by a 'supplement' which relates particularly to the choice of conditions.

SAMPLES AND MATERIALS

Plasma is obtained from blood, collected in disposable plastic tubes (Sarstedt) containing EDTA, by centrifugation (6200 × **g**; <20°).   **Extraction** is performed in 10 ml glass-stoppered centrifuge tubes, and evaporations in conical tubes (~12 ml) made in-house.   For cleaning glassware, steam-sterilization is followed by treatments with basic and acidic agents and repeated rinsing (at 90° initially) with pure water (Millipore).

Reagents and chemicals are Merck 'z.A' or similar; for HPLC they are of appropriate purity.   Buffer solutions (in Millipore Milli-Q Plus water) are passed through 0.45 µm filters. **Standards** are kept in methanol (working solution 100 µg/ml) at room temperature in the dark. They show good stability, as do plasma samples containing the drug (up to 7 days/6 mths. at 4°/-30°).

DETERMINATION BY HPLC

To 1 ml plasma, or 200 µl urine diluted with 800 µl water, 400 ng i.s. (an analogue of BAY u 3405 with, in position 6, fluoro) are added.   Two extractions are performed with shaking, each with 4 ml diethyl ether/hexane (1:1).   The organic phase is evaporated at 40° (N_2 stream), and the residue reconstituted in 300 µl 0.02 M pH 5.5 phosphate buffer; 50 µl is loaded by autosampler into an HPLC system (Spectra Physics) with an SP 8480 detector (at 230 nm) and SP 4200 computing integrator.

The RP-18 (5 µm) columns, run at 40°, are from Merck: main column LiChrospher 100, 125 mm; guard column suitably LiChroCART, 40 mm.   The mobile phase is 0.02 M pH 5.5 phosphate buffer/acetonitrile (68:32 by vol.), with flow-rate 0.85 ml/min for 9 min and then 1.8 ml/min for 3.5 min. The retention times are 5.9 min for BAY u 3405 and 7.1 min for the i.s. (see later for illustrative chromatograms).

Analytical parameters.- The LOD in plasma was 2 ng/ml (5 ng/ml for quantification); urine 5 ng/ml (10 ng/ml). Within-day precisions (with n) at ng/ml values as stated were:-
5, 19.2% (10); 10, 6.35% (5); 20, 2.63% (5); 30, 1.37% (5); 50, 1.88% (10) and 1.01% (10); 300, 1.00% (10); 500, 1.91% (10).
Between-day precisions were: 50, 0.96% (7) and 500, 1.39% (7); (for deep-frozen samples) 50, 3.81% (14) and 500, 2.73% (14); 50, 3.57% (17) and 500, 2.64% (23).
Inaccuracies (ng/ml) were 1.4 at 50 ng/ml (n = 22) and 8.2 at 500 (n = 25). The detector response is linear from 5 to 5000 ng/ml.

DETERMINATION BY GC-ECD

The i.s. is an analogue of BAY u 3405 with 4'-chloro replacing 4'-fluoro. To 1 ml plasma or urine, 500 ng are added, and 2 ml of pH 4.0 citrate buffer (Titrisol, Merck). After extraction by shaking with 3 ml DCM, the organic phase is evaporated at 40° (N_2 stream). The residue is dissolved in 100 µl acetone, and 20 µl 1% diazomethane in ether (ampoules stored at -30°) is added. After 5 min at room temperature, the acetone and excess diazomethane are evaporated under N_2; residue is dissolved in 150 µl toluene and 1µl loaded for GC. The instrument (Dani #6500 HR) has a programmed temperature vaporizer injection system (set at 110°), autosampler and ECD #655, at 325°.

The DB-1 (0.1 µm) column (30 m x 0.32 mm i.d.) is run with H_2 (1.2 bar), N_2 (as auxiliary gas; 0.6 bar) and, after 1 min at 200°, 30° rise per min. up to 240°, then 5°/min to 300° and kept there for 1 min. The retention times are 7.3 min for BAY u 3405 and 9.1 min for i.s. (see later for illustrative chromatograms).

Analytical parameters.- The LOD is 1 ng/ml in plasma and 2 ng/ml in urine. Within-day precisions (with n) at ng/ml values as stated were:-
10: 9.35% (5); 50: 1.10%, 1.68% & 3.79% (each 5); 200: 2.35%, 2.60% & 5.60% (each 5).
Inaccuracies (ng/ml) were: for 10, 0.40 [= 4.0%] and 0.80 [= 8.0%] (each n = 5); for 50, 0.65 [= 1.30%] (n = 18); for 200, 2.70 [= 1.25%] (n = 20).

COMPARISON OF HPLC AND GC

Both assay procedures for BAY u 3405 in plasma and urine have been shown to be specific and reliable. HPLC is the preferred method when many samples have to be analyzed, e.g.

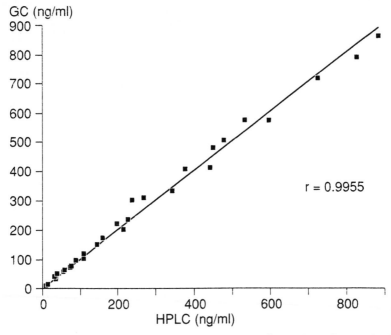

Fig. 1. Correlation of HPLC and GC (peak-height ratio basis) for assaying BAY u 3405 in plasma (30 spiked samples).

in pharmacokinetic studies in volunteers or patients [6, 7] and during drug monitoring in patients. GC is less robust than HPLC, and sample throughput is generally lower. The HPLC method has already been successfully set up in laboratories in the U.S.A., South Africa and Japan.

In a few instances when analyzing patients' plasma, peaks from endogenous components or former medication interfered with BAY u 3405 assay in plasma. In these cases the HPLC results were cross-checked by GC. From this experience, GC-ECD of BAY u 3405 in body fluids can be regarded as more selective than the HPLC assay with UV detection.

The two assay procedures, however, proved to be equally precise and accurate, giving near-identical values for spiked human plasma samples in the concentration range 10–900 ng/ml (Fig. 1). Both have served well for studies in the dosage range 6–650 μg/kg, in healthy volunteers and in patients [7, 8].

===========

CHROMATOGRAPHIC PATTERNS

Representative chromatograms for plasma are shown in Fig. 2 for HPLC and in Fig. 3 for GC.

Fig. 2. HPLC
chromatograms
for plasma (200 ng
i.s./ml). BAY u 3405
spiked in (**A**: 200 ng
/ml; found: 197.1) or
ingested 12 h before
(**B**; 77 mg controlled
release tablet;
found, 10.8 ng/ml).
Drug peak marked *,
at 5.8 min (i.s. at
6.75 min). **C** = pre-
dose blank; *arrow*
shows 'window' in
drug position. (See
Fig. 3 legend for
comment on R_T's.)

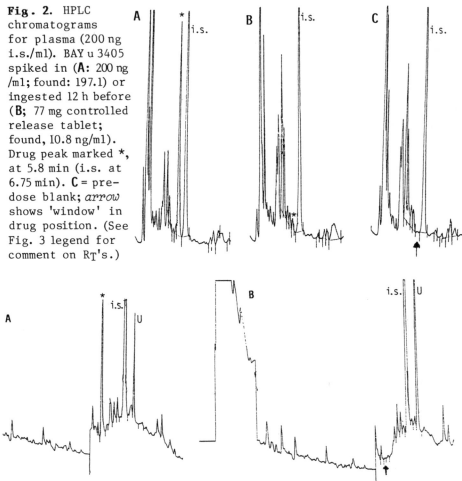

Fig. 3. GC chromatograms for plasma, spiked with i.s. and in **A**,
BAY U 3405 (30 ng/ml: peak *, at 9.3 min; i.s. at 10.5 min). **B**, blank:
arrow shows 'window' in drug position. U, unidentified plasma peak.
(R_T's differ from those stated in text, since conditions different.)

METHOD-DEVELOPMENT CONSIDERATIONS

Since there seem to be no metabolites in plasma (bile
and faeces contain glucuronide), only the parent drug had
to be analyzed.   It is an acid, water-soluble even at neutral
pH.   **Solvent extraction** was tried at different pH's (2-6.5)
and with diverse solvents (or mixtures; ranging from heptane
to ethyl acetate) using, for **HPLC**, an optimized mobile phase
with detection at 230 nm (the 285 nm alternative gives fewer
interferences but a worse LOD).

As usual, a compromise was adopted; thus for HPLC it was more advantageous to 'live with' a low extraction yield from plasma (47±1%) and achieve 2 ng/ml LOD than to obtain 95% recovery (with diethyl ether) but a worse LOD (10 ng/ml).

Solid–phase extraction (SPE) was also considered and investigated (C-8 and C-18 from two manufacturers, 40 µm; 6 ml/500 mg columns). Conditioning with methanol (2 ml) and water (2 × 2 ml), and loading the plasma (1 ml), was followed by washing (3 × 2 ml water) and elution with 500 µl methanol. Extraction efficiency (90±2%) was much better than with liquid–liquid extraction conditions that gave a comparable LOD. However, within-day and between-day imprecision (variability) was much greater; moreover, SPE was much more time-consuming than solvent extraction. An especially critical feature of SPE, in our experience, is the meticulous attention needed to ensure that the columns do not dry out during conditioning and washing. In view of the unacceptable imprecision, the SPE approach was finally abandoned.

References

1. Rosentreter, U., Böshagen, H., Seuter, F., Perzborn, E. & Fiedler, V.B. (1989) *Arz.-Forsch./Drug Res. 39*, 1519–1521.
2. Perzborn, E., Boberg, M., Bühner, K., Böshagen, V.B., Fiedler, V.B., Gardiner, P.J., Müller, U., Norman, P., Ritter, W., Rosentreter, U., Seuter, F. & Weber, H. (1991) *Drugs of the Future 16*, 701–705.
3. Perzborn, E., Seuter, F., Fiedler, V.B., Rosentreter, U. & Böshagen, H. (1989) *Arz.-Forsch./Drug Res. 39*, 1522–1525.
4. Seuter, F., Perzborn, E., Rosentreter, U., Böshagen, H. & Fiedler, V.B. (1989) *Arz.-Forsch./Drug Res. 39*, 1525–1527.
5. Fiedler, V.B., Perzborn, E., Seuter, F. & Rosentreter, U. & Böshagen, H. (1989) *Arz.-Forsch./Drug Res. 39*, 1527–1530.
6. Francis, H.P., Morris, T.G., Thompson, A.M., Patel, U. & Gardiner, P.J. (1991) *Ann. N.Y. Acad. Sci. 629*, 399–401.
7. Ritter, W., Weber, H. & Kuhlmann, J. (1990) *Arch. Pharm. 323*, 692.
8. Ritter, W., Weber, H. & Kuhlmann, J. (1990) *J. Clin. Pharmacol. 30*, 849.

#C-7

METHODOLOGICAL PROBLEMS IN THE STUDY OF THE METABOLISM OF THE CHIRAL THROMBOXANE ANTAGONIST ICI 185,282

L. Witherow, A. Warrander and I.D. Wilson[⊕]

Safety of Medicines Department, ICI Pharmaceuticals,
Mereside, Alderley Park, Macclesfield SK10 4TG, U.K.

The metabolism and disposition of racemic [14]*C-ICI 185,282
has been investigated in male and female rats following oral
administration at 20 mg/kg. The compound was administered
either as a racemate with both enantiomers radiolabelled or
as a pseudoracemate with only the pharmacologically active enantiomer labelled. Differences were observed in both the
routes of elimination and the* [14]*C-metabolite profiles between
male and female animals dosed with the same radiolabelled
form of the compound. Similarly the routes of elimination
and* [14]*C-metabolite profiles were observed to vary between
animals of the same sex dosed with the same radiolabels.
These differences, coupled with the complexity of the metabolite
profiles themselves (irrespective of sex or radiolabel) made
the interpretation of the results difficult.*

The last decade has witnessed a dramatic increase in
the awareness of the importance of chirality as a factor
to be considered in the design of the drug metabolism studies
to be undertaken during drug development. Indeed, where
the desired pharmacology resides predominantly in one enantiomer,
the usual decision nowadays is not to develop the racemate
but to progress only the active enantiomer. However, there
are many racemic compounds already at an advanced stage of
development for which the adoption of such an approach would
cause profound difficulties. For such compounds, studies
must be designed which provide information on the metabolism,
pharmacokinetics and disposition of the racemate and the
individual enantiomers both in the species used for toxicological
evaluation and in man. Our concern here is with ICI 185,282, i.e.
5(Z)-7-([2RS,4RS,5SR]-4-o-hydroxyphenyl-2-trifluoromethyl)-1,3-dioxan
– a racemate. This compound,
being a thromboxane antagonist,
is potentially useful as, e.g.,
a bronchodilator [cf. W. Ritter's
art., #C-6].

[⊕]to whom any correspondence
should be addressed.

This article describes metabolism studies on ICI 185,282, performed in male and female rats. The studies were undertaken using racemic forms of the drug differing in that only one enantiomer (the pharmacologically active ICI 188,434) or both of the enantiomers were radiolabelled.

EXPERIMENTAL

Compounds.- Racemic ^{14}C-ICI 185,282 was synthesized in the Radiochemical Laboratory at ICI Pharmaceuticals with a radio-chemical purity of 98%. This material was mixed with unlabelled ICI 185,282 to give material with a specific activity of 5 µCi/mg. The radiolabelled active enantiomer of ICI 185,282 (ICI 188,434) was prepared from the racemic ^{14}C-radiolabelled compound by fractional crystallization with a chiral amine. A pseudoracemate was then prepared by mixing ^{14}C-ICI 188,434 with an equal amount of the unlabelled, inactive enantiomer.

Animals.- Male and female albino rats (alpk:APFSD Wistar-derived; ~200 g) were obtained from the Alderley Park animal breeding unit. Rats were acclimatized for 3 days prior to dosing in glass metabowls at room temperature (23°) with light-and-dark cycles every 12 h. Food and water were available *ad libitum.*

Treatment.- Four groups of 3 animals were dosed orally at a dose level of 20 mg/kg (20 µCi/animal). Thus 3 male and 3 female rats (Groups 1 & 2) each received a single oral dose of ^{14}C-ICI 185,282 racemically radiolabelled. A further 3 male and 3 female rats (groups 3 & 4) each received a single oral dose of the pseudoracemic ^{14}C-ICI 185,282 in which only the active enantiomer (ICI 188,434) was radiolabelled.

Sample material.- Urine, faeces and cage-wash were collected for the periods 0-24, 24-48, 48-72, 72-96 and 96-120 h post-dose. The rats were killed 120 h after dosing (by anaesthetic inhalation) and the carcasses retained. All samples except cage-wash were stored frozen at -20° until analyzed. Cage-wash was stored at 0-4°.

Sample analysis for [14]C

Urine and cage-wash aliquots were taken (4 x 500 µl or, for 0-24 h urine, 4 x 200 µl) and diluted to 1 ml with distilled water. Beckman 'Ready Value' LSC* cocktail (10 ml) was added to each aliquot. Counting was done (Packard 1900C or equivalent counter) for 10 min or until 10^4 counts had accumulated. Quench correction was calculated using the 'H number' (Beckman) or 'ISIE' (Packard) method.

Faecal samples mixed with an equal weight of water were homogenized. From the homogenate 4 aliquots were taken (0.5 g
*LSC *denotes* liquid scintillation counting.

or, for 0-24 h samples, 0.2 g) and combusted in a sample
oxidizer (Packard 306). The combustion products were absorbed in
LKB Optisorb '1' (8 ml) which was then mixed with LKB Optisorb 'S'
scintillant (12 ml) for LSC as above.

Carcasses were dissolved overnight at 50° in 2M methanolic
NaOH containing detergent. From each sample 4 aliquots (~200 µl)
were taken by weighing, neutralized with 4.4 M nitric acid,
diluted to 1 ml with water, mixed with Beckman 'Ready Value'
scintillant (10 ml), and assayed by LSC.

Separation and quantification of metabolites

Sample preparation.- Faecal homogenates (0-24 h, 2 g) were
extracted with methanol (30 ml). Urines (0-24 h, 1 ml) were
extracted and concentrated using RP cartridges (C-18 Bond Elut).

Normal phase TLC was performed on silica gel glass-backed
plates (20 × 20 cm; E. Merck, Poole, U.K.) which incorporated
a fluorescent indicator. Samples of urine and concentrates
of the methanolic faecal extracts were applied as 2 cm bands
using an automatic applicator (Camag Linomat IV). Ascending
chromatography was performed in glass TLC tanks using the
solvent system chloroform/ethyl acetate/formic acid (5:4:1 by
vol.). The plates were allowed to develop to 17 cm. Labelled
(^{14}C) metabolites were located by autoradiography and quantified using a linear analyzer — LabLogic (Sheffield, U.K.)
RITA, AMBIS [1], or Berthold (St. Albans, U.K.) LB 2842 — or
a Phosphor Imager (Molecular Dynamics, Sevenoaks, U.K.) [2].

Reversed phase TLC was performed on C-18-bonded glass-backed
plates (10 × 20 cm; E. Merck) which incorporated a fluorescent
indicator. The urine and faecal extracts were applied as
above (2 cm bands), with similar ascending chromatography.
The solvent was water/acetonitrile/trifluoroacetic acid (60:40:0.1).
After development to 17 cm, radioactivity was detected and
quantified as above.

RESULTS

Elimination was mainly in the faeces, whatever the nature
of the radiolabelled material employed and irrespective of
the sex of the animals. The detailed excretion-balance
results appear in Table 1. Excretion was generally rapid,
the majority of the dose being recovered in the 0-48 h period
after administration. Differences in the routes of elimination
of the different radiolabelled forms were most pronounced
for the female animals. Thus, ~63% of the dose was eliminated
in the faeces of female rats dosed with the racemically
radiolabelled ICI 185,282 whilst for the pseudoracemate only

Table 1.
Recovery of the dose following oral administration of ^{14}C-ICI 185282 to male and female rats. Values are means ±S.E. (n = 3).

	Time	Percentage dosed radioactivity recovered				
		Urine	Cagewash	Faeces	Carcasses	Total
racemate						
Male	0 - 24	15.27 ± 0.81	0.92 ± 0.17	55.47 ± 6.24		
	24 - 48	1.38 ± 0.35	0.25 ± 0.07	13.86 ± 2.68		
	48 - 72	0.21 ± 0.02	0.31 ± 0.28	2.25 ± 0.49		
	72 - 96	0.07 ± 0.00	0.03 ± 0.01	0.52 ± 0.10		
	96 - 120	0.04 ± 0.01	0.02 ± 0.00	0.41 ± 0.17		
	0 - 120	16.97 ± 1.16	1.29 ± 0.71	72.51 ± 3.08	0.50 ± 0.01	91.52 ± 2.26
Female	0 - 24	22.05 ± 0.97	3.86 ± 0.55	56.39 ± 2.68		
	24 - 48	1.51 ± 0.26	0.55 ± 0.14	5.44 ± 0.58		
	48 - 72	0.37 ± 0.05	0.14 ± 0.01	0.83 ± 0.08		
	72 - 96	0.14 ± 0.00	0.07 ± 0.01	0.48 ± 0.18		
	96 - 120	0.06 ± 0.00	0.03 ± 0.00	0.13 ± 0.02		
	0 - 120	24.14 ± 1.25	4.66 ± 0.64	63.26 ± 2.13	0.28 ± 0.02	92.33 ± 0.38
pseudoracemate						
Male	0 - 24	17.96 ± 1.04	1.19 ± 0.42	56.20 ± 2.60		
	24 - 48	1.19 ± 0.13	0.21 ± 0.03	12.01 ± 1.62		
	48 - 72	0.26 ± 0.03	0.06 ± 0.01	2.97 ± 0.57		
	72 - 96	0.11 ± 0.01	0.03 ± 0.01	0.70 ± 0.07		
	96 - 120	0.05 ± 0.01	0.30 ± 0.28	0.36 ± 0.08		
	0 - 120	19.65 ± 0.71	1.78 ± 0.32	74.06 ± 3.06	0.67 ± 0.08	96.17 ± 3.44
Female	0 - 24	30.00 ± 2.00	4.80 ± 1.65	44.37 ± 1.08		
	24 - 48	2.29 ± 0.20	1.23 ± 0.41	6.55 ± 0.68		
	48 - 72	0.56 ± 0.08	0.41 ± 0.18	1.19 ± 0.18		
	72 - 96	0.24 ± 0.05	0.07 ± 0.00	0.44 ± 0.06		
	96 - 120	0.14 ± 0.03	0.08 ± 0.03	0.18 ± 0.03		
	0 - 120	33.04 ± 1.74	6.62 ± 2.29	52.73 ± 0.75	0.39 ± 0.04	92.96 ± 0.26

~53% was eliminated by this route. There was a corresponding increase in the excretion of radioactivity in the urine of females dosed with the pseudoracemate (33% compared to 24% for racemic radiolabel-dosed females). Much smaller differences were noted for male rats with a similar, but not statistically significant, trend towards the greater urinary excretion of the pseudoracemate (Table 1). There were also differences in the routes of elimination of radioactivity for male and female rats dosed with the same radiolabelled form of ICI 185,282 (Table 1).

Metabolite profiles

Following the elimination studies, metabolite profiles were obtained using normal- and reversed-phase TLC for urine and faecal extracts, for the period 0-24 h post-dose. Complex, multi-component, metabolite patterns were observed in all cases. Differences in metabolite patterns were noted between the two sexes dosed with the same radiolabelled form of the compound and between animals of the same sex dosed with the different radiolabelled forms of the drug. Metabolite patterns within dose groups were, however, in good accord. For the sake of clarity and ease of discussion, only differences between animals of the same sex but dosed with different forms of the radiolabelled ICI 185,282 will be considered here. Typical ^{14}C-metabolite profiles for the urine of male rats dosed with either the racemic or the pseudoracemic forms of ICI 185,282 are shown in Fig. 1A & 1B respectively.

These chromatograms show two main groups of metabolites, one at or near the origin and the second, less polar, group centered at R_f ~0.5. The most striking difference between the two chromatograms was that in samples from animals given the ^{14}C-racemic radiolabel there was present a prominent peak at R_f 0.1 which was absent in chromatograms for rats given the ^{14}C-pseudoracemate. This peak must therefore have arisen from stereoselective metabolism of the inactive enantiomer of ICI 185,282. Similar results were obtained for the urine of female rats (Fig. 2, A & B).

^{14}C-metabolite profiles for faecal extracts are shown in Figs. 3 (male) and 4 (female). A complex pattern of radiolabelled metabolites was obtained, with between 9 and 11 'major' and a number of minor metabolites detectable by autoradiography (depending upon the chromatographic system employed). However, differences between the profiles were apparent, depending upon the radiolabelled form of the compound administered. In particular, animals receiving the pseudo-racemate showed a significant enhancement in the size of a peak of R_f 0.3-0.5 compared to rats dosed with the racemate.

Figs. 1–4. Radio-TLC after dosing (**A**) racemic, (**B**) pseudo-racemic ICI 185,282. Load-band front at o.

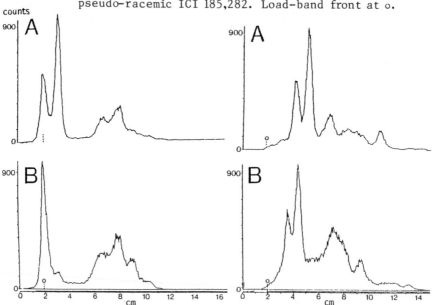

Fig. 1. Male-rat urine; normal-phase TLC.

Fig. 2. Female-rat urine; reversed-phase TLC.

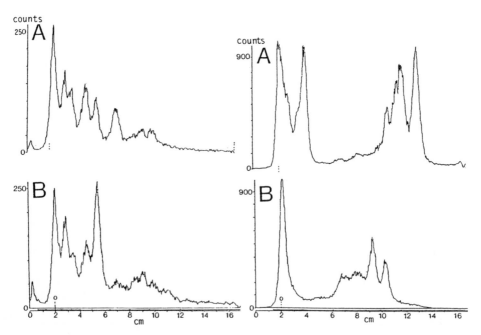

Fig. 3. Male-rat faeces; normal-phase TLC.

Fig. 4. Female-rat faeces; normal-phase TLC.

Similarly normal-phase TLC (data not shown) revealed the presence of a peak (R_f ~0.1) which was present only in extracts of faeces from animals receiving the racemic radiolabel.

Profile complexity.- A particular problem encountered in this work resulted from the complexity of the metabolite profile. This made interpretation of the results difficult but also resulted in technical difficulties in terms of the resolution of the radio-TLC detectors. Thus, although the individual components were sufficiently well resolved for auto-radiography to provide a good qualitative picture, the bands of radioactivity were often too close to be resolved by the linear analyzers, making quantification impracticable. We therefore briefly investigated the use of two devices with much better spatial resolution to determine whether they had any advantages in situations such as this.

Fig. 5 shows results for the same sample analyzed with either a conventional linear analyzer, or the AMBIS 2-D Linear Analyzer [1], or a Phosphor Imager [2]. The improvement in resolution afforded by the AMBIS and the Phosphor Imager is clear, and manifests the benefits of these devices when dealing with complex multi-component samples.

CONCLUSIONS

Based on the results of this study, a number of general conclusions can be drawn. Firstly, it is quite clear that by comparing the excretion of racemic and non-racemic labelled material (dosed as the racemate) it is possible to demonstrate that enantioselective elimination has taken place. Further, by comparing ^{14}C-metabolite profiles for extracts of excreta it is also possible to demonstrate that enantioselective metabolism has occurred. The absence of peaks where the extracts were from the study where only the active enantiomer was radiolabelled, as distinct from the study with radiolabelled racemate, also provides good evidence that the metabolite was produced from the inactive enantiomer.

Similar conclusions can be drawn from significant enhancement of radiolabel peaks in the study with labelled active enantiomer compared with the study using racemic radiolabel, although of course in this case the radiolabelled metabolite will be present in both extracts. However, it is much more difficult to be confident in the case of small changes where both inactive and active forms of the compound might be converted to the same metabolite in slightly different proportions. To unambiguously deal with this difficulty, a further study where the radiolabelled inactive enantiomer is dosed as a pseudoracemate is clearly indicated.

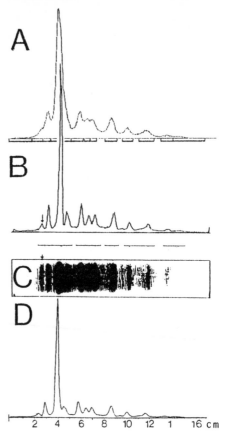

Fig. 5. Comparison of peak
resolution on three different
instruments for radioactivity
detection on TLC plates: ^{14}C
in male-rat urine after dosing
the pseudoracemate (cf. Fig. 1B).
A, the Berthold LB 503 as used
for the majority of analyses in
this study;
B, the chromatogram, and **C,** the
'autoradiograph' produced by
the AMBIS 2-D instrument;
D, the 'autoradiograph'
produced by the Phosphor Imager.
The TLC system was of NP type.

Probably the best course of action where the investigation
of the metabolism and excretion of racemic compounds is to
be investigated would therefore be to use two pseudoracemic
radiolabelled forms of the compound.   This would readily
show gross differences in metabolic fate, and would allow
common metabolites to be detected.   Having established that
enantioselective elimination and metabolism has occurred, the
isolation, identification and determination of the enantiomeric
composition (e.g. by NMR or chiroptical techniques) of the
individual metabolites could be undertaken.   In the light
of the daunting amount of additional work that would be
required for an extensively metabolized compound such as ICI 185,282
(whatever the final strategy used to determine its chiral
metabolism and disposition), the adoption of a policy where
only the active enantiomer of a racemic compound is developed
is clearly to be welcomed.

References

1. Smith, I. & Furst, V. (1989) *J. Planar Chromatog. 2,* 233-237.
2. Johnson, R.F., Pickett, S.C. & Barker, D.L. (1990)
   Electrophoresis 11, 335-360.

#ncC

NOTES and COMMENTS relating to
ANTI-ASTHMATICS AND KINDRED AGENTS

Forum comments relating to the preceding main arts., and
to the 'Notes' that follow, appear (not in exact order)
on pp. 247-249.

Eicosanoids guide

from Vol. 21 of this series (ed. E. Reid et al.; same publisher)

- pertinent especially to arts. #C-1, #C-6, #C-7 and #ncC-1;
  cf. **Arachdonic** entry in Analyte Index, #Ia; standard
  abbreviations include LT = leukotriene, TX = thromboxane


```
5-HETE                      PL-A₂ action
   ↑                          furnishes
5-HPETE ←——————————— arachidonic acid, aa → 15-HPETE——→ LX's
   ↓       lipoxygenase                │cyclo-oxygenase
 LT's                         endoperoxides, e.g. PGG₂
[& routes                 ———————————————|↓L——————————————→ TX's
via 8- or   PG's ←
12-HPETE]                         PGI₂
                            (prostacyclin)
```

*NOTE: HPETE = hydroperoxyeicosatetraenoic acid (isomers!).
Further guidance, with a diagram, appears on p. 63 of Vol. 18*
('Bioanalysis of Drugs, Especially Anti-inflammatory and Cardio-
Vascular', ed. E. Reid *et al.*; Plenum, 1988), *which has arts. on
eicosanoid assay.*

#ncC-1

A Note on

HPLC METHODS DEVELOPED TO AID IN THE ASSESSMENT OF THREE ARACHIDONATE 5-LIPOXYGENASE INHIBITORS

G.S. Land, J.P. Sharpe, G.G. Lovell, R.J. Mason and †J.A. Salmon

Department of Bioanalytical Sciences/†Research Directorate, Wellcome Research Laboratories, Langley Court, Beckenham, Kent BR3 3BS, U.K.

Require-ment	*A bioanalytical method, applicable to plasma from different animal species, to aid in establishing the most suitable candidate from a programme of research into inhibitors of arachidonate 5-lipoxygenase, particularly BWA4C, BWB70C and BWA137C (formulae overleaf, Figs. 1 & 2).⊕ Once a 'general' method had been devised, it was optimized for each analyte and for metabolites.*
End-step	*RP-HPLC (C-18 column, pre-conditioned), with detection by UV (at 226 nm in the general method). Advantageous wavelength variation and fluorescence in later work. Mobile phase: water/THF/TFA* (e.g. 40:6:0.1), with 0.5 mM oxalic acid except when, later, a PRP-1 column was used.*
Sample treat-ment	*The sample (+ i.s. if applicable) was freed from protein by acetone addition and centrifugation, and the super-natant was extracted with chloroform. The residue obtained by drying down was dissolved in aqueous THF for HPLC (Scheme 1; unaffected by method modification).*
Comments	*The general method is versatile and quite sensitive (calibration range 0.2-50 µg/ml). Advantages of variant methods for particular analytes may include better sensitivity. Conditioning unnecessary with PRPl column.*

The background to investigation of potential inhibitors of the lipoxygenase has been given elsewhere [1]. For the general method that was developed in this investigation, details are given above and later, particularly in Scheme 1 and Fig. legends. When applied to the analysis of plasma samples from several animal species, the method had to be adapted to embrace metabolites such as those shown in Fig. 1 for BWA4C. The emphasis of the investigation shifted from a

*Abbreviations.- TFA, trifluoroacetic acid; THF, tetrahydro-furan; i.s., internal standard. % values are on a vol. basis.
⊕Hydroxyureas or acetohydroxamic acids. Lipoxygenase role: p. 225.

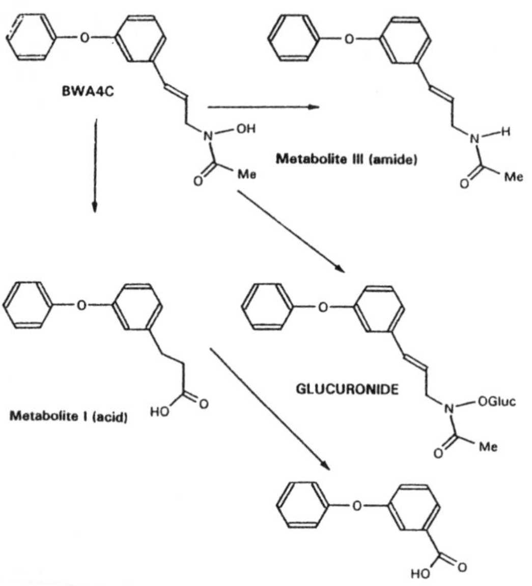

Fig. 1. Structure and metabolism of BWA4C.

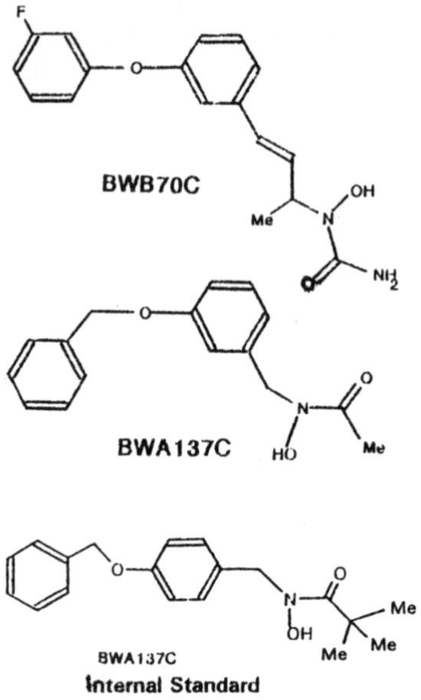

Fig. 2. Structures of BWB70C and of BWA137C and its i.s.

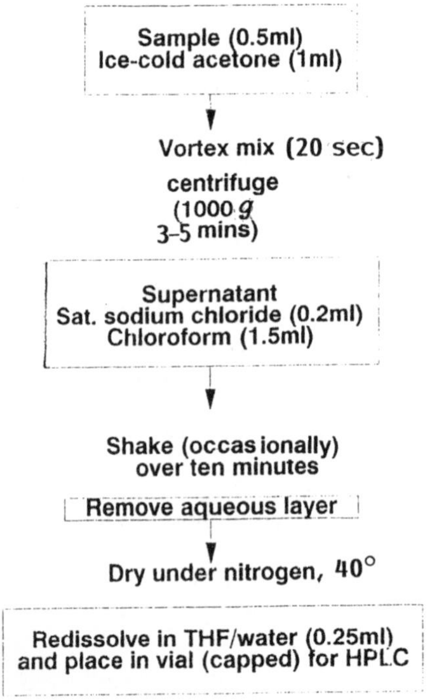

Scheme 1. General extraction method.

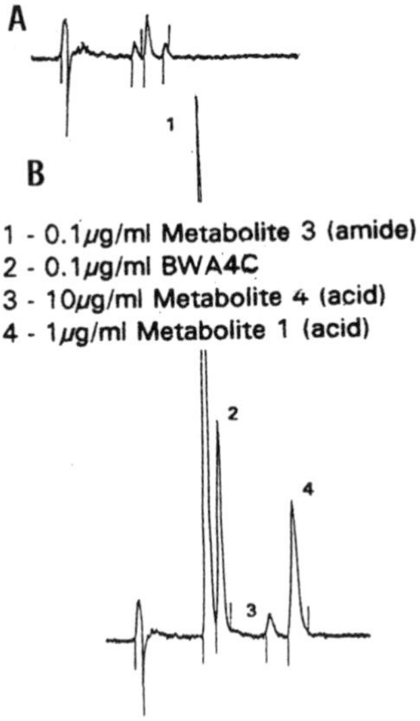

Sample (0.5ml)
Ice-cold acetone (1ml)

Vortex mix (20 sec)

centrifuge
(1000 *g*
3–5 mins)

Supernatant
Sat. sodium chloride (0.2ml)
Chloroform (1.5ml)

Shake (occasionally)
over ten minutes

Remove aqueous layer

Dry under nitrogen, 40°

Redissolve in THF/water (0.25ml)
and place in vial (capped) for HPLC

BWA4C

Metabolite III (amide)

Metabolite I (acid)

GLUCURONIDE

BWB70C

BWA137C

BWA137C
Internal Standard

1 - 0.1μg/ml Metabolite 3 (amide)
2 - 0.1μg/ml BWA4C
3 - 10μg/ml Metabolite 4 (acid)
4 - 1μg/ml Metabolite 1 (acid)

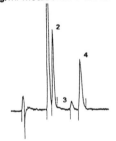

Fig. 3. HPLC of BWA4C and metabolites in plasma (50 μl; rabbit) with fluorescence detection (ex 235. em 320 nm; if UV instead, 240 nm). Column at 40°; 40:60:0.1 H_2O/THF/TFA. **A**, blank; **B**, spiked.

comparative assessment of the bioavailability/exposure of candidate drugs to a more detailed metabolic and pharmacokinetic assessment. Therefore the bioanalytical objectives became the provision of a selective, sensitive method capable of analyzing the parent drug and relevant metabolites in several different animal species and then (cf. Figs. 4 & 5) in man.

The general method

Usually the eluent specified above (ratio 47.5:52.5:0.1; 1 ml/min) was applied to a Spherisorb ODS 5 μm column (250 x 4.6 mm). conditioned, when new, with water (1 h), 3% EDTA (30 h), then water (5 h). **Advantages** of the method include the following:
- With an i.s. added, the method is reasonably easy to operate.
- A number of analogues similar to the primary analyte can be assessed without changing the assay.
- Run times are acceptable: 10-30 min, depending on the analyte.
- The assay can serve for a number of the metabolites (amine, acid, glucuronide), semi-quantitatively and quantitatively.
- Sensitivity is reasonably good for 0.5 ml samples, the calibration range being typically 0.2-50 μg ml.
Disadvantages include the following.
- The assay may not be optimal for a particular analyte.
- The chosen i.s. (cf. Fig. 2) may not always be appropriate.
- The column has to be pre-conditioned to ensure consistency.
- Use of THF is not a favoured practice.
- For some analytes the limit of quantification needs improvement.

Variant assays for particular analytes including metabolites

Use of multiple detectors, with UV wavelength switching and fluorescence, reduced the need to compromise in multi-analyte assays and, helped by reduced background interference, gave 10-fold improvement in sensitivity. This allowed <500 μl (even 50 μl) samples to be assayed. Parameters such as eluent composition, temperature and column type were varied to optimize separations. Trial of solid-phase instead of solvent extraction showed troublesome resistance of some analytes to elution procedures.

Representative chromatograms are shown in Fig. 3 for BWA4C and 3 metabolites, in Figs. 4 and 5 for BWB70C and metabolites, and in Fig. 6 for BWA137C. The separation shown in Fig. 5 (BWB70C) is noteworthy because individual enantiomers from the administered racemate were separated, using a coupled-column system [cf. Lindberg's art., #C-3 - *Ed.*]: column #1 was C-18 (room temp.; 40 x 6 mm), and #2 was Chiracel OD (0°; 250 x 4.6 mm; polar eluent).

Concluding comments. - An especially advantageous improvement was the use of polymer (spherical particle) in place of silica-based columns, removing the need for a pre-conditioning and having oxalic acid in the eluent (THF/TFA still needed). But eventually - maybe when drug development moves to another compound - one has to 'call-it-a-day' on method refinement.

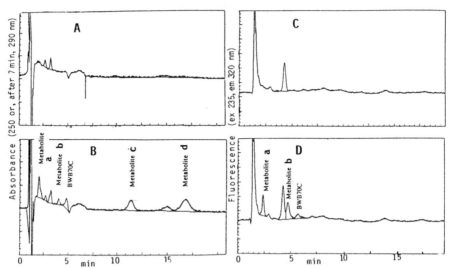

Fig. 4. Blank plasma extracts (**A, C**) and spiked plasma (BWB70C and metabolites, mid-range; **B, D**). Detection as in axis caption: **A** & **B**, dual UV; **C** & **D**, fluorescence (poor for 2 metabolites). Column: Hamilton PRP-1 10 μm, 250 × 4.2 mm, at 35°; 100 μl injected; 38.5:61.4:0.1 water/THF/TFA.

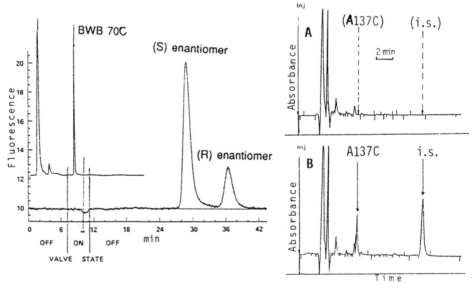

Fig. 5. Enantiomer separation of heart-cut BWB70C peak from plasma after *in vivo* dosing with the racemate. Chiral column received slightly more polar eluent than the C-18 column.

Fig. 6. BWA137C (**B**; **A** = blank) in plasma from a dosed rat. 'General' method.

Reference.- 1. Salmon, J.A., Jackson, W.P. & Garland, L.G. (1989) in *Therapeutic Approaches to Inflammatory Disease* (Lewis, A.J., ed.), Elsevier, Amsterdam, pp. 137-146.

#ncC-2

A Note on

TRACE ANALYSIS OF CLENBUTEROL IN BIOLOGICAL FLUIDS BY AUTOMATED CAPILLARY GC-MS

H.J. Förster, K.L. Rominger, E.M. Ecker, H. Peil and A. Wittrock

Division of Research, Boehringer Ingelheim KG,
D-6507 Ingelheim, Germany

Clenbuterol, 4-amino-α-[(*tert*-butylamino)methyl]-3,5-dichlorobenzyl alcohol hydrochloride (Ventipulmin®, Planipart®), is a potent sympathomimetic drug used in man (e.g. for asthma therapy) and domesticated animals.

The common therapeutic dose is rather low, only 20 µg/day. This, together with a large volume of distribution, results in plasma concentrations of the analyte in the low pg/ml range. We have described [1] a stable isotope dilution assay with clenbuterol-D9 as internal standard. The sample preparation comprises liquid-liquid extraction of alkalinized plasma with methyl *tert*-butyl ether, purification of the extract by an additional analyte transfer (acid pH/alkaline pH), followed by derivatization with hexamethyldisilazane. The resulting O-trimethylsilyl ethers of clenbuterol and the internal standard are analyzed by ammonia-CI GC-MS with selective monitoring of the $(M+H)^+$ ions at m/z 349 and 358 (Fig. 1).

The C.V. was 11% at the 25 pg/ml level and 4% at the 100 pg/ml level. The limit of detection was 15 pg/ml in plasma. Accuracy checks with analyte-supplemented plasma samples (0-150 pg/ml) showed a bias of +3.8%. The overall extraction yield was 67%. Analysis of variance done on the data for 100 pg/ml showed that the sample work-up contributed 20% to the total variance; 80% was due to the GC-MS analysis. Calibration curves derived from plasma containing clenbuterol-HCl (15 to 200 pg/ml) and 2.5 ng/ml clenbuterol-D9-HCl were linear and passed through the origin.

For successful application of the assay, several potential troubles had to be guarded against:-

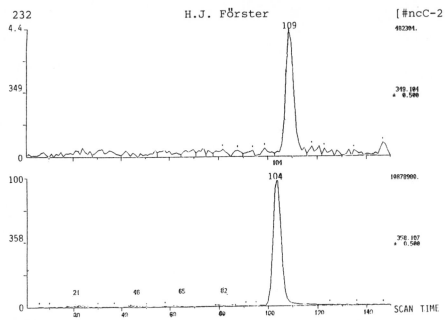

Fig. 1. GC-MS pattern for a human plasma sample (monitoring at e/z 349 and, for ^{2}D-drug, 358) 2 h after administration of 20 µg clenbuterol hydrochloride (found: 106 pg/ml).

(1) impurities in the extraction solvent which may react with the analyte to form compounds of different molecular mass;

(2) sample loss and memory effects associated with the glassware;

(3) hydrolytic instability of the clenbuterol derivatives;

(4) GC-MS conditions not optimally adjusted for maximum mass flow into the ion source of the spectrometer.

The remedy for (1) was use of bunitrolol, which has a side-chain structurally similar to that of the analyte, as a scavenger compound to remove impurities from the methyl-*tert*-butyl ether.

(2) Solvent evaporation and derivatization were done in a mini-version of a Kuderna-Danish flask to avoid sample losses. With the following treatment of the glassware to eliminate memory effects from previous sample work-up, silanization was unnecessary:- after soaking in CrO_3/H_2SO_4, washing with distilled water and drying, the glassware was annealed at 500°C in the glass-blower's workshop.

Reference

1. *Present authors* (1988) *Biomed. Environ. Mass Spectrom. 17,* 417-420.

#ncC-3

A Note on

HPLC DETERMINATION OF SALMETEROL IN PLASMA

P.V. Colthup and G.C. Young

Glaxo Group Research, Ware, Herts. SG12 0DP, U.K.

The development of salmeterol has required methods for its determination in plasma, sufficiently sensitive, accurate and precise to allow pharmacokinetic profiles in animals to be established. Salmeterol in plasma can be determined by GC-MS with high accuracy and precision (see Higton's art., #ncC-4; structural formula shown), but the equipment required is very expensive and sample throughput is limited. As outlined in Scheme 1, assay of animal plasma can be performed by HPLC with fluorescence detection, after initial solid-phase extraction. Fig. 1 shows typical chromatograms.

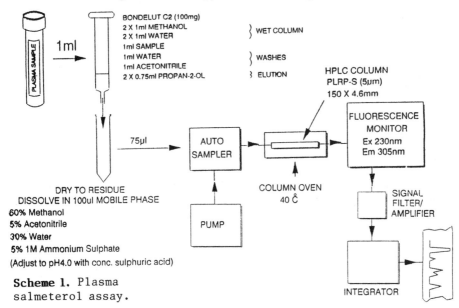

Scheme 1. Plasma salmeterol assay.

The method is quantitative down to 1 ng/ml, and has been used successfully with assay runs of up to 50 samples besides standards and quality-control samples. For spiked samples in the range 1-20 ng/ml, the differences between theoretical and observed means were 5.3% at 3.0 ng/ml but otherwise <3.0%. The mean C.V. was 7.0% (range 3.7-16.3%). Over 6 independent assays the differences between theoretical and observed

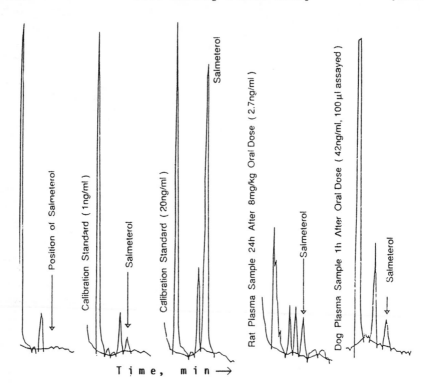

Fig. 1. Representative chromatograms. See Scheme 1 for % composition of mobile phase (v/v basis).

(3 quality-control samples) were 14.6% for 2.0 ng/ml (C.V. 10.4%) but otherwise <6.0% (C.V. <6%). The method is specific for salmeterol with respect to endogenous plasma constituents and identified metabolites, and has been applied successfully to plasma samples obtained during pharmacokinetic studies in rats and dogs.

#ncC-4

A Note on

GC-MS DETERMINATION OF SALMETEROL IN HUMAN PLASMA FOLLOWING INHALED ADMINISTRATION

D. Higton, C. Clegg and J. Oxford

Biochemical Pharmacology Department,
Glaxo Group Research, Park Road, Ware, Herts. SG12 0DP, U.K.

Salmeterol (formula overleaf) is a long-acting B_2 adreno-receptor agonist used in asthma treatment [1], administered (as xinafoate) by inhalation at low dosage (50 µg b.d.*). It acts locally in the lung and hence its therapeutic action is not related to circulating plasma concentrations. However, a specific assay was required to determine them, to allow comparison with the concentrations found in animals during toxicity testing.

The extraction procedure was developed from the HPLC method described in an accompanying article (#ncC-3). [$^2H_3$]-salmeterol was added, as internal standard, to 2 ml portions of plasma samples and standards. Bond Elut C-2 solid-phase extraction columns (100 mg; large reservoir capacity) were conditioned using methanol (15 ml) and water (2 ml). After sample loading the columns were washed with water (1 ml) and acetonitrile (1 ml), and then eluted with isopropanol (1.5 ml). The eluates were evaporated to dryness and then reacted at 70° with N,O-bis(TMS)trifluoroacetamide (10 µl, 5 min) followed by N-methyl-bis(trifluoroacetamide) (10 µl, 60 min) to form the tris-TMS mono-TFA (TMS/TFA) derivative. Analysis was by capillary column GC-MS on a Pye 304/VG 7070E instrument. Injection (2 µl) was in the cool on-column mode onto a retention gap (2 m × 0.53 mm uncoated, deactivated fused silica). The column was J&W DB1, 15 m × 0.25 mm, with 0.1 µm film thickness. The GC instrument, with helium as carrier gas (8 psig), was temperature-programmed with the peak of interest eluting at ~290°. The MS was operated at an instrumental resolving power of 5000, in the EI mode (70 eV electron energy, 200 µA trap current; ion source at 250°).

EI-MS on the TMS/TFA derivative (Fig. 1) showed a major ion at m/z 369, the ^2H internal standard giving an ion at m/z 372. With selected ion monitoring of these ions, the method had a 25 pg/ml limit of quantification for 2 ml of plasma.

Abbreviations.- EI-MS, electron-impact mass spectrometry; TFA, trifluoroacetyl; TMS, trimethylsilyl; b.d., *bis in die.*

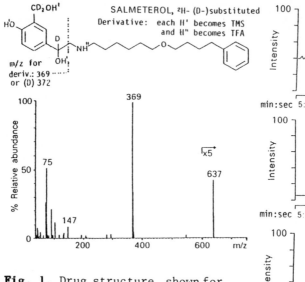

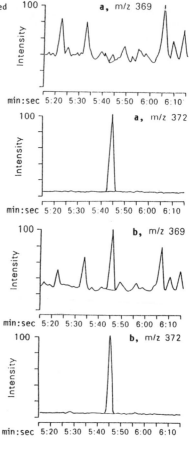

Fig. 1. Drug structure, shown for $^2H_3$-salmeterol, and EI-MS spectrum for the TMS/TFA derivative of ^1H-salmeterol. Mol. wt. of derivative 727 (^1H) or 730 ($^2H_3$).

Fig. 2 *(right)*. GC-MS chromatograms, for the ions indicated, from plasma extracts containing 0.5 ng/ml $^2H_3$-salmeterol and, in **b** but not **a**, 25 pg/ml ^1H-salmeterol.

Fig. 2 shows chromatograms for normal plasma spiked with internal standard and, except in the blanks, salmeterol. The on-column injection cannot be automated in this instrument, and therefore this assay has not come into routine use. Hence it has not been fully validated; but the results from quality control samples analyzed alongside clinical samples showed adequate accuracy and precision (Fig. 3). A time curve is shown in Fig. 4.

During the development of this assay, several technical difficulties were encountered.-
(1) At a concentration of 100 ng/ml or less in water, salmeterol adsorbed onto glassware, the loss being ~30% in the first 15 min. This problem disappeared when all solutions below 1 mg/ml were made up in 25% (v/v) aqueous plasma.
(2) Components were found to be eluting from non-conditioned Bond Elut columns and causing interference in the peak assignment in the chromatogram. These components were eluting

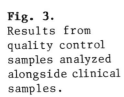

Fig. 3.
Results from
quality control
samples analyzed
alongside clinical
samples.

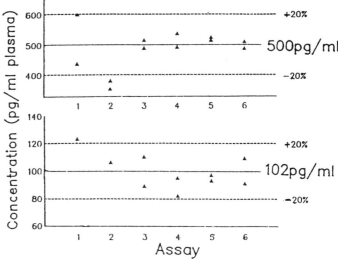

Fig. 4. Concentration-time
curve for salmeterol in
plasma samples from a patient
receiving 50 µg salmeterol
(as xinafoate) b.d. by
metered dose inhaler.

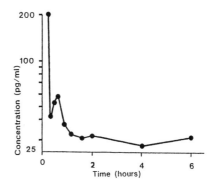

at or near the peak of interest and caused difficulties
in the detection of the derivatized salmeterol peak. This
interference was reduced by conditioning the columns with
15 ml of methanol rather than 1 ml.
(3) When using on-column injection, non-volatile material from
the sample accumulated in the retention gap, leading to broadening
of the chromatographic peak. When the gap was replaced,
the peak shape was restored.
(4) With MS resolving power set at 1000, large chromatographic
peaks arising from septum material were observed. These
interfered with the assignment of the peak derived from salmeterol.
The mass specificity of the assay was improved by increasing
the MS instrument's resolving power to 5000, thereby reducing
the effect of the co-chromatographing components.

   In summary, a procedure has been developed for the analysis
of salmeterol in human plasma. This procedure, whilst unsuited
to routine use, has been used for the analysis of clinical
samples following administration of 50 µg of salmeterol (as
xinafoate) b.d.

Reference

1. Ullman, A. & Svedmyr, N. (1988) *Thorax 43*, 674-678.

#ncC-5

A Note on

IgG AND IgE RESPONSES IN MICE FOLLOWING INJECTION OF mPEG-MODIFIED ALDER POLLEN ALLERGENS

Hogne Vik[⊕], Erik Florvaag, Elisabeth Holen and Said Elsayed

Allergy Research Group, Clinical Biochemistry Laboratory, University Hospital, Bergen, Norway

Alder, birch and hazel, belonging to the same genus *(Betulacea)*, share most of their antigens, whilst distantly related plants have a few cross-reacting antigens [1, 2]. The sequence is known for the major **bp**[*] antigen (Bet v I) and also, in part, for the major allergens of **ap** (Aln i I) and of **hp** (Cor a I) [3–5]. Based on optimal hydropathicity and the helicity assignment of Bet v I, the N-terminal region of the molecule has been postulated to represent an IgE-binding epitope [6].

Structurally modified allergens (allergoids) have been shown to be safer for immunotherapy than native allergens [7]. Elevated IgG synthesis but virtually unchanged IgE synthesis has been shown in mice [8] after administration of low doses of **bp** allergens, both native and conjugated to monomethoxy polyethylene glycol (mPEG) in an 'activated' form, 2-O-mPEG-4,6-dichloro-5-triazine (M_r 6 kDa).

For specific immunotherapy of **tp** allergies, mPEG-mod. **tp** extracts have been proposed, being able to enhance specific IgG Ab synthesis against the respective allergens. Native compared with mod. **tp** allergens gave weaker IgE elevations in pollen-allergic patients [9]. Since these observations had not been borne out by our studies in mice [10] with mPEG-mod. **bp** extracts, we have now repeated them using **ap** extracts.

MATERIALS AND METHODS

Antigen preparation.- From the aqueous crude **ap** extract, fraction AI 34 was isolated by gel filtration as 2 combined fractions [11]; it included 24 of the 27 antigenic precipitates, with all 3 of the **ap** allergenic precipitates. It was conjugated [8]

[⊕]addressee for any correspondence; now at Nycomed AS, P.O. Box 4220 Torshov, N-0401 Oslo 4, Norway.

[*]*Abbreviations.*- Pollens: tree, **tp**; **ap**, **bp**, **hp**: alder, birch and hazel (*Alnus incana, Betula verrucosa, Coryllus avellana*; for AI 34 and Aln i I *see text*); mod. = modified, generally by mPEG (*see above*), the *prefix* mPEG- being omitted where clearly implied. Saline, 0.9% NaCl. Adjuvants: Al.hyd., aluminium hydroxide gel; FIA, Freund's incomplete. Ab, antibody; anti-mouse sera: **GAM**, goat anti-IgE; SAM, sheep anti-IgG. NC, nitrocellulose.

by adding 300 mg 'activated' mPEG to 50 mg AI 34 (5 µmol). RAST-
inhibition [cf. #ncC-6, *this vol.*] showed retention of allergenicity.

Animals and immunization.- The mice were CBA/Ca females
(~15 g, pellet-fed), 7 per cage [8]. Sensitization and immunization were as in the **bp** studies [8]. For the 4 experiments
performed, Table 1 along with Figs. 1-3 gives the conditions,
varied in respect of adjuvant (none in expt. 3) and dosage
schedule. The primary comparison was between native AI 34
(Aln i I being the effective antigenic component) and mPEG-AI 34
(mPEG-Aln i I). Blood sampling, from 7 mice per group, was by
heart puncture under anaesthesia. (Ethical regulations precluded
repeated sampling from each mouse for longitudinal studies.)

Assay of Ig's.- Triplicate assays on each blood sample
were done with ^{125}I-Ig's (NE 1600 gamma counter, correcting
for non-specific binding and background counts) [12]. For
IgE [8, 13] the Ig fraction was isolated from freeze-dried
GAM (Nordic Immuno. Labs., The Netherlands) and tagged using
Bolton-Hunter reagent, giving 300,000 cpm/ml; for IgG, ^{125}I-SAM
was purchased (Amersham Internat'l.). The procedure for IgG
([8]; cf. [14]) entailed coupling Aln i I to nylon balls (6.4 mm
diam.) which, after incubation with 1:5 test serum (200 µl
per ball), were treated with ^{125}I-SAM to label the bound
IgG. For IgE, AI 34 was bound to NC discs (6 mm diam.; 0.2 mm
pores), and ^{125}I-GAM was applied (225×10^3 cpm per disc).

RESULTS (Table 1 &, for expts. 2-4, Figs. 1-3)

IgG responses.- In expt. 1 [illustration excluded - *Ed.*],
tests on days 79 and 86 showed elevated IgG, the highest
values being seen with mPEG-mod.-AI 34 on day 86. For all
3 treatments in expt. 2, IgG was maximal on day 142, the
rise persisting slightly longer with native compared with
mod.-AI 34 (Fig. 1). The greatest increase was found where

Figs. 1 and *(opposite page)*
2 & 3. Ig levels in mouse
serum after allergen administration (see Table 1 for doses
and adjuvant, and text for
other conditions). In each Fig.
the left and right axes are,
respectively, IgG and IgE as cpm,
represented ●——● and ■--■ for
administration of native AI 34
allergen. Referring to the mod.-
AI 34 regimens in Table 1, the
second in expt. 2 is shown in Fig. 1
as ☐—☐, △--△; the first in Figs. 2
& 3 is shown ▲—▲, △--△, and the
second ■—■, ◇—◇. *Broken* line = IgE.

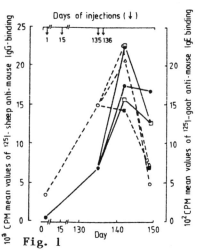

Fig. 1

Table 1. Immunization conditions, and Ig responses (– denotes no change). Injections (generally s.c.), concluding on the day denoted [ ], were spaced as shown in the Figs.; in Expt. 1 the second was on day 15. Agent: **ap** AI 34; 'Mod.' = mPEG-treated.

Expt. & adjuvant	Native: doses, µg	IgG	IgE	Mod.: doses, µg	IgG	IgE
1; FIA [72]	10, 1; (i.p.) 10	+	–	10, 1; (i.p.) 10	++	+
2, Fig. 1; FIA [136]	10, 1; (i.p.) 10, 1	++	+	10, 1; (i.p.) 1, 1	++	+
				10, 1; (i.p.) 10, 1	++	+
3, Fig. 2; NaCl [47]	0.1×10, 0.1, 0.1×10	(+)	–	0.1×10, 10, 0.1×10	+	–
				0.1×10, 100, 0.1×10	+	–
4, Fig. 3; Al.hyd. [47]	0.1×10, 0.1, 0.1×10	++	–	0.1×10, 10, 0.1×10	++	–
				0.1×10, 100, 0.1×10	++	–

the day–135 dose was 10 µg of mod.–AI 34.   In expt. 3 (Fig. 2) there were IgG rises, near-identical, on day 33.   The greatest increase was seen on day 50 where the pre-treatment was with mod.–AI 34 on day 33.   In expt. 4 (Fig. 3), IgG was notably increased on day 50 in all groups, especially where the day-33 pre-treatment was with 100 µg of mod.–AI 34.

IgE responses.- In expt. 1, rises were found with native but not mod. AI 34 on days 72 and, transiently, 79.   In expt. 2 IgE showed a transient rise, similar in all 3 groups, on day 142. In expt. 3 IgE showed decreases on day 33, notably with native AI 34 which, in expt. 4 gave on day 50 the lowest IgE.

DISCUSSION

Hyposensitization therapy with various extracts and regimens is known to give relief of symptoms, not clearly correlated with Ig titres or allergen dosage.   The occurrence of anti-IgG Ab's is uncorrelated with clinical effect or degree of side-effects [15, 16].   The therapy therefore lacks understanding

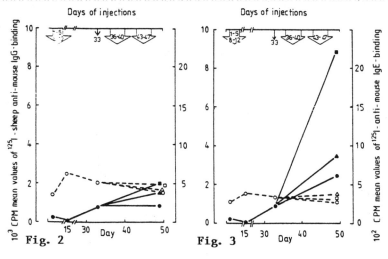

Fig. 2   Fig. 3

of how the Ab responses are regulated. During hyposensitization therapy, IgE often shows an initial rise, then a fall. With chemically modified extracts the IgG response is sought whilst avoiding an IgE rise. We reported [8] that with mod. rather than native **bp** extracts, specific **bp** IgG-type Ab's rose markedly but IgE did not fall. Now, with native and especially with mPEG-mod. AI 34, we have found IgG-titre rises, lasting for at least a week. Only in some instances did IgE rise.

An mPEG-mod. version of a mixed **tp** extract has been introduced for immunotherapy [9]. A mod. compared with native **bp** extract had no clear advantage in mice: IgG responded more strongly but IgE was unaffected. Both immunological and clinical improvements are requisite in mod. extracts for hyposensitization. Our **ap** results show a rise in IgG and only moderate rises in **ap**-directed IgE. Whilst our animal findings call for more extended studies in animals and humans with various hyposensitization regimes, evidently mod. **ap** allergens may have potential for combatting seasonal **tp** allergy.

Acknowledgement.- This study was supported by the Research Fund of the Norwegian Asthma and Allergy Federation.

References[⊕] (*Al* = *Allergy*; *Ap* = *Applied*; *Im* = *Immunol.*; *IA* = *Int. Arch.*)

1. Weber, R.W. (1981) *Ann. Al 46*, 208-215.
2. Ipsen, H., Bøwadt, H., Janniche, H., Petersen, B.N., Munch, E.P., Wihl, J-Å. & Lowenstein, H. (1985) *Al 40*, 510-518.
3. Borch, S.M. (1979) Thesis, Univ. of Oslo.
4. Elsayed, S. & Vik, H. (1990) *IA AlApIm 93*, 378-384.
5. Valenta, R., Breiteneder, H., Pettenburger, K., Breitenbach, M., Rumpold, H., Kraft, D. & Scheiner, O. (1991) *J. Al Clin. Im 87*, 677-682.
6. Elsayed, S., Eriksen, J. & Stavseng, L. (1991) *Schw. Med. Wochen. suppl. 1.40*, 10.
7. Sehon, A.H. & Lee, W.Y. (1981) *IA AlApIm 66 suppl. 1*, 39-42.
8. Vik, H. & Elsayed, S. (1986) *IA AlApIm 81*, 315-321.
9. Åkerblom, E., Annerhed, A., Berglund, A., Boo, E., Einarsson, R., Malmquist, M., Ponten, S. & Steringer, I. (1985) *Ann. Al 55*, 381.
10. Vik, H., Sørnes, S. & Elsayed, S. (1984) *IA AlApIm 74*, 55-62.
11. Florvaag, E., Elsayed, S. & Hammer, A.S.E. (1986) *IA AlApIm 80*,
12. Baldo, B.A. (1983) *Al 38*, 535-546.   [26-32.
13. Harbo, M. & Ingild, A. (1973) *Scand. J. Im suppl. 1*, 161-169.
14. Djurup, R., Søndergaard, I., Minuva, U. & Weeke, B. (1983) *Im Meth. 62*, 283-296.
15. Nickelsen, J.A., Gearitis, J.W. & Reisman, R.E. (1986) *J. Al Clin. Im 77*. 48.
16. Østerballe, O., Egeskjold, Johansen, A.S. & Skov, P. (1982) *Al 37*, 209-216.

[⊕] *Editor excised some non-vital refs. (and some text wording)*

#ncC-6

A Note on

ANTIGENIC AND ALLERGENIC ACTIVITIES OF DIALYZABLE
LOW-M$_r$ PEPTIDES FROM ALDER AND BIRCH POLLEN EXTRACTS

Hogne Vik[⊕], Erik Florvaag and Said Elsayed

Allergy Research Group, Clinical Biochemistry Laboratory,
University Hospital, Bergen, Norway

Hypersensitivity towards various **tp**'s[*] represents a major
clinical health problem. During the last 2 decades several
tp allergens have been identified and purified to homogeneity,
and primary amino acid sequences determined [e.g. 1-3]. During
natural exposure, **tp**-allergic individuals are challenged by
whole pollen grains and, after their deposition onto the
mucosal membranes, by proteins released from the grains.
When producing allergen extracts it is of interest to know
whether allergenic molecules of potential importance to the
individual patient are lost during the extraction procedures.
This could, for example, arise from use of dialysis tubing
with a mol. wt. (M$_r$) cut-off of 8 kDa.

In 1978 (see [4]) the antigenic and allergenic content
of *Paretaria officinalis* pollen was described by Garcia &
co-workers. Among others, pollen components of M$_r$ <8 kDa
were isolated. These low-M$_r$ components were characterized
by their IgG- and IgE-binding capacities, and were demonstrated
to represent haptenic molecules [4]. The present investigation
has been performed to verify the antigenic and allergenic
activity of low-M$_r$ components from **ap** and **bp**.

MATERIALS AND METHODS

Preparation of low-M$_r$ pollen fractions.- Non-defatted
ap and **bp** grains, both containing <0.5% foreign protein,
were purchased (Allergon, Engelholm, Sweden). Dry pollen
in 10 g amounts was suspended in 100 ml pH 7.2 phosphate
buffer (75 mM phosphate/375 mM NaCl) and gently shaken for 3 h
at 4°. After centrifugation at 4°, as in the production
of ordinary crude extracts, the supernatants were dialyzed

[⊕]now at Nycomed AS, P.O. Box 4220 Torshov, N-0401, Oslo 4.
[*]*Abbreviations.-* Pollens: **tp**, tree; **ap**, alder (*Alnus incana*,
AI); **bp**, birch (*Betula verrucosa*, BV). (C)IE, (crossed)
immunoelectrophoresis; RAST, radio-allergosorbent test; SPT,
skin prick test.

in bags (Spectrum Medical Industries, Los Angeles) with a cut-off of 8 kDa, against 750 ml distilled water for 2, 6, 12 and 24 h at 4°. The 24-h dialysate was freeze-dried and stored at -20° until used [5] ('AlniD' from AI, 'BetvD' from BV).

Gel exclusion (filtration) chromatography was done with polyacrylamide gel of 6 kDa cut-off (Bio-Rad, Richmond, CA; P-6, 'fine' grade). The columns were 30 × 1.5 cm. The eluent was 50 mM NH_4HCO_3, pH 8.2.

Immunoelectrophoresis.- A CIE procedure [6] was employed to investigate the antigenicity of the 24-h dialysis extracts. In rocket-line IE a precipitation line was obtained using 1 mg of each crude sample, evenly distributed in a gel section (5 × 1 × 0.15 cm) next to the cathode. The dialysates (40 mg/ml; 10 µl) were put into 2.5 mm diam. wells in an intermediate gel (5 × 1.5 × 0.15 cm) close to the gel containing crude extract; 0.9% NaCl was put in control wells. Electrophoresis was performed at 2 V/cm overnight into gels (5 × 4.5 × 0.15 cm) containing 300 µl (13 µl/cm^2) appropriate rabbit antibodies, anti-AI-crude or anti-BV-crude ([7]; immunization as in [5]).

Human sera, and tests (RAST as in [5]).- Individual sera as tested for RAST inhibition were from patients with a clinical history of **tp** allergy and class 2-4 serum RAST to **ap** and **bp**. A pool was also prepared with equal amounts of sera from 10 individuals reactive to both **ap** and **bp** allergens[5]. The SPT [8] was done, with informed consent, in a **ap**-allergic patient (E.F.N.) and a healthy control (H.V.), with the crude AI and BV dialysates and the 4th UV-absorbing peak from the P 6 column.

RESULTS

UV spectra and peaks obtained.- Aln i D and Bet v D gave near-identical peptide-type spectra, with maxima at 265 and 200 nm. P-6 gel chromatography on Aln i D gave peaks 'I-IV' (Fig. 1A); the 4th (highest) gave 2 peaks when run on P-2 (Fig. 1B). Bet v D gave identical results (not shown).

Precipitation lines.- CIE gave no visible lines with AlniD or Bet v D up to 100 mg/ml. Rocket-line IE showed a visible deflection of the Bet v crude line for Bet v D and, slightly, of the Aln i crude line for Aln i D.

Specific IgE binding.- In RAST, Aln i D at 100 mg/ml gave 13% *vs.* Aln i crude extract (10 mg/ml) binding, falling to 3.5% at 3.125 mg/ml. For Bet v D the values for 100 and 12.5 mg/ml were 8.5% and 4.0% (Fig. 2A). In inhibition titration (Fig. 2B) the corresponding values for Bet v D were 65% and near-nil; Aln i gave 33% and 10% inhibitions at 100 and 6.25 mg/ml.

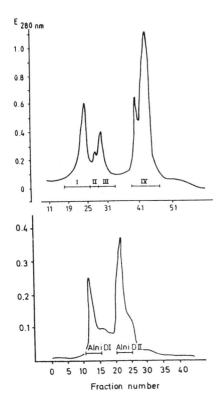

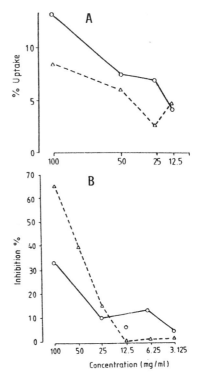

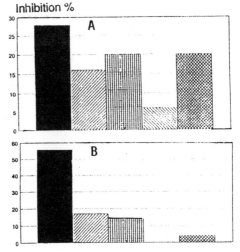

Inhibition %

Fig. 1, *above left.* Gel 'filtration' separations, by 50 mM NH₄HCO₃. **(A)** Aln i D, on Bio Rad P6; 105 mg of material, giving 4 peaks; the predominant ones were re-run on Bio Rad P2 – Bet v DIV, and **(B)** Aln i DIV, giving mainly Aln i Di & Aln i DII.

Fig. 2, *above right.* RAST curves for Aln i D (——) & Bet v D (---), as % of crude extract: **(A)** direct, **(B)** inhibition titration; 10 mg/ml.

Fig. 3, *right.* RAST inhibition diagrams for P 6 fractions (uptakes as % of total cpm). *See text.*

KEY:
3 A   ▮ Aln i D 100   ▨ Aln i D I 100   ▦ Aln i D II 4   ▧ Aln i D III 4   ▨ Aln i D IV 8 mg/ml

3B   ▮ Bet v D 100   ▨ Bet v D I 100   ▦ Bet v D II 25 mg/ml
     ▧ Bet v D III 4   ▨ Bet v D IV 2 mg/ml

Further study of the dialysis products for IgE-binding capacity by RAST inhibition is shown, with the concentrations used, in Fig. 3. All 4 peaks from each dialysate manifested IgE binding, least for the 3rd peak. The predominant peaks (1st and 4th) were further investigated.

SPT observations.- Aln iD or Bet v D at 100 mg/ml gave a wheal reaction   equivalent to 10 mg histamine/ml, and half this response was obtained with Bet v DIV at 2 mg/ml. No other dialysis fraction gave a reaction. No reactions were evoked by any of the fractions in one control individual.

CONCLUSIONS

Both AI- and BV-dialysates contain low-M_r material that is usually excluded during extract-preparation procedures. However, compared with the total antigens and allergens in conventional crude extracts, such proteins (< 8 kDa) seem to be of minor importance.   These observations accord closely with data reported [9] for pollen from *Parietaria juidaica*.

References (Al = Allergy; Ap = Applied; Im = Immunol.; IA = Int. Arch.)

1. Elsayed, S. & Vik, H. (1990) *IA Al Ap Im 93*, 378-384.
2. Borch, S.M. (1979) Thesis, Univ. of Oslo.
3. Roebber, M.G. (1975) PhD Thesis, McGill Univ., Montreal.
[Complementing these 3 refs.: Valenta *et al.*, cited in #ncC-5, *this vol.]*
4. Giallongo, A.,Oreste, U., Cocchiara, R., Ruffilli, A. & Geraci, D. (1980) *Mol. Im 17*, 1019-1024.
5. Vik, H. & Elsayed, S. (1982) *IA Al Ap Im 68*, 70-78.
6. Weeke, B. (1973) *Scand. J. Im 2, suppl. 1*, 47-48.
7. Aukrust, L., Apoid, J., Elsayed, S. & Aas, K. (1978) *IA Al Ap Im 57*,   253-262.
8. Florvaag, E. & Elsayed, S. (1984) *IA Al Ap Im 75*, 300-308.
9. Feo, S., Cocchiara, R. & Geraci, D. (1984) *Mol. Im 21.1*, 25-36.

Comments on #C-1, I.F. Skidmore - THE ANTI-ASTHMATIC DRUG SCENE
            #C-2, J.J. Gardner - ASSAY OF SOME ANTI-ASTHMATICS
            #C-3, C. Lindberg - HPLC APPROACHES
            #C-7, W. Ritter - THROMBOXANE ANTAGONIST ASSAY

Skidmore, replying to M. Doig.- IgE molecules indeed have beneficial physiological activity; they play an important role in removing or killing parasitic infections. Regulatory dilemma: **Ritter** and **Lindberg asked** how authorities reacted if they were told, in respect of a drug given by inhalation (e.g. to asthma patients), that it was not feasible to furnish pharmacokinetic data or even to measure plasma levels. **Skidmore's response.**- Most authorities, even the FDA, would accept a statement such as "<0.1 ng/ml" because the general circulation lies beyond the organ (lung) where the drug acts; one would argue that the drug had no oral bioavailability, being completely metabolized.

Commenting (from the Chair) on the talk by **Gardner, Ritter commended it** as illustrating well how progress in drug monitoring and pharmacokinetics depends entirely on progress in methodology and technical facilities. [This dependence is the theme of a 1988 survey by Ritter which deals with nitrofurantoin and other nitrofurans, glyceryl trinitrate, and synthetic prostaglandins such as arbaprostil (pp. 139-144 in 'Metabolism of Xenobiotics', ed. J.W. Gorrod *et al.*, Taylor & Francis, London).]

C.A. James asked Ritter about the reaction of regulatory authorities if indeed each of two techniques (GC, HPLC) was a substantial source of data within a single clinical study. **Reply.**- Almost all the data were obtained by HPLC. GC was used only in situations, usually in patients, where with plasma levels <20 ng/ml there could be interferences, and in any case the excellent correlation between HPLC and GC should reassure the authorities. **Ritter, answering Skidmore** concerning the notably large $t_{\frac{1}{2}}$: it seems to be concentration-dependent, not appearing merely when there is a measurable concentration in plasma at the time when this elimination phase becomes significant. **D. Schmidt queried** the use of a capillary GC column 30 m long (**Ritter** said there was no special reason for this): it is better to use shorter columns, giving faster analysis at lower temperatures with higher sensitivity.

Re #C-3: **Lindberg, replying to A. Beerahee.**- In routine use of TSP LC-MS, high throughput is maintained, even with derivatized products, but it depends on sample-preparation time in the case of cortisol.

[Note by Ed.- COMMENTS not in exact sequence of art. ref. nos.]

Comments on #C-5, D. Higton - SALMETEROL BY GC-MS
              #C-6, K.D. Jenkins - RIA OF FLUTICASONE PROPIONATE
              #ncC-2, H.J. Förster - CLENBUTEROL ASSAY

Higton, replying to questions.- (**A. Beerahee**) A push-type
connector was used to attach the retention gap to the main
column. (**Mira Doig**) Attachment presents no problem: one opens
the GC oven door and, wearing gloves, makes the connection
as quickly as possible. The retention gap is deactivated
by phenylmethylsilyl treatment. (**C. Lindberg**) The choice of
Bond Elut columns for extraction was based on work in a companion
contribution (#ncC-3).

Jenkins, replying to questions.- (**D. Chapman**) Thyroglobulin
was used for the immunogen merely because it had been used
successfully for previous steroid assays. (**B. Law**) Non-specific
binding was as high with polystyrene tubes as with the glass
ones we used (**Law** suggested that polypropylene or polyethylene
tubes might give lower binding). (**H. de Bree**) We could not
correlate the sensitive RIA with an instrumental method, as
none is available; we believe regulatory authorities will
accept our arguement that an instrumental assay as sensitive
as the RIA is technically not feasible.

Mira Doig asked Förster about his opinion that the large
increases in C.V. at high concentrations were due to the
dwell times adopted for monitoring the drug-standard and internal-
standard peaks: did you check this by doing repeat injections
using single-ion monitoring of your standard? **Reply:** not
tried. **S. Westwood:** have you tried NI-CI to increase the
sensitivity of your assay? **Reply.-** Yes, but NC-CI is ineffective
for TMS ethers, although published work has shown it to
be good for PFB-Br derivatives. **Reply to I.F. Skidmore:** For
assay of clenbuterol in tissues a method is available, with
sensitivity 100 pg/g tissue. [See entry on p. 151.]

Comments on #ncC-1, G. Land - LIPOXYGENASE INHIBITORS
              ncC-3, G.C. Young - SALMETEROL BY HPLC
              ncC-4 & 5, H. Vik - POLLEN ANTIGENS

Replies by Land.- (Query by **Simmonds**) The oxalic acid put
into the mobile phase serves to improve chromatography of the
acetohydroxamic acids and hydroxyureas. (Query by Ed., **E. Reid**)
Whilst disfavouring THF because of its smell and toxicity,
we often find it highly useful for achieving good peak shape.
Young, answering H. de Bree.- Measurement of metabolite is
not a feature of our HPLC method.

Replies by H. Vik.- (**Doig**) Coating of the antigens onto
nylon and nitrocellulose balls was done not chemically but
passively, merely adding the balls to antigen solution and mixing.

(**Skidmore**) The native antigen produces more of a reaction than mPEG in the nose, but we are unsure whether they differ in skin tests.

Editor's interpolation: précis of a Forum abstract (no text)
- M. Guerret (Sandoz, Rueil-Malmaison): KETOTIFEN ASSAY

Ketotifen, a cycloheptathiophenone derivative, has anti-anaphylactic properties and is efficacious in the treatment of asthma. Due to the low dosages (<2 mg/day), a large distribution volume and an extensive metabolism in man, plasma concentrations of unchanged drug are ~0.5 ng/ml. Non-specific assay can be done by RIA, and specific assay by GC-NPD or GC-MS. Each technique has played a role in pharmacokinetic work. Derivatization is detrimental to 10-OH-ketotifen.

Guerret, answering Skidmore.- Whilst ketotifen is marketed in Japan as well as in Europe, we don't know whether any differences exist in kinetics or metabolism.

Citations contributed by Senior Editor
besides those in Sect. #2 of #ABC (Compendium of assays)

Asthma - a disorder of adrenergic receptors? - Insel, P. & Wasserman, S.I. (1990) FASEB J. **4**, 2732-2736.

Drug disposition (anti-allergics and anti-asthmatics) in infancy and childhood.- Sweeney, G.D. & McLeod, S.M. (1989) Clin. Pharmacokinet. **17** (Suppl. 1), 156-168.

Possible prophylaxis of asthma, by isoenzyme-selective cyclic nucleotide phosphodiesterase (PDE) inhibitors that act mainly on cell types responsible for promoting airways inflammation (the nature of which is discussed).- Giembycz, M.A. (1992) Biochem. Pharmacol. **43**, 2041-2051. This informative and cogent 'Commentary' argues that the need for bronchodilator symptomatic therapy could shrink through trial of such inhibitors. Examples include vinpocetine, rolipram and zaprinast.

Section #D

BIOANALYTICALLY EXPLOITABLE TECHNIQUES

#D-1

RECENT ADVANCES IN THE BIOANALYTICAL POTENTIAL
OF MASS SPECTROMETRY, ESPECIALLY LC-COUPLED

W.M.A. Niessen, U.R. Tjaden and   J. van der Greef

Division of Analytical Chemistry,
Center for Bio-Pharmaceutical Sciences,
P.O. Box 9502,   2300 RA Leiden,   The Netherlands

Nearly 20 years of investigation has led to considerable progress in on-line LC-MS* [1, 2].   Expectations that LC-MS would have a large impact on bioanalysis have been fulfilled, largely through new MS techniques comprising a spin-off of the LC-MS research efforts [2].   LC-MS can now embrace polar and even ionic analytes, including drugs and their phase I and phase II metabolites,   and bioconstituents such as proteins. LC eluents with involatile components can also be coped with. Some recent developments are now critically reviewed, with examples of trace-level analytes in biological matrices. Approaches such as CE to circumvent the LC step are also considered.

INTERFACING LC AND MS

For the interfacing, which is not as straightforward as GC-MS, ingenious devices have been developed [2].   The following approaches (Fig. 1) have been most successful: (a) nebulization of the column effluent, removal of the mobile-phase constituents, vaporization of the analyte and subsequent ionization, as with the moving-belt interface$^{\oplus}$ and the PBI; (b) direct ionization from the effluent stream, as in CF-FAB; or (c), as with TSP$^{\oplus}$, ESP, ISP and HNI: nebulization of the effluent into either an atmospheric-pressure region [M.V. Doig, #ncD-1, this vol.- *Ed.*] or a reduced-pressure region, desolvation of the droplets, then either gas-phase CI or ion-evaporation. The various interfaces have different applicability constraints, and are generally limited to a narrow range of flow-rates and mobile-phase compositions [2].

Abbreviations.- LC, *prefixing* MS, implies HPLC, on-line; CE, capillary electrophoresis; ITP, isotachophoresis. CF-FAB, continuous-flow fast atom bombardment; CI, chemical ionization; EI, electron impact; IT, ion-trap. *Interfaces:*- heated nebulizer, HNI; particle-beam, HNI;   spray *(denoted* SP):- electro-, ESP; ion-, ISP; thermo-, TSP.   Continuous-flow dialysis, CFD; phase-system switching, PSS.   [$^{\oplus}$See end of refs. list.- *Ed.*

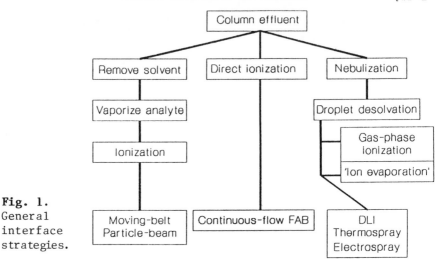

Fig. 1.
General
interface
strategies.

PRESENT-DAY SCOPE OF LC-MS

With accelerating developments, use of LC-MS has given
a significant impetus to some researches. Aspects of these
developments are now briefly reviewed, particularly routine
application of LC-MS in qualitative and quantitative bioanalysis,
ESP-MS for sensitive M_r determinations (especially of biomacro-
molecules), and external ion sources for IT-MS.

LC-MS in biological-sample assays and metabolite identification

The most important development in LC-MS of the last
few years is the ability to use LC-MS in routine applications,
especially in semi-automated, quantitative assays. Generally
there is a sample pre-treatment step (liquid-liquid extraction
or solid-phase isolation), LC-MS usually with the TSP interface,
and use of isotopically labelled internal standards. For
plasma in particular, analyte examples include:- the phenothia-
zine derived from the antiarrhythmic agent moricizine [3],
the enantiomers of the asthma agent terbutaline using coupled-
column chromatography [4], labetalol [5] and sumatriptan
succinate [6] using an automated sample processor, and,
with full automation, budesonide [7]. These methods have
limits of quantification ranging from high pg/ml to mid ng/ml.
Whilst assay data are likewise obtainable by HNI or ISP,
reports on assays using these interfaces are scant compared
with those based on TSP. Yet ISP and especially TSP and
HNI represent important tools in quantitative bioanalysis.

LC-MS has also been widely used in qualitative analysis,
e.g. for identifying drug metabolites or degradation artefacts.
Tandem MS (MS-MS) with TSP is typically used. Examples
are the identification of metabolites of the H_1-antagonist

temelastine [8] and pyrilamine ([9]; urine) and of a metabolite
and sample work-up artefacts of heptabarbital [10]. PBI
is of growing importance in this field, exemplified by the
identification of oxodipine metabolites [11], especially since
PBI can be used to obtain EI and solvent-independent CI
mass spectra.

Protein analysis by electrospray LC-MS

Another development of potential importance is the ability
to perform accurate (±1 Da) mol. mass determinations on bio-
macromolecules, especially proteins, at the low pmol/μl level
using ESP and ISP [12, 13]. The ESP mass spectra of proteins
consist of a series of multiply-charged ions. With knowledge
of the charging ions, e.g. by a proton or by a sodium ion,
and of the m/z difference between two adjacent peaks in
the spectrum, the mol. mass of the protein can be calculated
[13, 14]. Computer programs are available that can do the
calculations, even when the spectra are obscured by the presence
of more than one protein in a mixture [14]. At present,
mol. mass determinations on large proteins are usually performed
off-line by direct infusion of the protein solution into
the ESP interface. The ability to determine mol. mass accurately
could allow detection and possibly identification of small
differences between proteins, e.g. of only 1 amino acid. The
method is a powerful tool in characterizing proteins, e.g.
recombinant porcine somatotropins [15] and outer and inner
core proteins from HIV-1 and HIV-2 [16]. ESP-MS is applicable
to the clinical diagnosis of variant haemoglobins [17]. Obviously,
ESP-MS or ISP-MS can be used in combination with enzymic
protein digestion and MS-MS to achieve protein sequencing
[18], similarly to the use of FAB or CF-FAB [19].

With smaller proteins and with peptides, ESP-MS is applicable
in combination with LC or CE. An example of on-line CE-MS
using an ESP interface is given in Fig. 2, which shows the
electropherogram of a separation of a series of ACTH-related
fragments ranging in mol. mass from 400 to 1000; the fragments
are detected as either singly or doubly charged ions. The
amount of compound injected electrokinetically into the CE
system was only 110 fmol (~2 nl injection volume).

Ion trap technology

The analytical potential of ITMS has been demonstrated
in a manufactured (Finnigan MAT) GC-MS combination system. ITMS
allows excellent sensitivities, high resolution, a broad range
of m/z ratios or multiple stages of MS-MS (MS^n). Obviously
these results cannot be achieved simultaneously; thus, high
resolution incurs the penalty of longer analysis time and
poorer sensitivity.

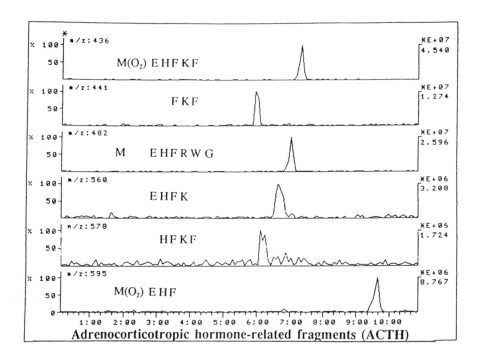

Fig. 2. CE-MS with ESP of ACTH-related fragments. CE: electro-kinetic 2 nl injection of 55 fmol/nl into a fused-silica capillary (800 mm × i.d. 75 μm) filled with 5 mM ammonium acetate (pH 8.2), operated at ~22 kV. ESP: interfaced to Finnigan MAT TSQ-70 operated at 3.5 kV, 2 L/min countercurrent N_2 and a 1 μl/min sheath flow of 50% methanol in water containing 1% acetic acid.

The use on an ITMS of external ion sources, e.g. for TSP [20] or ESP [21], significantly broadens its analytical potential. Whether the ion-trap characteristics, e.g. its sensitivity, can actually be used in on-line LC-MS applications awaits further investigation [21]. The problems of on-line LC-ITMS, wherein ionization and scanning are performed in alter-nation, arise from the fact that in the ion trap the analysis is performed as a sequence of events, which is even more complicated in MS-MS which entails ionization, parent selection and isolation, dissociation and scanning the product ions.* Filling up the trap with ions, requisite for obtaining good-quality mass spectra, takes some time, especially with an external ion source such as ESP. So as to keep the number of ions in the trap relatively constant, the ion injection time should change with the sample concentration, which can result in difficulties in accurately describing the chromato-graphic peak and with the analysis of minor components in mixtures. Furthermore, additional time for ion desolation

*Here and elsewhere, W.M.A.N. disfavours clarifications made by Eds.

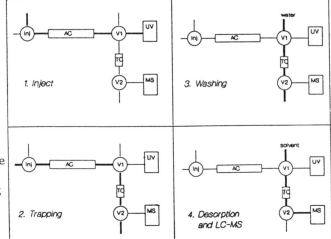

Fig. 3. Sequence in the phase-system switching approach (diagrammatic).

appears to be needed in ESP [21]. Other ITMS features, e.g. high-resolution MS and MS , are even more time-demanding and therefore more difficult to perform within the time frame of the chromatographic analysis. Some of these problems may be solved by applying more powerful computers and software for instrument control and data acquisition. The potential of a sample-saving, pulsed sample introduction in tune with the actual ionization period within the duty cycle of the analysis has not yet been investigated. While such an approach is not readily applicable in LC-MS, there are considerable possibilities in combination with electromigration techiques such as CE (see later).

POST-LC PHASE-SYSTEM SWITCHING

One topic in our LC-MS researches is the development of approaches to solve the incompatibility in LC-MS coupling with respect to mobile phase composition and flow-rate [22]. Most LC-MS interfaces can handle mixtures of water and organic solvents containing volatile buffers. However, many analytical LC separations entail use of non-volatile mobile constituents, e.g. phosphates and ion-pairing agents. Amongst remedial approaches, PSS is widely applicable although we reckon that its primary role is in assay of already identified analytes.

PSS is based on valve-switching techniques [22-25]. As outlined in Fig. 3, the sample is injected onto an LC column, operating under conditions of flow-rate and mobile-phase composition favourable for the separation. The peak of interest is heart-cut from the column effluent and enriched on a short trapping column. When necessary, post-column water addition can be done to increase the capacity ratio of the analyte on the packing material of the trapping column, resulting in a peak compression onto it. Hydrophilic compounds are removed from it with water. Finally the analyte is eluted from it to the LC-MS interface with a solvent and flow-rate favourable to the MS analysis.

The flexibility of this approach is evident from consideration of some of its applications. Mitomycin C has been analyzed using PSS with a reversed-phase LC system containing a phosphate buffer at 1 ml/min flow-rate coupled to a moving-belt interface running with methanol at 0.2 ml/min [23]. With a TSP interface, running with 1.2 ml/min 20% (v/v) methanol in 50 mM ammonium acetate, metoprolol enantiomers have been determined using PSS combined with a chiral α_1-acid glycoprotein column using 0.25% 2-propanol in 20 mM phosphate buffer at 0.8 ml/min [24]. PSS has also been applied in order to avoid splitting in CF-FAB. Erythromycin ethylsuccinate was assayed in plasma by reversed-phase LC with 1.0 ml/min 70% methanol in 50 mM phosphate buffer, enrichment on a trapping column after the addition of water (2 ml/min) and desorption and CF-FAB analysis using 8% (v/v) glycerol in methanol at 15 µl/min [25]. The variety of applications clearly demonstrates the versatility of PSS in circumventing incompatibilities in LC-MS coupling.

CONTINUOUS-FLOW DIALYSIS, WITH NO ANALYTICAL COLUMN

The PSS approach is in fact an adaptation and simplification of coupled column chromatography. Enrichment through trapping on and subsequent elution from short columns is widely used in sample pre-treatment, e.g. in solid-phase extraction and pre-concentration. The technique can be combined with other on-line sample pre-treatment techniques, e.g. CFD [26]. The combination of CFD, trapping column and LC-MS interface (Fig. 4; no analytical column) can be used as a sample inlet system for MS or MS-MS in the screening for analytes in samples manifesting macromolecular interferences, e.g. for drug monitoring in plasma. This approach has been used, with MS-MS in the selective reaction monitoring mode, to assay the anti-cancer drug pyridoglutethimide (with glutethimide as internal standard) in plasma from patients [27]. The limit of quantification was 5 ng/ml plasma. The analysis time was 14 min per sample. Faster analysis should be feasible using a smaller dialysis system.

ISOTACHOPHORESIS AND CAPILLARY ELECTROPHORESIS

CE-MS developments are to some extent handicapped insofar as MS is a mass-flow sensitive detector, i.e. a certain number of molecules rather than a concentration must be offered to the MS to obtain a signal. Due to the small internal diameter of the CE capillary, the intrinsic properties of

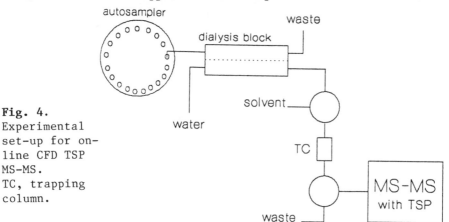

Fig. 4.
Experimental
set-up for on-
line CFD TSP
MS-MS.
TC, trapping
column.

the electrophoretic separation, and the present detection limits
of MS, the dynamic range of CE-MS in respect of concentration
is rather narrow. Improving the MS detection limits by use
of a 'PATRIC' array detector is a feasible but costly solution
[28]. Capillary ITP and CE might be combined on-line [29].
ITP can be applied to samples to obtain analyte concentration
from dilute solution to allow UV or (laser-induced) fluorescence
detection. For a detection technique with a narrow dynamic
range such as the MS, ITP can also be used to adapt the
sample concentration to the dynamic range of the CE-MS combination, as we are investigating.

CONCLUSIONS AND PERSPECTIVES

Some recent developments in the LC-MS and related fields
have been briefly reviewed. Although significant progress
has been achieved, there is still ample room for improvements,
especially in respect of concentration detection limits.
However, in evaluating what limits of quantification are achievable in LC-MS, it should also be kept in mind that the
confirmation of molecular mass and structural identity has
only recently become possible for most of the labile and
highly polar compounds that are now amenable to routine application of LC-MS. Despite its limitations, LC-MS has earned
its place in the analytical laboratory and has become indispens-
able in some fields of application. Recent progress in
the more biochemically oriented applications of LC-MS will
certainly have considerable impact in (bio)pharmaceutical applications, e.g. in the characterization of receptors and in
peptide drug research.

References (*BmEv* = *Biomed. Environ.*; *MS* = *Mass Spectrom.*; *JChr* = *J.*
[*Chromatog.*)
1. Tomer, K.B. & Parker, C.E. (1989) *JChr 492*, 189-221.
2. Niessen, W.M.A., Tjaden, U.R. & van der Greef, J. (1991) *JChr 554*, 3-   [26.
3. Pieniaszek, Jr., H.J., Shen, H-S.L., Garner, D.M., Page, G.O.,
   Shalaby, L.M., Isensee, R.K. & Whitney, Jr., C.C. (1989) *JChr 493*,
   79-92.

4. Edholm, L-E., Lindberg, C., Paulson, J. & Walhagen, J. (1988)*JChr 424*,
5. Lant, M.S., Oxford, J. & Martin, L.E. (1987) *JChr 394*, 223- 230. [61-72.
6. Oxford, J. & Lant, M.S. (1989) *JChr 496*, 137-146.
7. Lindberg, C., Paulson, J. & Blomqvist, A. (1991)*JChr 554*, 215-226.
8. Beattie, I.G. & Blake, T.J.A. (1989) *BmEv MS 18*, 860-866.
9. Korfmacher, W.A., Freeman, J.P., Getek, T.A., Bloom, J. &
    Holder, C.L. (1990) *BmEv MS 19*, 191-201.
10. Heeremans, C.E.M., Stijnen, R.A.M., van der Hoeven, R.A.M.,
    Danhof, M. & van der Greef, J. (1991) *JChr 554*, 205-214.
11. Julien-Larose, C., Voirin, P., Mas-Chamberlin, C. &
    Dufour, A. (1991) *JChr 562*, 39-45.
12. Fenn, J.B., Mann, M., Meng, C.K., Wong, S.P. &
    Whitehouse, C.M. (1990) *MS Rev. 9*, 37-70.
13. Smith, R.D., Loo, J.A., Edmonds, C.G., Barinaga, C.J. &
    Udseth, H.R. (1990) *Anal. Chem. 62*, 882-899.
14. Mann, M., Meng, C.K. & Fenn, J.B. (1989) *Anal. Chem. 61*, 1702-1708.
15. Baczynskyj, L. & Bronson, G.E. (1990) *Rapid Comm. MS 4*, 533-535.
16. van Dorsselaer, A., Bitsch, F., Green, B.N., Jarvis, S., Lepage, R.,
    Bischoff, R., Kolbe, H.V.J. & Roitsch, C. (1990) *BmEv MS 19*, 692-704.
17. Shackleton, C.H.L., Falick, A.M., Green, B.N. &
    Witkowska, H.E. (1991) *JChr 562*, 175-190.
18. Huang, E.C. & Henion, J.D. (1990) *J. Am. Soc. MS 1*, 158-165.
19. Ackermann, B.L., Coutant, J.E. & Chen, T-M.(1991) *Biol. MS 20*, 431-440.
20. Kaiser, R.E., Jr., Williams, J.D., Lammert, S.A.
    & Cooks, R.G. (1991) *JChr 562*, 3-11.
21. McLuckey, S.A., van Berkel, G.J., Glish, G.L., Huang, E.C.
    & Henion, J.D. (1991) *Anal. Chem. 63*, 375-383.
22. van der Greef, J., Niessen, W.M.A. & Tjaden, U.R. (1989) *JChr 474*,
    5-19.
23. Verheij, E.R., Reeuwijk, H.J.E.M., LaVos, G.F., Niessen, W.M.A.,
    Tjaden, U.R. & van der Greef, J. (1988)*BmEv MS 16*,
    393-397.
24. Walhagen, A., Edholm, L-E., Heeremans, C.E.M.,
    van der Hoeven, R.A.M., Niessen, W.M.A., Tjaden, U.R. &
    van der Greef, J. (1989) *JChr 474*, 257-263.
25. Kokkonen, P.S., Niessen, W.M.A., Tjaden, U.R. & van der Greef, J.
    (1991) *JChr 565*, 265-275.
26. Tjaden, U.R., de Bruijn, E.A., van der Hoeven, R.A.M., Jol, C.,
    van der Greef, J. & Lingeman, H. (1987) *JChr 420*, 53-62.
27. van Baakergem, E., van der Hoeven, R.A.M., Niessen, W.M.A.,
    Tjaden, U.R., Poon, G.K., McCague, R. & van der Greef, J.
    (1992) *J Chr*, in press.
28. Reinhoud, N.J., Schröder, E., Tjaden, U.R., Niessen, W.M.A.,
    ten Noever de Brauw, M.C. & van der Greef, J. (1990)*JChr 516*, 147-155.
29. Stegehuis, D.S., Irth, H., Tjaden, U.R. & van der Greef, J.
    (1991) *JChr 538*, 393-402.

Note by Editor.- Cited authors with arts. in present book:
Lindberg (#C-3); Oxford (#A-2). Interfacing also features in
earlier vols. (Plenum; ed. E. Reid *et al.*): 14 & 16 (Oxford/Martin;
esp. moving-belt), & 18 (Blake; Bruins). Vol.20 (R. Soc. Chem.) has
MS-MS and CE studies. In *all* vols., look up Index entry 'MS'.

#D-2

PEROXYOXALATE CHEMILUMINESCENCE DETECTION IN HPLC[⊕]

[1]P.J.M. Kwakman and [2]G.J. de Jong

[1] Department of Analytical Chemistry, Free University,
De Boelelaan 1083, 1081 HV Amsterdam, The Netherlands

[2] [formerly at [1]] Analytical Development Department,
Solvay Duphar B.V., 1380 DA Weesp, The Netherlands

As was expounded in an earlier contribution to this series [B, V. 18][⊕] and is now amplified, POO-CL[*] has progressed to become one of the most sensitive detection methods available for HPLC, with on-column detection limits in the fg-pg range. The mechanism of the POO-CL reaction and several detection systems are briefly discussed. Derivatization reactions enable labels with good CL characteristics to be introduced into various analytes. Biomedical and related applications including detection of trace amounts of H_2O_2 are surveyed. The general scope of the detection system is indicated, with mention of FIA.

Even detection by fluorescence (conferrable by a derivatization step) is often too insensitive for biomedical and environmental analyses in the low-ppb and ppt range, largely because of stray light detrimental to the background signal and noise. CL^*, commonly POO-CL, circumvents this HPLC-detection problem, giving a 10- to 100-fold improvement in sensitivity; application to HPLC dates back to 1980 [B35][⊕].

Following a 1963 report by E.A. Chandross of the reaction of oxalylchloride with H_2O_2 in the presence of a fluorescent compound, the POO-CL reaction was studied with aryl oxalates

---

[⊕]Text as furnished was condensed by Editor; it mirrored the broad survey given at the Forum (by Dr de Jong) and did not take account of the survey in Vol. 18 (this series) by Brinkman and Amsterdam co-authors - now *cited as* 'B, V.18'.
  This volume is listed at end of refs. list, where there is some 'fill-out' in connection with an Editor's stratagem:- where the present MS. had refs. in common with refs. in 'B, V.18', only the citation no. in the latter is now given, prefixed by 'B' or, if from the Amsterdam group, '**B**'. Thus 'B35' signifies ref. 35 in B, V.18.- *See end of article.*
[*] *Abbreviations (the first due to Ed.).-* POO, peroxyoxalate; CL, chemiluminescence; FIA, flow-injection analysis. *Others listed at foot of next p.*

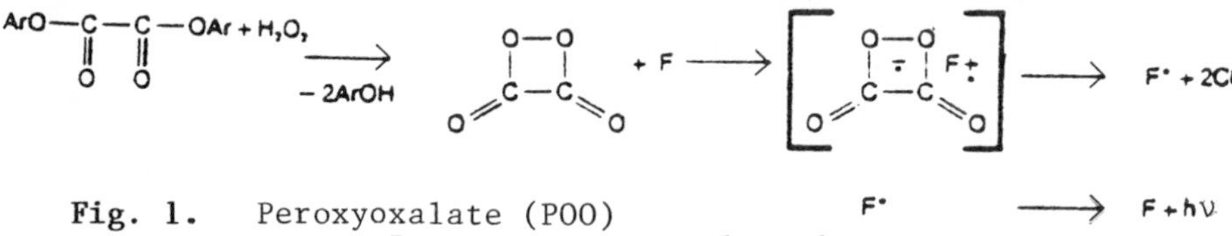

Fig. 1. Peroxyoxalate (POO) reaction scheme. F = fluorophore. *Adapted from [B, V. 18] by Ed.*

and a mechanism proposed [1]:- a supposed energy-rich 1,2-dioxe-tanedione intermediate serves to chemically excite the fluoro-phore, producing its singlet state and hence fluorescence emission (Fig. 1). CL efficiency values in the range 1-23% have been reported [1]. Electronically excited states of energy up to ~430 kJ/moles can be generated [2]. The CL decomposition of dioxetanes has been studied [3, 4], and the CIEEL* concept introduced [5] and explored for POO-CL [6]. The chemi-excitation step is the electron transfer from the intermediate back to the fluorphore resulting in its excited state, implying that its oxidation potential should correlate well with the CL efficiency. CL intensity was indeed found to depend strongly on the electronegativity of the aryl group of the oxalate ester. It was proposed that there was not a common intermediate as above, but rather an intermediate retain-ing one aryl group [6]. A study of the TEA-catalyzed reaction of H_2O_2 with TCPO [B40] showed a biphasic intensity/time profile, and a multiple intermediate mechanism was proposed in which more light-producing pathways are possible.

POO-CL REACTION SYSTEM

The oxalate and H_2O_2 are mixed with the column eluate in a coil just before the photomultiplier (diagram in [B49]), with benefit to sensitivity. All oxalates investigated are a compromise, none having all the desired features - solubility in common HPLC solvents, stability in the presence of H_2O_2, and good CL intensity [7, 8, B42]. TCPO and 2-NPO, in the range 1-10 mM, are favoured, because of stability with H_2O_2 present. DNPO is more soluble in common HPLC solvents such as acetonitrile and methanol, but has limited applicability because of very fast reaction kinetics. TDPO is soluble

---

*Abbreviations (ctd. from previous p.).- CIEEL, chemically initi-ated electron exchange (concept); PAH, polycyclic aromatic hydro-carbon; TEA, triethylamine. DCIA, 7-(diethylamino)-3-[4-(iodo-acetyl)aminophenyl]-4-methylcoumarin; NBD-F, 4-fluoro-7-nitro-benzoxadiazole; NDA, naphthalene-2,3-dialdehyde; 2-NPO, bis(2-nitrophenyl)oxalate; TCPO, bis(2,4,6-trichlorophenyl)oxalate; TDPO, bis[4-nitro-2-(3,6,9-trioxadecyloxycarbonyl)-phenyl]oxal-ate. Variants on the Cl moiety of **dansyl**-Cl, viz. 5-dimethyl-aminonaphthalene-1-sulphonyl chloride: -BAP, *N*-bromoacetamide-piperazide; -Hy, hydrazine; -OH, hydroxide.

in acetonitrile up to even 1 M [7], but cannot be used in such high concentrations due to CL quenching by the phenolic reaction product [8]. The H_2O_2 concentration is typically 10-500 mM, higher than the oxalate concentration.

DERIVATIZATION APPROACHES AND APPLICATIONS

Some analytes have an intrinsic CL capability, notably PAH's which can be detected in sub-pg amounts after conversion, advantageously pre-column, to amino-PAH's ([B57]: a Fig. there-from re-appears in [B, V.18]; [B42]). Most analytes have to be derivatized to give a fluorophore [9, 10], commonly by dansyl tagging. Entries in the following list include the detection limit, (); some refer only to pure compounds.-

Dansyl-Cl.- [B35, B46, B48-50]: amino acids (1-10 fmol; illustrated in [B, V.18]). [11]: bradykinin. [B48]: catechol-amines (5-10 fmol). [12]: amphetamines in urine (10 fmol; **NDA** & **NBD-F** as alternatives to dansyl-Cl). [13]: a sec.-amine drug in serum (1-10 pg). [14]: substituted phenols in river water (15-30 fmol; **laryl-Cl** an alternative). [15]: oestradiol in serum (50 pg). [16]: same (50 fmol).

Dansyl-Hy.- [B51]: flucortin butyl in plasma (7.5 pg). [17]: an immunosuppressive macrolide antibiotic, FK-506, in serum (12 pg). [18]: oxo-steroids & -bile acids in urine (5-10 fmol).

Dansyl-OH.- [B53] a tert.-amine drug, secoverine (500 pg).

Dansyl-BAP.- [19]: carboxylic acids (10-25 fmol).

3-Aminofluoranthene.- [B54]: carbonyls (100 fmol).

Luminarin 1.- [20]: prim./sec. amines (1-6 fmol).

Fluorescamine.- [B56]: catecholamines in urine (25 fmol).

NDA.- [21]: catecholamines in urine (1 fmol). [22]: a prim.-amine drug in serum (5 fmol).

DCIA.- [23]: fluoropyrimidines in serum (30 fmol).

Coumarins.- [B55]: carboxylic acids (50-75 fmol).

3-Aminoperylene.- [B42]: carboxylic acids (1-10 fmol).

Comments (each preceded by the ref. no.; e.g. [11] connotes bradykinin).- #[B50].- Imai *et al.* could measure pure amino acids at 10^{-9} M (20 µl injections) using a microbore column and gradient elution without major shift of the baseline (Fig. 10 in [B, V.18]). #[11], N-terminal group; microbore column.- Recovery was low (~10%) due to incomplete separation of the dansyl-bradykinin derivative and reagent by-products and to the peptide-hydrolysis step (110°, 20 h). #[13].- See Fig. 2. In 1987 [B46] we achieved similar detection limits (~100 fg; 0.5 µl injection) with a packed capillary column and, as illustrated in [B, V.18], a special mixing device. #[12].- Detection of the amines (primary or, exemplified by methamphetamine, secondary) was worse (~5-fold) with NBD-F and better (~5- to 10-fold) with NDA compared with dansyl-Cl. #[14] & #[16].- The derivatization was by a novel procedure we developed, using

Fig. 2 *(from [13], by permission).*
Determination of a dansylated secondary
amine by HPLC with POO-CL detection.
Serum sample spiked with 4.8 ng/ml (24 pg
injected) of the drug. C-18 column; THF/
imidazole buffer pH 7 (1:2 by vol.).

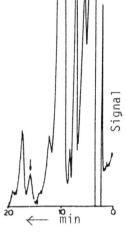

Signal

Fig. 3, *below (from [15], by permission).*
Normal-phase HPLC of dansylated steroids
with POO-CL detection (see text): I, dansyl-
17α-oestradiol; II, dansyl-17β-oestradiol.
A: authentics, 150 pg each; B, blank serum
(500 μl); C, serum pool spiked with 200 pg I
and 300 pg II per 500 μl serum.

20   10   0
←— min

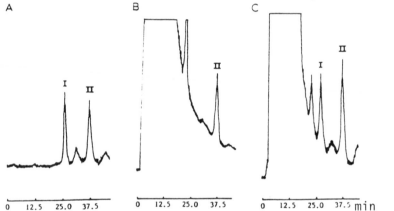

A   B   C

0   12.5   25.0   37.5   0   12.5   25.0   37.5   0   12.5   25.0   37.5 min

a 2-phase liquid-liquid extraction system.   (A different post-
column system [**B**53] was used for secoverine [**B**,V.18].)   #[15].-
A novel feature of the analysis of serum for dansylated
oestrogens (with ~90% recovery) reported in 1988 by O. Nozaki
et al. was the use, now illustrated (Fig. 3), of a normal-phase
(silica) HPLC column, using hexane/chloroform/ethanol (70:30:0.1,
by vol.) as eluent; the solutions added post-column were TCPO
and TEA in chloroform and H_2O_2 in methanol.

Further comments: agents other than dansyl-Cl.-   #[B51].-
For the 3α-ketosteroid flucortin butyl, dansyl-Hy was used
(Fig. 4), and as little as 100 pg/ml was measurable in plasma.
The excess dansyl-Hy caused large interferences; other authors
removed it by column switching [17] or gel-permeation chromatography [18].   #[B54].- The C=O label 3-aminofluoranthene,
reported [B57] to be very apt for CL detection, unfortunately
has limited applicability because of the unfavourable derivatization conditions of the reductive amination reaction (14 h

Fig. 4 *(adapted from a 1984 paper, T. Koziol et al., J. Chromatog. [B51]).* RP-HPLC of flucortin butyl in dog plasma by POO-CL after dansyl-Hy treatment, which gave 2 dimers *(arrows).*

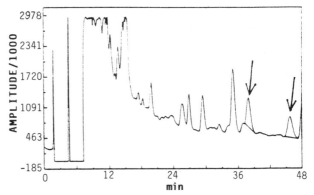

for ketones) and of interference, causing poor detection limits, by a long tailing peak. #[B53].- We adapted a stratagem that circumvents the non-susceptibility of tertiary amines to derivatization, entailing ion-pair formation and post-column extraction (by stream segmentation) into an organic solvent; we used dansyl-OH as the CL counter-ion, and added H_2O_2 by pumping an apolar solvent (1,2-dichloroethane) through a column packed with perhydrit (H_2O_2 held on a urea support). Detectability is limited by the background signal associated with extraction of the counter-ion. #[20].- With the novel agent luminarine 1, non-aqueous derivatization conditions were 20 min at 60° for n-pentylamine and 30 min at 80° for dipropylamine. #[B56].- For noradrenaline and dopamine in human urine (17.5 µl injected), Kobayashi *et al.* found the detection limit with fluorescamine to be comparable with that listed [#B48] for dansyl-Cl. #[21], catecholamines, & #[22], fluvoxamine.- In using NDA, a primary amine label developed for specific fluorescent detection of amino acids and peptides, side-product interferences induced by the co-reagent cyanide can be minimized by multidimensional HPLC (O.S. Wong) or by liquid-liquid extraction [22].

Comments on concluding entries in the list.-#[B55].- In the well-known 4-bromomethyl-7-methoxycoumarin, replacement of methoxy by diethylamino dramatically improved the CL sensitivity for fatty acids or prostaglandins (cf. luminarine [20]); with DCIA the strong electron-donating influence of the diethylamino group lowered the oxidation potential and so facilitated the energy transfer (see above, CIEEL mechanism). #[23].- The rate of reaction of the acidic pyrimidine -NH group, using DCIA in acetone containing freshly powdered K_2CO_3, was significantly enhanced by sonication in an ultrasonic water-bath (30 min, room temp.). #[B42].- In the use, for carboxylic acids, of 3-aminoperylene (formed from perylene), the derivatization conditions are rather unfavourable (50°, 2 h). 'Real' samples have yet to be studied, and for prostaglandins in

urine and serum the sensitivity need is challenging. #[19].-
Using dansyl-BAP, wherein a piperazine spacer links the dansyl
and bromoacetamide moieties, we derivatized several aliphatic
and aromatic amino acids and achieved trace-level assay of
retinoic acid.

DETERMINATION of H_2O_2 GENERATED BY ENZYME REACTIONS

Trace amounts of H_2O_2 such as some enzymes generate
can be determined by POO-CL in the presence of excess fluorophore,
at pH 7 such that there is good activity - which is not
the case with other CL reactions (e.g. with luminol, lucigenin
or acridinium esters). There have been diverse applications,
mainly for bioconstituents. FIA rather than HPLC was used
in certain studies cited in our previous survey [B, V.18]:
[B63-66] - e.g. cholesterol, using immobilized cholesterol
oxidase (0.1 pmol detectable), L-amino acids using L-amino
oxidase (2 pmol) and glucose, by β-galactosidase (12 ng); [B70]
- glucose, by glucose oxidase (2 pmol detectable). HPLC
was employed in estimating acetylcholine and choline with
immobilized choline oxidase and cholinesterase, by K. Honda
et al. ([B67]; TCPO and perylene used) and by Amsterdam
colleagues [B68] who used, for urine and serum (1 pmol detectable), a special flow-cell (depicted in [B, V.18]). The Amsterdam
approach with post-column reactors has also been applied to
glucose [24] and L-amino acids [25]. 8-Hydroxyquinoline and
Rhodamine B (FIA) have also been used, immobilized in a flow-
cell, for H_2O_2 detection [26]. For glucose with FIA ([B70];
mentioned above) Rhodamine B was used, and a water-soluble
oxalate which, however, had to be dissolved in acetonitrile
because of limited stability in water. HPLC and putrescine/poly-
amine oxidases were used in assay of hydrolyzed urine for
polyamines [27]; 5 pmol was detectable (interferences trivial).

OTHER DETECTION SYSTEMS BASED ON PEROXYOXALATE

Detection of compounds having a C-H bond strength <95 kcal/mol
was investigated [28] as a follow-up to the observation that
quinones in the presence of oxygen photocatalyze H_2O_2 production,
detectable by POO-CL, offering better selectivity than UV
detection in determining anthraquinones. Excited anthraquinone
abstracts a proton from susceptible compounds, forming the
semiquinone QH* which, by donating its electron to O_2, forms
the hydroperoxy radical, giving H_2O_2 by disproportionation:

$$QH^* + O_2 \longrightarrow Q + HO_2\cdot; \qquad HO_2\cdot + HO_2\cdot \longrightarrow H_2O_2 + O_2$$

Thereby compounds such as sugars and alcohols that lack a
UV chromophore can be detected selectively, without derivatization, in the low-ng range [28]. This approach offers promise.

Another interesting approach that circumvents H_2O_2 addition
is photo-initiated POO-CL detection. Irradiation of TCPO with

dissolved O_2 present yields reactive intermediates that transfer energy to fluorophores. This system was applied to analysis of air for carbonyl compounds [29], which were derivatized by drawing air through cartridges packed with porous glass particles impregnated with dansyl-Hy. For several analytes this approach is ~5-50 times less sensitive than H_2O_2-initiated CL and >10 times less sensitive than fluorescence detection; yet it is simpler and easier to operate and could be apt for automated procedures.

Two other approaches warrant follow-up [**B**, V.18]. Urea or ouabain, although non-fluorogenic, enhances the CL-background generated by H_2O_2 and DNPO, by an unknown mechanism, and can thereby be estimated [B60]. For HPLC detection, quenching of the CL signal by analytes such as some anilines or organo-sulphur compounds or anions such as I^- or NO_2^- can be exploited [30, 31]; the high selectivity of this approach was shown by assay of unprocessed urine for methimazole in the low-ng range ([B59]; P. van Zoonen *et al.*).

CONCLUDING COMMENTS

Of the various CL approaches applied for highly sensitive (fg-pg range) HPLC detection, POO-CL seems the most versatile. The complex mechanism and, governing the part of the CL decay curve measured in a dynamic flow system, the kinetics of the reaction are still under investigation. The scope of POO-CL hinges on derivatizing analytes lacking a fluorophore. POO-CL has notable biomedical applicability for determining low amounts of analytes from which H_2O_2 can be produced enzymatically. Whilst HPLC detection by POO-CL is notably sensitive, there is still a need for systematic optimization and for purchasable CL detectors, with a view to routine use of CL for complex samples in many laboratories.

References (*JChr* = J. *Chromatog.*)

The foregoing survey (condensed editorially) complements and, *as denoted by prefix* 'B' or '**B**', cites refs. and Figs. from the following art., *cited in text as* '**B**, V.18':-
   Brinkman, U.A.Th., de Jong, G.J. & Gooijet, C. (1988) in *Bioanalysis of Drugs and Metabolites, Especially Anti-inflammatory and Cardiovascular* (Reid, E., Robinson, J.D. & I.D. Wilson, eds.) [Vol.18 of present series], Plenum, New York, pp. 321-338: particularly p. 331 onwards.

Refs. therefrom are now listed (in approx. order of citation) with publication year, facilitating a search if V.18 is not at hand; firstly '**B**' (cf. 'B') refs., all from our Amsterdam group:-
   #Brinkman, U.A.Th. & co-authors: **B**53, 1987. #de Jong, G.J., *et al.*: **B**46, 1987. #Frei, R.W., & co-authors: **B**68 & **B**59, 1987; **B**69, 1985.
'B' refs., listed **either** by first author *(....et al.)* **or**, to indicate the research group, by a 'key' name:-
   #Alvarez, F.J., *et al.*: B40, 1986. #Birks, J.W.: B57, 1984.

Capomacchia, A.C., *et al.* : B60, 1986.   #Grayeski, M.L.: B51,
1984; B54 & B55, 1987; B70, 1986.   #Imai, K.: B42, 1985;
B47, 1987; B35, 1980; B49 & B50, 1984; B56, 1981; B67,
1986. #Melbin, G.: B48, 1983. #Rigin, V.I.: B63-B66, 1978-83.
========

1. Rauhut, M.M., Bollyky, L.J., Roberts, B.G., Loy, M., Whitman, R.H., Ianotta, A.V., Semsel, A.M. & Clarke, R.A. (1967) *J. Am. Chem. Soc. 89,* [6515-6522.

2. Lechtken, P. & Turro, N.J. (1974) *Mol. Photochem. 6,* 95-99.   [6515-6522.

3. McCapra, F., Beheshti, I., Burford, A., Hann, R.A. & Zaklika, K. (1977) *J. Chem. Soc. Chem. Comm.* 944-946.

4. Adam, W. & Liu, J-C. (1972) *J. Am. Chem. Soc. 99,* 2894-2895.

5. Schuster, G. (1979) *Acc. Chem. Res. 12,* 366-373.

6. Catherall, C.L.R., Palmer, T.F. & Cundall, R.B. (1984) *J. Chem. Soc. Farad. Trans. 80,* 823-836 & 837-849.

7. Imai, K., Nawa, H., Tanaka, M. & Ogata, H. (1986) *Analyst 111,* 209-211.

8. Imai, K., Nishitani, A. & Tsukamoto, Y. (1987) *Chromatographia 24,* 77-81.

9. Imai, K. & Weinberger, R. (1985) *Trends Anal. Chem. 4,* 170-175.

10. Imai, K. (1987) Ch. 10 in *Chromatog. Sci. Series 48 (Detection Oriented Techniques ....),* Dekker, New York, p. 359.   [260.

11. Miyaguchi, K., Honda, K., Toyo'oka, T. & Imai, K. (1986) *JChr 352,* 255-

12. Hayakawa, K., Imaizumi, N., Ishikura, E., Minogawa, E., Takayama, N., Kobayashi, H. & Miyazaki, M. (1990) *JChr 515,* 459-466; *also 464,* 343-352.

13. de Jong, G.J., Lammers, N., Spruit, F.J., Brinkman, U.A.Th. & Frei, R.W. (1984) *Chromatographia 18,* 129-133.

14. Kwakman, P.J.M., Kamminga, D.A., Brinkman, U.A.Th. & de Jong, G.J. (1992) *JChr 553,* 345-356. *[cf.* (1988) *459,* 139-149].

15. Nozaki, O., Ohba, Y. & Imai, K. (1988) *Anal. Chim. Acta 205,* 255-260.

17. Takada, K., Oh-Hashi, M., Yoshikawa, H., Muranishi, S., Nishiyama, M., Yoshida, H., Hata, T. & Tanaka, H. (1990) *JChr 530,* 212-218.

18. Higashidate, S., Hibi, K. Senda, M., Kanda, S. & Imai, K. (1990) *JChr 515,* 577-584.

16. *As for* 14. *J. Pharm. Biomed. Anal.,* in press.

19. Kwakman, P.J.M., van Schaik, H.P., Brinkman, U.A.Th., & de Jong, G.J. (1991) *Analyst 116,* 1385-1391.

20. Tod, M., Prevot, M., Poulou, M., Farinotti, R., Chalom, J. & Mahuzier, G. (1989) *Anal. Chim. Acta 223,* 309-317.

21. Kawasaki, T., Imai, K., Higuchi, T. & Wong, O.S. (1990) *Biomed. Chr 4,* [113-117.

22. Kwakman, P.J.M., Koelewijn, H., Kool, I., Brinkman, U.A.Th. & de Jong, G.J. (1990) *JChr 511,* 155-166.

23. Yoshida, S., Urakami, K., Kito, M., Yakeshima, S. & Hirose, S. (1990) *JChr 530,* 57-64.

24. van Zoonen, P., de Herder, I., Gooijer, C., Velthorst, N.H. & Frei, R.W. (1986) *Anal. Lett. 19,* 1949-1961.

25. Jansen, H., Brinkman, U.A.Th. & Frei, R.W. (1988) *JChr 440,* 217-223.

26. Ding, X., Wang, P. & Liu, G. (1988) *J. Luminescence 40 &41,* 844- 845.

27. Kamei, S., Ohkubo, A., Saito, S. & Takagi, S. (1989) *Anal. Chem. 61,* 1921-1924.

28. Aichinger, I., Gübitz, G. & Birks, J.W. (1990) *JChr 523,* 163-172.

29. Nondek, L., Milofsky, R.E. & Birks, J.W. (1991) *Chromatographia 32,* 33-39.

30. van Zoonen, P., Kamminga, D.A., Gooijer, C., Velthorst, N.H., Frei, R.W. & Gübitz, G. (1986) *Anal. Chem. 58,* 1245-1248.

31. Gooijer, C. & Velthorst, N.H. (1990) *Biomed. Chr 4,* 92-95.

#D-3

CAPILLARY ELECTROPHORESIS OF DRUGS IN BIOLOGICAL FLUIDS

David Perrett and Gordon Ross

Department of Medicine, St. Bartholomew's Hospital
Medical College, West Smithfield, London EC1A 7BE, U.K.

There are many reports on CE* separation of drugs in standard solutions but few on drug assay by CE in biofluids. Consideration is given here to the development of MECC or CZE assays for both drugs and endogenous molecules in biological samples, and to problems encountered.

We have assembled a CE apparatus having a fused silica capillary connecting the two chambers, with either UV or fluorescence detection. Using plasma and urine samples we have satisfactorily separated paracetamol and its metabolites by MECC, and urinary aspirin metabolites by CZE or, obviating sample variability, by MECC. MECC could separate nucleotides in cells, including human tumour cells in fluorouracil-treated rats.

Whilst sensitivity is adequate, quantitation is less good than by HPLC, and migration times may lack reproducibility. Whilst CE separations are efficient, they are relatively slow; but sample preparation can be minimal. With approaches that are considered, CE could become a sensitive, robust and useful technique for drugs in biofluids. Protein-bound drug can be distinguished from free drug. Research applications include the measurement of neurochemicals in single cells[⊕].

Recently CE has received much attention as a 'new' means of analyzing a wide variety of biomolecules [1-3][#] including some drugs [4, 5]. The first description of this instrumental approach to electrophoresis is usually credited to four classical papers from Jorgenson's laboratory published 10 years ago [e.g. 6]. CE is notable for its ability to separate and quantify, with analytical precision, the components of complex aqueous samples with very high resolution (N >100,000 theoretical plates)

---

*Abbreviations.- CE, capillary electrophoresis (CZE if in free solution); DTAB, dodecyltrimethylammonium bromide; LIF, laser-induced fluorescence; MECC, micellar electrokinetic capillary chromatography; RP, reversed phase; TCA, trichloroacetic acid.

[⊕]This published example (dopamine) is illustrated, and 'MECC' briefly explained, in Dolphin's Vol. 20 art. [1990; eds. and publisher as for present vol.]. [#]Apparatus diagram appears in [2].

using <10 nl of sample. It can separate cations, anions and uncharged molecules simultaneously. Additionally, it can separate any charged species ranging from simple molecules to macromolecules and even charged particles.

Although many reports of the separation of drugs in standard solution by CE have been published [see 4, 5], actual assays of drugs in biofluids are still relatively few in number. From a comprehensive computer database maintained by us [7], a list of substantive publications (up to December 1991) on drug analytes has now been compiled, with listing of the matrix [S = serum, P = plasma, U = urine; detection mode indicated ()]:-
- lithium: S (conductivity) [8]
- methotrexate: S (LIF) [9]
- cefpiramide: P (UV) [10]
- aspoxicillin: P (UV) [11]
- S-carboxymethyl-L-cysteine: U (UV) [12]
- barbiturates: S, P, U (UV) [13]
- thiopental: S (UV) [14]
- cimetidine: S, P (UV) [15]
- cytosine-β-D-arabinoside: S (UV) [16]
- glyphosate: S (UV) [17]
- paracetamol: P, U (UV) [this art.]
- salicylate: U (fluorescence) [this art.].
This list comprises only 10 publications prior to the present article. However, numerous separations of pharmaceuticals in standard mixtures have been demonstrated. The question may therefore be asked why a technique with such attractions to the analyst and a literature base of >750 publications has produced only some 10 papers on actual assays of drugs in biofluids.

Initially this lack of method descriptions was explained by the fact that most early reports came from analytical groups more interested in the technique and/or its instrumentation than in its applications. However, the recent availability of commercial instrumentation for CE has not been accompanied by a substantial increase in the numbers of 'real-life' applications reported. Equally it must be remembered that the majority of present-day pharmaceuticals are relatively uncharged, although their metabolites may be ionic, and since, moreover, RP-HPLC has proved excellent for their analysis both in pharmaceutical formulations and in biofluids, analysts may have seen no reason to change. Accordingly, the question has to be asked: "is the dearth of CE applications to biofluids due to the conservative attitudes of bioanalysts or do more fundamental problems remain to be overcome?"

Here we discuss some of the problems in developing assays for drugs and endogenous molecules in biological samples and attempt to answer the above question.

MATERIALS AND METHODS

Initially we assembled a CE system using a 30 kV HV source (230R; Bertan, Hicksville, NY, U.S.A.) mains-powered *via* a safety interlock connected to microswitches on a perspex isolation box.   Only when the lid was closed was HV applied to the platinum electrodes dipping into two electrolyte chambers. The chambers were connected by an electrolyte-filled   fused silica capillary (typically 50 µm i.d. × 70 cm; SGE, U.K.). A detection window 50 cm from the loading end was created by removing the polyimide coating.   This window was aligned in a variable wavelength detector (CV[4]; Isco, Lincoln, NE, U.S.A.).

We have also used with this CE system a Jasco 821-FP fluorescence HPLC detector able to scan in both the excitation and the emission modes which we equipped with a laboratory-made flow cell.   The fused silica capillary employed was as above (50 cm to detector) with a UV-transparent coating (Polymicro Technologies, Phoenix, AZ).   The output from both detectors was to a HP3396 integrator.

More recently we have employed an automated SpectraPHORESIS 1000 system (Spectra-Physics, Hemel Hempstead, U.K.) with rapid scanning detection and controlled by an IBM PS/2 computer.

Paracetamol and aspirin given to humans have been used as test drugs.   The metabolism of 5-fluorouracil has been studied in a cell line carried in rats.

RESULTS AND DISCUSSION

Because biofluids abound in endogenous molecules a prime requirement of most bioassays is good specificity. In analyses involving separation processes this is in part achieved by the use of high-resolution techniques. Additional specificity can also be achieved by the use of appropriate detectors.

Paracetamol is rapidly metabolized, particularly forming glucuronide and sulphate conjugates. These conjugates, because of their polarity, are often difficult to analyze by standard RP-HPLC techniques.   Paracetamol is uncharged under normal pH conditions and therefore has to be separated by MECC. In MECC the analytes undergo partition between the aqueous phase moving under electroosmotic flow and detergent micelles migrating under the influence of the electric field in the oppposite direction.   Hydrophobic interactions are therefore dominant in the separation, and non-polar molecules can be separated in what is in effect a form of chromatography with electroosmotic flow comprising the pump.

Following an oral dose (500 mg), paracetamol and its metabolites were readily separated from endogenous compounds

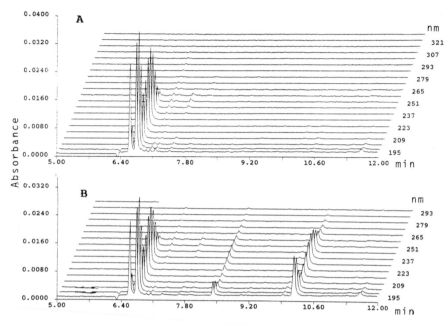

Fig. 1. MECC of urine up to (**A**) and during 2 h after (**B**) ingestion of paracetamol. Capillary: 50 μm × 70 cm (63 cm to detector); buffer: 50 mM phosphate/30 mM borate pH 8.6 + SDS to 50 mM; temp. 35°; loading at 5 kV, 5 sec; run at 20 kV. Peaks include: ~8.5 min, paracetamol; ~10.2 min, glucuronide.

in urine, and the appearance of the metabolites could be followed by UV detection with rapid scanning (Fig. 1). Sample introduction was electrokinetic, and gave a retention reproducibility of 4.9% (n = 6) with an efficiency for paracetamol of 254,200 theoretical plates. Such a CE separation, giving good resolution of the glucuronide, compares very favourably with established HPLC methods in terms of separation efficiency. Typically, in these studies, MECC resolved 50% more components from urine than gradient RP-HPLC with columns of standard dimensions (4.6 × 100-150 mm) although the new peaks observed were often minor components. The sensitivity of CE is often concentration-limited due to the very short pathlength in the detector combined with the low sample loadings. Therefore relatively high sample concentrations are required before detection is possible using a UV detector.

It was considered that the concentration of paracetamol would be high enough to permit its determination in plasma. Accordingly, plasma was prepared from heparinized blood which was collected at 30 min intervals following administration of a single 400 mg tablet of paracetamol to a normal individual. Plasma proteins were precipitated with 1 volume of TCA (12%

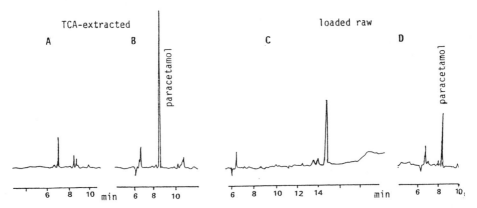

Fig. 2.   MECC of plasma from a volunteer, pre-dose **(A, C)** and
1 h after a paracetamol dose **(B,D)**, TCA-extracted **(A,B)** or un-
processsed **(C,D)** [see text].   Conditions as for Fig. 1 but note
differing time axes; detection at 242 nm.

w/v), and the excess TCA removed by multiple extractions
with water-saturated diethyl ether.   The same MECC separation
technique as for urines was used.   As shown in Fig. 2A (blank)
and 2B, the TCA extracts were clean, such that paracetamol
could readily be determined with high efficiency.

     With electrokinetic sample introduction CE can be highly
selective, since only those species with a high mobility
are injected during the brief time in which the loading
voltage is applied.   This can lead to a sample bias [18]
or can be exploited for the selective loading of samples
of interest from complex matrices.   Nishi & co-authors [11]
showed that aspoxicillin (a penicillin) in human plasma could
be determined by MECC with direct sample injection since
the drug was selectively loaded relative to plasma proteins.
In that example detection was at the relatively non-selective
wavelength of 210 nm and the proteins appeared as a large
broad peak on the electropherogram but with a migration time
about twice that of the aspoxicillin. Using the same rationale
we have been able to determine paracetamol directly in plasma
without any sample preparation (Fig. 2, C & D).   Here protein
peaks were near-absent since proteins absorb poorly at the
paracetamol detection wavelength (242 nm) besides differing in
migration from paracetamol.   The peaks migrating at >13 min
and the rise in baseline at 18 min may represent the proteins
(Fig. 2C).   Paracetamol could readily be determined in raw
plasma (Fig. 2D).   Thus, provided that analyte concentrations
are sufficiently high, the reduction in sample preparation
that CE with electrokinetic loading offers could well be
an important advance in bioanalysis.   However, more work
with a wider variety of drugs is required to show that the
requisite precision and reproducibility are routinely achievable.

Fig. 3. MECC of
(**A**) a mixture of
salicylate (SAL),
gentisic acid (GA)
and o-hydroxyhipp-
uric acid (oHA),
each 100 µM; (**B**)
pre-dose and (**C**) 1 h
post-dose urine
after 250 mg aspirin.
Capillary: 50 µm ×
51 cm (37 cm to detec-
tor); buffer: 25 mM
Tris pH 9.5 + SDS to
50 mM + methanol to
10% v/v; ambient temp.;
load 2 sec hydrodyna-
mic; run 30 kV.

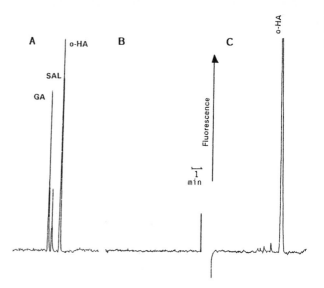

Aspirin is a model analyte that can be studied under
conditions of free-solution CE (CZE) since salicylate is ionized
under normal pH conditions. However, one problem that we
encountered with CZE was that variations in the salt concentration in the sample can lead to variations in the migration
times of the compound of interest. This was particularly
so when separating metabolites in urine where the osmolarity
of the urine can vary 10-fold. Migration times for salicylate
and its main metabolites were more constant under MECC conditions,
and the anticipated good resolution was obtained, thus aiding
peak identification in complex samples. However, because
of the presence of numerous endogenous compounds, UV measurement
on urine samples at 300 nm gave poor selectivity, and even
with continuous scanning peak identification was difficult.

With fluorescence detection (ex 310, em 400 nm) the selectivity of the system was greatly enhanced, and salicylate
and its main metabolites could, as shown in Fig. 3, be readily
separated and detected. Adequate sensitivity for their assay
in urine was achievable. One advantage of CE over HPLC
is that when the applied voltage is turned off, peaks can
be stopped dead in the flow cell. Thereby we were able
to confirm peak identities by stop-flow scanning (Fig. 4).
It is noteworthy that under stopped conditions the detector
appears to be at least 10-fold more sensitive than when
the peak is moving through the flow cell. Possibly the
linear velocity of the peak (~1.5 mm/sec) is too large to
permit maximum fluorescence efficiency.

Cellular nucleotides.- In order to analyze closely related
but diverse groups of compounds such as amino acids and

Fig. 4. Stopped-flow scanning of the principal fluorescent peak shown in Fig. 3C. The applied voltage was turned off at 8 min to hold the peak in the flow cell, and the emission spectrum from 310 to 500 nm scanned (note the maximum at 407 nm). Other conditions as in Fig. 3.

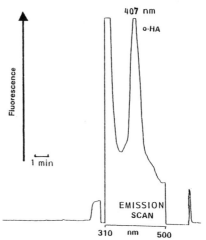

nucleosides in biological fluids by HPLC, it is usually necessary to employ gradient elution techniques in order to obtain adequate resolution. Recently we have established an MECC separation of the nucleotides found in cell extracts [19]. MECC with 100 mM DTAB present gave separation of 15 naturally occurring nucleotides. Their resolution was optimized through choice of conditions in respect of pH, DTAB concentration, applied potential, presence/absence of metal chelators and running temperature. In these studies the fused silica capillaries were 50 μm i.d. × 73 cm (63 cm to detector) and the base electrolyte was 25 mM Na_2HPO_4 pH 7, but the polarity was reversed from the normal mode, being maintained at -20 or -30 kV.

In order that many anti-cancer drugs can become pharmacologically active they must first be metabolized to their phosphorylated nucleotides. We have applied our system developed for intracellular nucleotides to extracts of tumour cells grown in rats dosed with fluorouracil, in order to ascertain the degree of phosphorylation of the drug. Fig. 5 shows a preliminary experiment revealing the complex nucleotide pattern for such extracts. Identifications so far made include adenosine nucleotides and fluorouracil itself. Preliminary identifications have been made for some fluorinated metabolites using the rapid scanning detection facility of the SpectraPHORESIS system, but further work is needed to settle peak identities firmly. The tumour is evidently ischaemic since the ATP/ADP ratio is low. Similar experiments are being performed in rats dosed with methotrexate.

CONCLUSIONS

The possibility of a resurgence of electrophoretic techniques for drug assays on biofluids is now real. A number

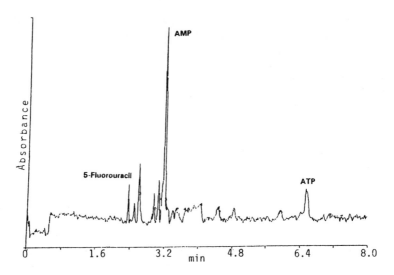

Fig. 5. MECC of an extract of a human tumour cell line grown
in a rat. Samples were kindly provided by Dr Paul McSheeny,
NMR Unit, St. George's Hospital, London; the cells (Ht29 cell
line) were extracted with 8% PCA (v/v; then neutralization, KOH).
Fused silica capillary: 50 µm × 73 cm (63 cm to detector);
buffer: 50 mM Na_2HPO_4/1 mM Na-EDTA pH 7 with 100 mM DTAB;
load: electrokinetic 5 sec. 5 kV; run: -25 kV; detection at
270 nm; temp. 35°.

of advantages of CE in this context are clear. CE is very
useful for the separation of both charged drugs using CZE
and uncharged drugs using MECC. In the CZE mode it may
prove very valuable for the study of charged metabolites
such as glucuronides. CE is excellent for chiral separations
of drugs, as demonstrated with standard mixtures and in quality-
control applications; but biofluids have yet to be studied
in this respect.

At present CE separations tend to be of high efficiency
but are relatively slow, and drug assays are no exception.
However, savings in time can be achieved by the reduction
in sample preparation that can be afforded by the selectivity
of CE with regard to both sample loading and system selectivity.
Paracetamol could be analyzed directly in plasma without any
need to remove proteins, although other assays have used
solid-phase extraction techniques to pre-fractionate the samples
before CE. Another advantage is that CE is capable of working
at low UV wavelengths, <215 nm, which enables the detection
of molecules lacking obvious chromophores. An important advantage of CE is that not only does it need little sample

preparation but also it consumes very little in the way of buffer and buffer additives such as chiral reagents. This aspect may have both cost and health-and-safety advantages.

However, much remains to be done to make CE a robust and useful technique for biofluids. In particular, CE still suffers from a general lack of concentration sensitivity. Many approaches to improving this are being explored. Some are generally applicable such as increased pathlength detectors and various means of increasing sample loading such as isotacho-phoretic sample introduction. Others, while achieving high sensitivity (e.g. LIF detection) have only limited applica-bility. At present there are no commercial modular fluorescence detectors available for CE work. The fluorescence detector used here was more versatile than an LIF detector since it had a xenon lamp, but the flow-cell design awaits much improvement. At present it gives only 5- to 10-fold better sensitivity than UV detection, whereas the supplied HPLC flow-cell gives a 500-fold improvement [20]. Matrix effects such as those found here with aspirin in urine need to be minimized by techniques such as isotachophoretic loading or solid-phase extraction of the sample prior to CE. At present quantitative accuracy and precision is less than that typically expected of HPLC assays [21], and migration-time reproducibility can also be low. Although the main area of dissatisfaction has usually been in quality-control analyses, the C.V.'s obtained (~5%) are generally quite acceptable to the bioanalyst.

CE also offers many interesting research possibilities. For instance, bound and free forms of a drug can be assayed simultaneously since the free drug and the protein-bound complex will have different mobilities in appropriate electrolytes. CE has already been shown to be compatible with *in vivo* microdialysis techniques. The small volume of ultrafiltrate collected by such devices either implanted in organs or sampling from peripheral veins is a constraint with respect to HPLC analysis but is ample for several CE runs. Taken to its limit, CE has been shown to be capable of measuring intracellular neurochemicals in a single cell ([22]; see ⊕ at start of art.). and there is no reason why the distribution of drugs within single cells could not also be studied, provided that a sufficient concentration is achieved.

References

1. Goodall, D.M., Lloyd, D.K. & Williams, S.J. (1990) *LC.GC International 3.6*, 28-40.
2. Karger, B.L., Cohen, A.S. & Guttman, A.J. (1989) *J. Chromatog. 492*, 585-614.
3. Deyl, Z. & Struzinsky, R. (1991) *J. Chromatog. 569*, 63-122.

4. Rahn, P.C. (1990) *Int. Lab. 20*, 44–50. *Also, see below.*
5. Wainwright, A. (1990) *J. Micro-col. Sep. 2*, 166–175.
6. Jorgenson, J.W. & Lukacs, K.D. (1981) *Anal. Chem. 53*, 1298–1302.
7. Perrett, D., Ross, G. & Goodall, D. (1991) *Capillary Electrophoresis (Bibliography/Database)*, The Chromatographic Society, Nottingham.
8. Huang, X.H., Gordon, M.J. & Zare, R.N. (1988) *J. Chromatog. 425*, 385–390.
9. Roach, M.C., Gozel, P. & Zare, R.N. (1988) *J. Chromatog. 426*, 129–140.
10. Nakagawa, T., Oda, Y., Shibukawa, A. & Tanaka, H. (1989) *Chem. Pharm. Bull. 37*, 707–711.
11. Nishi, H., Fukayama, T. & Matsuo, M. (1990) *J. Chromatog. 515*, 245–255.
12. Tanaka, Y. & Thormann, W. (1990) *Electrophoresis 11*, 760–764.
13. Thormann, W., Meier, P., Marcolli, C. & Binder, F. (1991) *J. Chromatog. 545*, 445–460.
14. Meier, P. & Thormann, W. (1991) *J. Chromatog. 559*, 505–513.
15. Soini, H., Tsuda, T. & Novotny, M.V. (1991) *J. Chromatog. 559*, 547–558.
16. Lloyd, D.K., Cypess, A.M. & Wainer, I.W. (1991) *J. Chromatog. 568*, 117–124.
17. Tomita, M., Okuyama, T., Nigo, Y., Uno, B. & Kawai, S. (1991) *J. Chromatog. 571*, 324–330.
18. Huang, S.H., Gordon, M.J. & Zare, R.N. (1988) *Anal. Chem. 60*, 375–377.
19. Perrett, D. & Ross, G. (1991) in *Purine and Pyrimidine Metabolism in Man VII* (Elion, G.B., Harkness, R.A. & Zollner, N., eds.), Plenum, New York,
20. Perrett, D., Faleye, J. & Ross, G. (1991) in *Monitoring Molecules in Neuroscience* (Rollema, H., Westerink, B. & Drifhout, W.J., eds.), Univ. Centre for Pharmacy, Groningen, 152–154.
21. Goodall, D.M., Williams, S.J. & Lloyd, D.K. (1991) *Trends Anal. Chem. 10*, 272–279.
22. Wallingford, R.A. & Ewing, A.G. (1988) *Anal. Chem. 60*, 1972–1975.

Added by Editor. - A 'near-duplicate' of ref. 4 appeared in *Am. Biotech. Lab. 8*, 24–29 (1990).

In Journals such as this (refs. 1 & 4 as examples), the page sequence may be broken by advertisements (as recognized in *Chemical Abstracts*).

#D-4

INSTRUMENTAL ANALYSIS IN BIOANALYSIS,
ESPECIALLY SUPERCRITICAL FLUID CHROMATOGRAPHY

Karin E. Markides

Institute of Chemistry, Uppsala University,
Box 532, S-75121 Uppsala, Sweden

Bioanalytical applications of SFC and SFE*, which were developed for lipophilic analytes, can be expected to widen with the introduction of more powerful mobile phases, solute derivatizations and controlled selectivity. SFE is attractive for sample preparation: no organic solvents are needed, temperature can be varied, and selectivity or trace enrichment can be achieved. Its success hinges on quantitative and reproducible extraction of diverse compounds, particularly trace amounts of polar solutes in a complex hydrophilic matrix.

For SFC, open tubular columns are efficient, inert and compatible with a wide spectrum of both GC and LC detectors, often without derivatization. On the other hand, packed columns have the advantages of speed, capacity and suitability for strong mobile phases that favour selective separation of the more polar solutes. Through temperature variation, SFC is notably suitable for chiral or other shape-dependent separations. State-of-the-art aspects are now considered in the bioanalytical context.

More than in any other field, the need for instrumental advances in bioanalysis seems insatiable. Complexity in the sample matrix, wide mol. wt. distribution, abundance of isomeric compounds, trace concentrations, presence of both hydrophilic and hydrophobic compounds, and molecular instability are some of the most important features in the analysis of biologically active compounds. In this perspective, the dimension of the problem presented by separation and identification of these compounds is readily understandable. The need for subtle separation of closely related compounds in complex matrices has focussed interest on highly efficient micro-column techniques.

---

*Abbreviations.- SF, supercritical fluid, *as in* SFC (chromatography) & SFE (extraction); SubF, sub-critical fluid. FTIR = Fourier transform infra-red. *'Ionization'(I) devices:* for MS, API = atmospheric pressure; for GC detectors, FID = flame, TID = thermionic. PG = prostaglandin.

COLUMNS AND OPERATING CONDITIONS

SFC represents an instrumental separation technique that has attracted increased interest amongst bioanalysts in recent years [1, 2]. The analytical column is the heart of any separation technique, and the choice of column type will largely govern utility and benefits with each particular system. SFC offers the options of using either open-tubular or packed columns. The strength of open-tubular SFC lies in high efficiency, inertness and a wide variety of detection methods, while packed-column SFC has advantages in speed of analysis and large stationary-phase surface area that give high sample and peak capacity. With both column types, a wide range of selective stationary phases is available. Several tailor-made stationary phases for open-tubular column SFC have been developed to give selectivity without increased retention. The polarizable biphenyl phase and the shape-selective liquid crystalline phase are good examples of this, while chiral-phase alternatives are still sought from manufacturers.

The two column types complement each other well, as in the two examples shown in Fig. 1. The first illustrates the separation, without derivatization, of closely related isomers of androstane hormones by high-resolution open tubular SFC [3]. In the second, a racemic mixture of substituted phosphine oxides was rapidly separated by packed column SFC, its high capacity enabling use of a strong mobile phase for the elution [4]. With the packed column the volume of organic solvents needed is only a fraction of that needed in HPLC; the saving in cost through the replacement (e.g. of hexane) by CO_2 is a growing attraction.

Mainly because of the surface-area difference, packed and open-tubular columns are not comparable when it comes to the chemical influence of the surface on the chromatography. This simple fact has, in many ways, resulted in a situation today where the packed column is used predominantly for highly polar solutes. Accepting the fact that CO_2 is the commonest mobile phase in SFC, it is well known that an organic modifier needs to be added to the CO_2 in order to dissolve highly polar solutes. This modifier will have the drawback of eliminating most of the detector options. Most polar solutes will, however, have chromophores for UV detection. Moreover, a modifier in the CO_2 will have the advantage of self-deactivating the particles in the packed column. This is extremely important since in no present-day packed columns with silica particles is the support truly inert.

The development of inert packings has posed a column-technology challenge for many years. The potential benefit to SFC is much greater than that to HPLC, in relation to

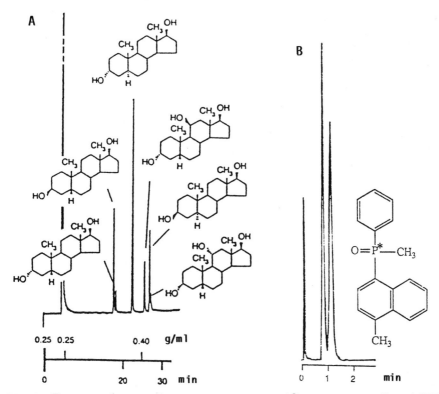

Fig. 1. Two examples: column open-tubular (**A**) or packed (**B**; subSFC).
(**A**) Androstane diols and triols. Liquid crystalline smectic
column, 6 m × 50 μm i.d.; 0.15 μm film thickness; CO_2; 120°;
density program from 0.25 to 0.50 g/ml at 0.0075 g/ml per min
after a 5 min isopycnic period. *From ref. [3], by permission.*
(**B**) Methyl-1-(4-methylnaphthyl)phenyl phosphine oxide. Column
10 cm × 4.6 mm i.d.; 10 μm aminopropyl Lichrosorb NH_2 silica
bonded with (R)-N-(3,5-dinitrobenzoyl)-phenylglycine; CO_2/
methanol/water at 20°; 271 atm inlet pressure, 227 atm outlet
pressure; detection at 280 nm. *From ref. [4], by permission.*

detectors. The problem of complete deactivation of the silanols
on the surface of silica particles has been discussed in
the literature [5, 6]. Other materials, such as carbon-based
polymers, have unacceptable property changes during density
programming. The open-tubular columns, on the other hand,
can be well deactivated. They are therefore mainly used
for non-polar to semi-polar analytes where high resolution
and/or other detector options than UV are wanted.

Examples of mobile-phase manipulation.- A liquid mobile
phase at room temperature and atmospheric pressure, e.g. a
gradient of 10→60% acetonitrile in CO_2, will dramatically change

Fig. 2. Separation of caprolactone diol oligomers under **(A)** SFC conditions, and **(B)** HPLC conditions, with a packed column (250 × 4.6 mm i.d.; 10 μm LiChrosorb silica particles). Mobile phase: acetonitrile in CO_2: **(A)** at 135°; **(B)** at 25°, with an acetonitrile gradient as shown and, at the start, 232 atm column exit pressure. Detection at 220 nm. *From ref. [7], by permission.*

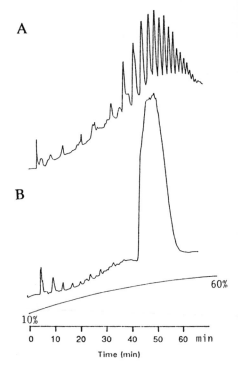

properties if subjected to supercritical temperature and pressure. Thus, with the same instrumental set-up, caprolactone diol oligomers were separated under liquid conditions (Fig. 2A) and SF-to-SubF conditions (Fig. 2B). Moreover, in Fig. 2B the transition from SFC to SubFC, in the last third of the chromatogram, did not noticeably impair the separation. The 100-fold enhancement in diffusion in the mobile phase, as it becomes supercritical, influences resolution and/or speed of analysis very effectively.

Among the neat supercritical fluids that could be used, only CO_2 is attracting enough attention today to become a practical choice. The fact that CO_2 is procurable in SFC/SFE purity grade is an unassailable advantage. In addition, the variable controls and the limits of the hardware as well as the materials are compatible with CO_2. Advances in hardware development are the key to performing SFC and SFE with new mobile phases such as NH_3 [8, 9]. An experimental separation of polar solutes with neat NH_3-SFC is shown in Fig. 3. Here, a non-polar packed column with polymer-based particles was used for chemical compatibility with the NH_3 and the chromatographic selectivity through large surface area. Routine use of NH_3 in open-tubular columns will not be possible until tailor-made stationary phases can be made with strong retention power and chemical stability.

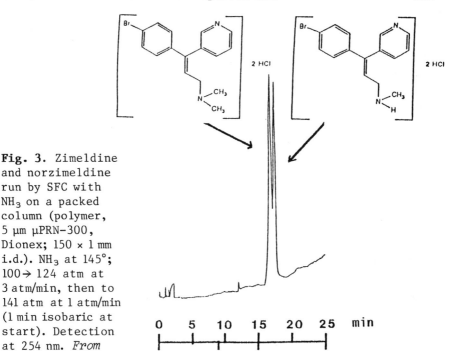

Fig. 3. Zimeldine and norzimeldine run by SFC with NH$_3$ on a packed column (polymer, 5 μm μPRN-300, Dionex; 150 × 1 mm i.d.). NH$_3$ at 145°; 100 → 124 atm at 3 atm/min, then to 141 atm at 1 atm/min (1 min isobaric at start). Detection at 254 nm. *From [9], by permission.*

Use of special SFC equipment.- Most published studies on packed column SFC have utilized instrumentation that represented more or less sophisticated modifications of HPLC equipment. An example of an exception is the use by K. Anton & co-authors [10] of equipment based on in-depth consideration. If capillary columns are used in SFC, i.e. <1 mm i.d., a syringe pump is normally utilized that can give a smooth, slow and well controlled flow. The packed capillary columns have many advantages over conventional packed columns. They are efficient and have good speed of analysis with moderate loadability, and can advantageously be coupled to an open-tubular column, with different selectivity, in a 2-dimensional chromatographic set-up (Fig. 4). The selectivity of the packed capillary column can be chosen to give a chemical class separation, and the class of interest is then separated into individual compounds by a highly efficient open-tubular column with a selectivity based on chemical structure.

SFC with assured inertness.- When high inertness is needed, well deactivated open-tubular columns are recommended for both chromatographic dimensions. This is illustrated in Fig. 5 for separation of an unknown natural compound from a complex biological matrix, a peritoneal dialysate. The columns used in the two SFC stages are specified (along with the authorship)

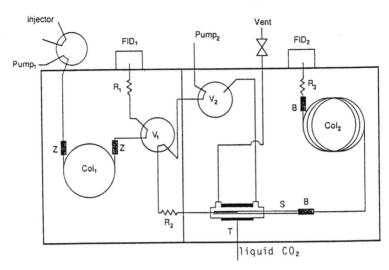

Fig. 4. Schematic diagram of a 2-dimensional SFC–SFC assembly. Col_1, primary packed capillary column; Col_2, secondary open-tubular column; V_1 and V_2, rotary valves; T, cold trap; S, capillary solute concentrator; R_1 and R_3, frit restrictors; R_2, linear restrictor; Z, zero dead volume union; B, butt connector.

in the Fig. legend. From chromatogram **B** the individual peaks revealed by the FID trace were collected and tested by a ^{86}Rb assay which, for the peak marked *, indicated high enough bioactivity to play a major role in causing salt-sensitive high blood pressure.

Sample injection and SFE

In the above example, injection of a 30 μl sample for trace analysis was achieved through a solvent-venting technique[⊕]. In reality, an injection volume of 100 nl is the limit in open-tubular columns of 50 μm i.d.; exceeding this limit causes >10% of the resolution to be lost through band-broadening during injection. Different types of solvent elimination methods have therefore been developed.

The ultimate solvent-free injector is, of course, the on-line SF extractor/injector. It is important to use and evaluate the available instrumentation in this area and to put demands on the next generation in respect of trace analysis and quantitation accuracy. The on-line SFE hardware niche is opening up the prospect that all micro-techniques will become routinely applicable in the future.

[⊕]I.J. Koski, K.E. Markides & M.L. Lee (submitted to *J. Microcol. Sep.*).

Fig. 5. Open-tubular column SFC-SFC chromatograms of a peritoneal dialysate extract. Conditions in the first dimension (**A**): 12 m × 50 μm column; oligoethylene oxide polysiloxane stationary phase; CO_2 at 90°; pressure program from 70 to 130 atm at 20 atm/min, then to 250 atm at 3 atm/min, after an initial 2 min isobaric period. Second dimension (**B**): 8 m × 50 μm column; liquid crystalline poly-siloxane stationary phase; CO_2 at 100°; pressure program from 70 to 130 atm at 20 atm/min, then to 330 atm at 6 atm/min after an initial 2 min isobaric period. See text for * (bio-active peak).
Acknowledgement: L.Q. Xie, K.E. Markides & M.L. Lee, submitted to *Anal. Chem.*; *permission in hand.*

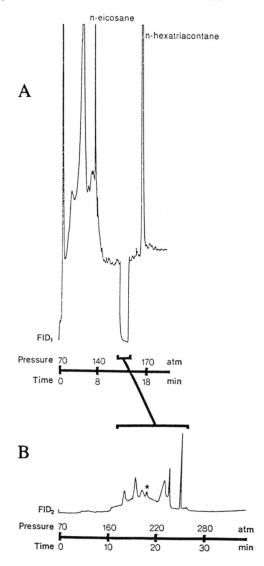

The on-line SFE injector makes it possible to use HPLC clean-up and to trap the compounds of interest on a solid-phase cartridge. An example is shown in Fig. 6 where underivatized PG's were analyzed after phase-transfer from water to CO_2 through on-line SFE injection and final detection by FID [11].

DETECTION

The need for derivatization is less and different in SFC than in the traditional HPLC, where detectability is often the reason, and in GC, where volatility and stability

Fig. 6. SFE/SFC chromatogram of aqueous
PG standards (13.3 or 20.0 ppm). SFE:
15 µl loaded onto XAD-2; extraction at
d = 0.80 g/ml and CO_2 at 45° for 45 min.
SFC: open tubular column (6.5 m × 50 µm);
oligoethylene oxide polysiloxane
stationary phase; CO_2 at 100°; density
program from 0.150 to 0.760 g/ml at
0.020 g/min after a 5 min isopycnic
period. *From ref. [11], by permission.*

Fig. 7, *below.* SFC chromatogram
of derivatized human urinary
α-keto acids. Open-tubular
column (10 m × 50 µm); poly-
methyl siloxane stationary
phase; CO_2; linear pressure
and temperature program; TID.
Peak identifications: 1, methyl
quinocalinol; 2, ethylquinoxal-
inol; 3, *p*-propylquinoxalinol;
4, 3-benzylquinoxalinol.
From ref. [12], by permission.

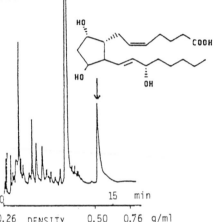

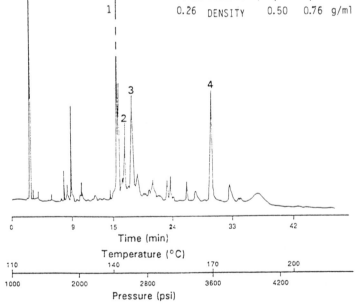

often is built into the analytes by the derivatization.   In
SFC, derivatization   is commonly a choice rather than a
necessity.   In the example in Fig. 7 [12], derivatization of
α-keto acids with *o*-phenylenediamine produced nitrogen-rich
quinoxalinols with enhanced TID sensitivity and selectivity.

It is well known that many detectors are compatible with SFC, whilst depending on the actual mobile phase used. The most commonly used detectors with the SFC technique are FID, TID, light detectors (UV, fluorescence), chemiluminescence, electron capture (ECD), light-scattering, FTIR and MS. All of these, and other, detectors can be important tools for the bioanalyst - which is propitious for the prospect that SFC will become accepted and often serve as a routine technique. The delicate balance and understanding of elution strength in the mobile phase *vs*. detector compatibility is an example of an area where more research in analytical chemistry is needed to find the true limits and capabilities of SFC, at present constrained by technical know-how.

Many of the above-mentioned detectors are in fact a focus for basic and applied research today, aimed at theoretical as well as practical progress. One example of an area where break-throughs have been made is SFC-FTIR [13]. Miniaturized high-pressure flow cells, sophisticated mathematical data-handling, fibre-optic technology and fast electronics are amongst notable advances that have rendered the FTIR detector suitable for SFC without detriment to the performance of either the SFC or the FTIR detector.

Mass-spectrophotometric (MS) detection

For MS detection it is likewise believed that basic knowledge could initiate a break-through. SFC-MS interfaces for several instruments with evacuated ion chambers are already commercially available. A mild separation technique such as SFC should be readily compatible with an identifying detector that likewise ensures intactness of the solutes of interest. API-MS [cf. M. Doig, art. #ncD-1- *Ed.*] is one example. The very mild ionization conditions normally give spectra dominated by the protonated molecular ion, such that MS-MS would be needed if structural information is sought.* The non-availability of SFC–API-MS interfaces has so far hampered developments in this area. It is reckoned, however, that a market will develop for API-MS linked to SFC and also SFE.

Use of a home-built SFC–API-MS interface is exemplified by the analysis of anabolic steroids shown in Fig. 8 [14]. It is commonly acknowledged, by investigators who possess both instrumental approaches, that SFC-MS is easier to perform successfully than HPLC-MS. This is due not only to the faster sample preparation procedure for SFC but also to its reproducibility and to the ease of keeping the ion source clean and of interpreting the data.

Note by Editor.- Use of SFC with EI-MS for identifying ibuprofen in horse urine (E.D. Lee *et al*.) is shown in Vol. 20 (art. by Dolphin).

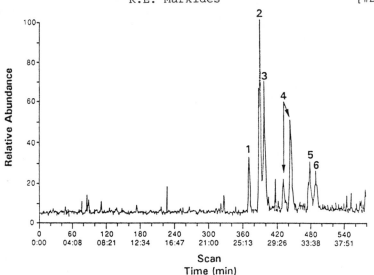

Fig. 8. SFC-MS chromatogram of a synthetic mixture containing anabolic and oestrogenic compounds (~20 ng each). Open-tubular column (5 m × 50 μm); poly(50% cyanopropyl)methylsiloxane stationary phase; CO_2; linear pressure program from 100 to 400 atm at 6.5 atm/min. Peak identifications: 1, melengestrol acetate; 2, medroxyprogesterone; 3, trenbolone; 4, diethylstilbestrol; 5, zeranol; 6, triamcinolone acetonide. *From [14], by permission.*

Mobile-phase choice in relation to detection

Commonly the choice of detector will determine the possible mobile phases that can be used for a separation. CO_2, for which detector incompatibilities are minimal, can cover a range of analyte polarity and basicity. Despite many reasons for choosing CO_2 wherever possible, other fluids such as NH_3, N_2O and Xe should not be overlooked. NH_3, which has a critical temperature of 132.4°C and a critical pressure of 111.3 atm, has given encouraging results in the separation of basic drugs, using present-day SFC equipment. However, the lack of a suitable column material poses a hardware limitation. Renewed interest is expected since NH_3 is a friendly solvent for biological samples and a good fluid to introduce into, for example, the MS detector.

Mixed mobile phases are seldom FID-compatible, except when formic acid or water is admixed with CO_2, N_2O or SF_6. With mixed mobile phases the goal of extending the SFC technique to more polar analytes ties in with the need to have universal detectors.

CONCLUDING COMMENTS

For the problems considered above it is evidently most important, through basic research, to continue to search for solutions not entailing compromises, but at the same time to use non-optimized solutions for the problems that need to be solved forthwith. Features of present-day SFC include diverse detectors, high separation power, ease of coupling techniques, and mild sample transport through extraction, fractionation, enrichment, separation and detection. Insofar as the hardware sets limits on the technique, we should have good expectations of future developments.

References

1. Lee, M.L. & Markides, K.E., eds. (1990) *Analytical Supercritical Fluid Chromatography and Extraction*, Chromatography Conferences Inc.,Provo, Utah: pp. 437-553.
2. Xie, L.Q., Markides, K.E. & Lee, M.L. (1992) *Anal. Biochem. 200*, 7-19.
3. Chang, H-C., Markides, K.E., Bradshaw, J.S. & Lee, M.L. (1989) (1989) *J. Microcol. Sep. 1*, 132-135.
4. Mourier, P., Sassiat, P., Caude, M. & Rosset, R. (1986) *J. Chromatog. 353*, 61-75.
5. Roumeliotis, P. & Unger, K.K. (1976) *J. Chromatog. 125*,115-127.
6. Payne, K.M., Tarbet, B.J., Bradshaw, J.S., Markides, K.E. & Lee, M.L. (1990) *Anal.Chem. 62*, 1379-1384.
7. Schmitz, F.P. & Gemmel, B. (1989) in *Progress in HPLC*, Vol. 4 (Yoshioka, M., Parvez,S.,Miyazaki, T. & Parvez, H., eds.), VSP, Utrecht, p. 73.
8. Kuei, J.C., Markides, K.E. & Lee, M.L. (1987) *J. High Resolu. Chromatog. Chromatog. Comm. 10*, 257-262.
9. Raynie, D.E., Payne, K.M., Markides, K.E. & Lee, M.L. (1992) *J. Chromatog.*, in press.
10. Anton, K., Bach, M. & Geiser, A. (1991) *J. Chromatog. 553*, 71-79.
11. Koski, I.J., Jansson, B.A., Markides, K.E. & Lee, M.L. (1991) *J. Pharm. Biomed. Anal. 9*, 281-290.
12. David, P.A. & Novotny, M. (1988) *J. Chromatog. 452*, 623-629.
13. Raynor, M.W., Bartle, K.D., Clifford, A.A., Chalmers, J.M., Katase, T., Rouse, C.A., Markides, K.E. & Lee, M.L. (1990) *J. Chromatog. 505*, 179-190.
14. Henion, J. (1990) in *Analytical Supercritical Fluid Chromatography and Extraction* (Lee, M.L. & Markides, K.E., eds.), Chromatography Conferences Inc., Provo, Utah, pp. 280-282.

#D-5

METHODOLOGICAL PROBLEMS IN THE ANALYSIS OF FLUOROQUINOLONES IN URINE BY ^{1}H AND ^{19}F NMR

[1]M. Tugnait, [1]F.Y. Ghauri, [1]J.K. Nicholson,
[2]K. Borner and [3]I.D. Wilson

[1]Chemistry Department, Birkbeck College,
29 Gordon Square, London WC1H OPP, U.K.

[2]Institute für Klinische Chemie und Klinische Biochemie,
Klinikum Steglitz der Freien Universität Berlin,
Hindenburgdamm 30, 1000 Berlin 45, Germany

[3]Safety of Medicines Department, ICI Pharmaceuticals,
Mereside, Alderley Park, Macclesfield SK10 4TG, U.K.

Orally administered ciprofloxacin and ofloxacin gave, for parent drug and metabolites, urinary ^{1}H- and ^{19}F-NMR spectroscopic profiles which showed very broad resonances. With EDTA added (at pH 7.4) in sufficient amount, the signals became sharpened, indicative of metal chelation which may relate to the biological action of fluoroquinolones. Evidence is presented for the view that in this metal-chelation phenomenon urinary Mg^{2+} plays a bigger role than Ca^{2+}. As exemplified by this study, NMR spectroscopy can give, besides analytical data, dynamic information on molecular interactions occurring in complex biofluids.

High-field NMR spectroscopy is continuing to excite interest as a means of studying the metabolic fate and disposition of xenobiotics, notably drugs, in animals and man [1, 2]§. In this connection we have made extensive use of the technique, particularly with ^{1}H and ^{19}F detection, and shown its value in obtaining both qualitative and quantitative information on the urinary excretion of drugs and their metabolites [1-4].

^{19}F-NMR spectroscopy has proved to be of particular interest in the study of the fate of fluorinated compounds because of the specificity provided by this nucleus for metabolite detection, there being no interference from endogenous compounds. Besides providing a specific NMR 'handle', ^{19}F-NMR is sensitive (~83% of ^{1}H) and, because of the large chemical shift range

§*Note by Ed.*- For other NMR arts., see #A-1 (Everett) in the present vol., and arts. involving the Birkbeck College laboratory in past vols.:- Vols. 16, 17 & (incl. ^{19}F-NMR) 18 [ed. by Reid *et al.*; Plenum Press].

of this nucleus in organic environments (>300 ppm), is a useful probe for structural changes due to metabolism (possibly >8 bonds distant from the site of fluorination [5, 6]). Structurally similar metabolites, e.g. transacylated glucuronides [6], can also be distinguished because of significant shift dispersion.

It was therefore of interest to us to examine the use of ^{19}F-NMR spectroscopy for the detection and analysis in urine of the fluoroquinolone antimicrobials ciprofloxacin and ofloxacin (structures shown in Figs.; HPLC studies appeared in Vol. 20 [7]) and to compare the results obtained with those for ^{1}H-NMR. Here we describe the results of these preliminary studies with particular emphasis on an unexpected methodological problem. This problem resulted from signal broadening mediated by chemical exchange and due to metal complexation by the analytes in urine, and was resolved by the use of a competing chelating agent.

EXPERIMENTAL

Drugs and samples.- Ciprofloxacin and ofloxacin were from Bayer AG, Wuppertal, and from Hoechst AG, Frankfurt, Germany. Urine samples were obtained from healthy volunteers receiving a normal therapeutic dose of the drug. Samples were stored frozen at -20° until analyzed. For analysis, samples (5-10 ml) were freeze-dried and (thereby becoming concentrated 5- to 10-fold) redissolved in 600 μl of D_2O to which 50 μl of TMS-PrS (2.5 mg/ml in D_2O) was added as internal standard.

^{1}H-NMR spectra were obtained using a Jeol GSX 500 NMR spectrometer (11.5 Tesla, 500 MHz proton frequency). Typically 256 free induction decays were collected for each sample into 35 data points following the application of 45° pulses, with a data acquisition time of 2.73 sec and a 6000 Hz spectral width. A pulse delay of 2.275 sec was used to ensure complete T_1 relaxation of the spectra. An exponential apodization corresponding to a line-broadening factor of 0.2 Hz was applied prior to FT*. A secondary gated irradiation field was applied at the water frequency in order to suppress the intense signal due to water protons. Chemical shifts were referenced to TMS-PrS* (2.5 mg/ml), and quantification performed by comparing the signal intensity of the analyte with that of the TMS-PrS.

[1]H-coupled ^{19}F-MR spectra were obtained on a Varian VXR 400 NMR spectrometer operating at 9.5 Tesla (376 MHz ^{19}F resonance frequency) using a 45° pulse (15 μsec) over a 60,000 MHz spectral width, with an acquisition time of 0.4 sec. Typically 500-600 free induction decays were collected into 8K data

*Abbreviations.- FT, Fourier transformation; TMS-PrS, Na 3-(trimethylsilyl-[2,2,3,3-^{2}H]-1-propionate (reference standard for chemical shifts; δ = 0 ppm).

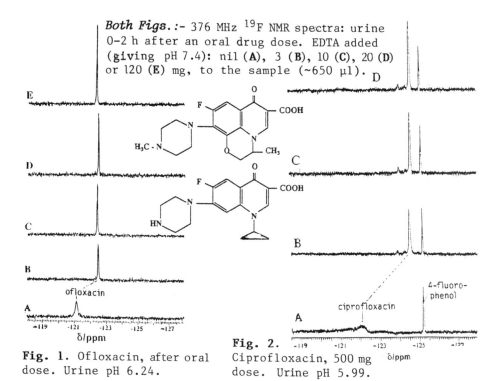

Both Figs.:- 376 MHz ^{19}F NMR spectra: urine 0-2 h after an oral drug dose. EDTA added (giving pH 7.4): nil (**A**), 3 (**B**), 10 (**C**), 20 (**D**) or 120 (**E**) mg, to the sample (~650 µl).

Fig. 1. Ofloxacin, after oral dose. Urine pH 6.24.

Fig. 2. Ciprofloxacin, 500 mg dose. Urine pH 5.99.

points with zero filling of the time domain spectrum to 64K data points prior to FT. An exponential weighting function, corresponding to a 1 Hz line-broadening, was applied prior to FT. Chemical shifts were referenced to trifluoroethanol (δ = -77 ppm).

RESULTS AND DISCUSSION

In our initial attempts to obtain ^{19}F NMR spectra of both drugs in urine, we unexpectedly encountered extremely broad peaks (>10 Hz) with poor signal-to-noise (Figs. 1A & 2A). Likewise the ^{1}H spectra exhibited broad and poorly defined peaks for the drugs when spiked into urine although peaks for endogenous substances were still sharp (Fig. 3A, ofloxacin). Such broadening is typical of the effects observed as a result of metal chelation and dynamic chemical exchange of metal ions between free-solution conditions and the complex where the rate of chemical exchange is intermediate on the NMR time-scale. Examination of the literature did indeed reveal a propensity of the quinolones to form chelates with divalent metal ions such as Ca^{2+} and Mg^{2+} (both present in mM concentrations in biofluids [8]). We therefore investigated the spectral effects of addition of EDTA, a very strong metal chelator; EDTA^{4-} has high association constants for Ca^{2+} and Mg^{2+}: log K = 11.0 and 8.7 respectively (20°, 0.1 M KNO$_3$).

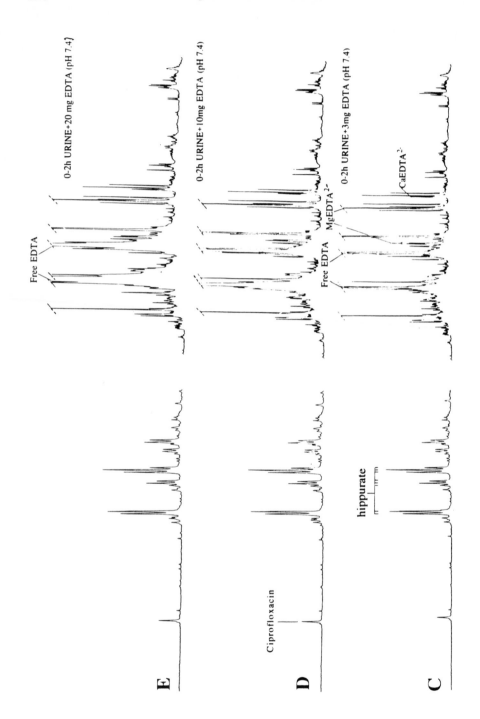

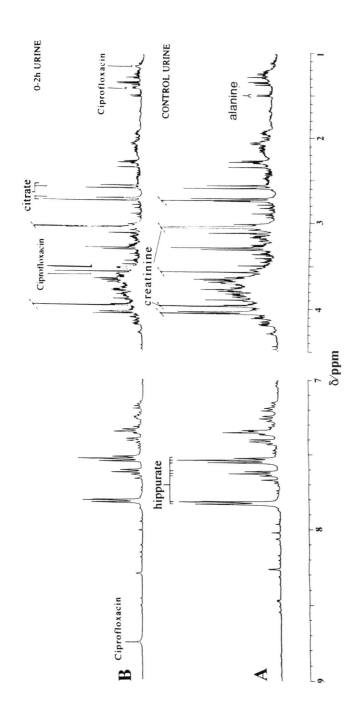

Fig. 3. 500 MHz ¹H NMR spectra from (**A**) control human urine, and (**B–E**) a 0–2 h urine from a healthy volunteer following oral administration of 500 mg of ciprofloxacin with, where indicated, addition of EDTA as in Fig. 2. Concomitantly with the sharpening by EDTA of the resonances for the drug, resonances are seen for EDTA complexes, Mg-EDTA^{2-} and Ca-EDTA^{2-}; these resonances have been previously assigned in NMR spectra of blood collected with an EDTA anticoagulant [9].

As shown in Fig.1, B-E, for the ofloxacin ^{19}F-NMR spectrum, even 3 mg EDTA significantly sharpened the signal with a shift to lower frequency. Increasing amounts of EDTA (up to 120 mg, ~185 mg/ml) gave further improvement, with progressive sharpening of the resonance lines revealing the underlying doublet structure (due to spin-spin coupling of the fluorine with an adjacent proton). The peak shape for ciprofloxacin similarly improved (Fig. 2, B-D) with EDTA up to 20 mg (~31 mg/ml). ^{1}H-NMR spectra showed similar effects on peak shape and, moreover, signals for EDTA complexes (Fig. 3, ciprofloxacin; comment in legend), as well as for organic bioconstituents.

We will be reporting the use of both ^{19}F- and ^{1}H-NMR for quantitative as well as qualitative analysis of urine from dosed subjects with EDTA added to the sample, necessarily in excess since high variability in Mg^{2+} and Ca^{2+} concentrations is to be expected. Assay down to ~100 µg/ml may be feasible.

CONCLUSIONS

From these findings it is clear that metal chelation, probably to Ca^{2+} and Mg^{2+}, was responsible for the broad, partial, resonances and dynamic chemical exchange observed in the ^{1}H- and ^{19}F-NMR spectra of ofloxacin and ciprofloxacin in urine. EDTA addition resulted in a significant improvement in spectral quality, offering the prospect of quantitative analysis. These studies provide a further demonstration of the ability of NMR to throw light on interactions between drugs, metabolites and endogenous components in biofluids.

References

1. Nicholson, J.K. & Wilson, I.D. (1987) *Prog. Drug. Res. 31*, 427-479.
2. Nicholson, J.K. & Wilson, I.D. (1989) *Prog. NMR Spectroscopy 21*, 449-501.
3. Wilson, I.D., Fromson, J., Ismail, I.M. & Nicholson, J.K. (1987) *J. Pharm. Biomed. Anal. 5*, 157-163.
4. Wade, K.E., Wilson, I.D., Troke, J.A. & Nicholson, J.K. (1990) *J. Pharm. Biomed. Anal. 8*, 401-410.
5. Everett, J.R., Tyler, R.C. & Woodnutt, G. (1989) *J. Pharm. Biomed. Anal. 7*, 379-403.
6. Ghauri, F.Y.K., Wilson, I.D. & Nicholson, J.K. (1990) in *Analysis for Drugs and Metabolites including Anti-Infective Agents* [Vol. 20, this series] (Reid, E. & Wilson, I.D., eds.), Royal Society of Chemistry, Cambridge, pp. 321-324.
7. Borner, K. (1990) *as for* 6., pp. 131-144.
8. Hoffgen, G., Borner, K., Glatzel, P.D. & Lode, H. (1985) *Eur. J. Clin. Microbiol. 4*, 345.
9. Nicholson, J.K., Buckingham M.J. & Sadler, P.J. (1983) *Biochem. J. 211*, 605-615.

#D-6

IMMUNOAFFINITY CHROMATOGRAPHY FOR THE EXTRACTION OF SALBUTAMOL FROM BIOFLUIDS PRIOR TO MS ANALYSIS

*L.F. Statham, P. Wright and C.P. Goddard

Biochemical Pharmacology Department,
Glaxo Group Research, Ware, Herts. SG12 0DP, U.K.

Immunoaffinity chromatography was investigated as a technique for extracting drugs from biological fluids. The test compound was salbutamol, but the methods could easily be adapted for any drug to which an Ab⊕ has been raised. A range of Ab purification methods, coupling reactions, affinity supports and eluting agents were investigated. A quick and efficient means of purifying antiserum prior to immobilization procedures was achieved, and a method developed for linking Ab's to an affinity support with the optimum orientaton for antigen binding. The affinity columns produced by this method were very stable, even when eluted with 100% methanol.

SPE⊕ is now widely used for sample preparation prior to analysis of drugs in biofluids. Unlike most SPE methods, which employ a stationary phase that binds the analyte through non-specific physicochemical interactions, in SPE methods based on the principle of affinity chromatography the binding is very specific. This is the case in immunoaffinity chromatography - where Ab's linked to a solid support provide the binding sites to which specific antigens, e.g. drugs, can attach and subsequently be displaced [1]. (Pre-column application to automated assay of hormonal steroids and aflatoxins was described in an earlier vol. [2].- *Ed.*)

The immobilized Ab's on an immunoaffinity stationary phase bind to analytes according to their 3-D structure. The degree of specificity with which Ab's bind to antigens can be made use of in a number of ways. Selective assays, such as GC-MS, can employ an immunoaffinity extraction method using Ab's that bind solely to the parent compound. Ab's that demonstrate some degree of cross-reactivity with metabolites can be used to develop extraction methods where the

---

*Any correspondence should be addressed to Dr I.F. Skidmore.
⊕*Abbreviations.*- Ab, antibody; BSTFA. *N,O*-bistrimethylsilyltrifluoroacetamide; EDC, ethyl dimethylaminopropyl carbodiimide; Ig, immunoglobulin; MS, mass spectrometry; SP(E), solid phase (extraction).

measurement of parent compound and metabolites is required[*].
Isolation of metabolites for structural elucidation can be
achieved in this way.   It is also possible to raise Ab's
against a structural component that is common to a group
of drugs and thus develop an affinity support that will extract
a series of structurally related compounds.

Salbutamol, a bronchodilator used in the treatment of
asthma, has now been used as a model compound to assess
the general usefulness of immunoaffinity chromatography in
the development of extraction procedures for drugs.   It has
been sought to develop practical immunoaffinity techniques
for extracting this drug from plasma prior to analysis by
an existing GC-MS procedure [3].   A suitable antiserum to
salbutamol, raised in rabbits, was available as a source
of Ab's.   Work could thus proceed on the three stages involved
in developing an immunoaffinity extraction method.-
1. Purification of Ab's, removing non-specific Ab's.
2. Coupling of Ab's to an affinity support - ideally leaving
antigen-binding properties unaffected and giving sufficient
stability to withstand the conditions used to elute the analyte
from the column.
3. Determination of the conditions for binding and eluting
the analyte.

PURIFICATION OF THE ANTIBODIES

Two different methods of Ab purification were investigated.

Affinity chromatography using salbutamol-linked gel

The salicylic acid derivative of salbutamol, which cross-
reacts with the antiserum, was linked to AH Sepharose 4B
using EDC, a water-soluble carbodiimide reagent (Fig. 1). This
salbutamol derivative (2.53 mg, 10 µmol) was dissolved in
dimethylformamide (50 µl), and water (1.5 ml) added.   This
solution was added to the gel (1 ml containing 6-10 µmol
of spacer groups to which the drug can be attached).   The
pH was adjusted to 4.5 using 0.1 M HCl.   A 10-fold excess
of EDC (19.2 mg, 100 µmol) was added, and the solution mixed
and left overnight at 4°.

The coupled gel was placed in a 6 ml disposable polystyrene
chromatography column and washed alternately with sodium citrate
buffer (pH 3.4, 0.1 M) and sodium borate buffer (pH 8.8, 0.2 M).
Finally the gel was washed with phosphate buffer (pH 7.4,
0.07 M).   Antiserum (100 µl) was diluted 1:10 in the phosphate
buffer, passed down the prepared immunoaffinity column, and
washed with 1 ml of the phosphate buffer.   The column was
finally washed with 2 ml of eluting reagent to remove the

[*]See Index entry 'Cross-reactivity' in earlier vols., especially Vol.
12 (which, as for 14, 16 & 18, was from Plenum Press; ed. E. Reid et al.).

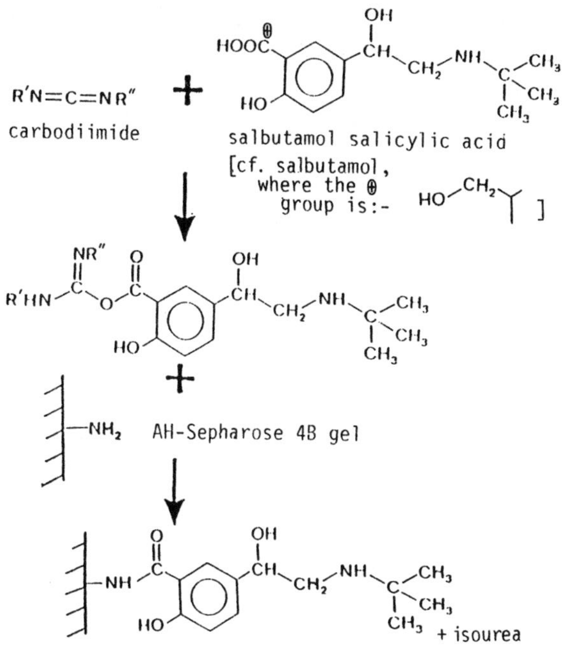

Fig. 1. The coupling of salbutamol salicylic acid to AH–Sepharose 4B gel.

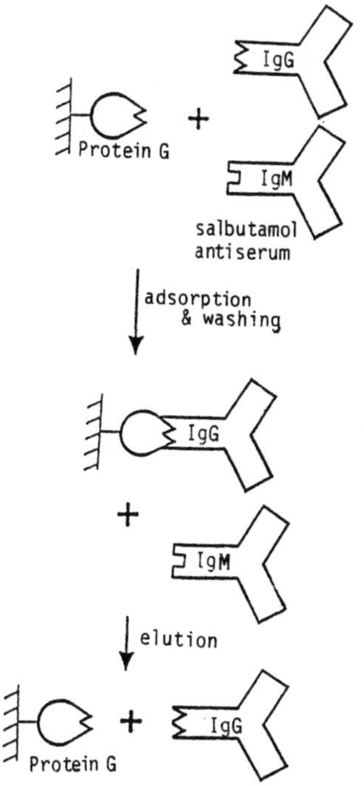

Fig. 2. Purification of salbutamol antisera using Protein G.

specifically bound Ab's. Amongst reagents tried for eluting the Ig's from the salbutamol column, the most effective was pH 3.4 citrate buffer.

Alternative approach: use of Protein G Sepharose 4 Fast Flow

This gel is produced (Pharmacia LKB) by immobilization on Sepharose 4 Fast Flow, with cyanogen bromide, of Protein G – a cell surface protein of Group G streptococci which binds in the Fc region of G class IgG's. This gel can bind, per ml, 19 mg of rabbit IgG and 18 mg of sheep IgG. The company's data sheet was followed for binding and eluting the salbutamol Ab's, as shown schematically in Fig. 2.

Into the column (as before) was placed pre-washed Protein G gel (0.5 µl). Rabbit antiserum (700 µl, ~7 mg IgG content) was diluted 1:10 with phosphate buffer and passed down the column, then re-passed to ensure that all of the IgG had bound to the Protein G. The column was washed with 7 ml of phosphate buffer and the IgG fraction eluted with 7 ml of pH 2.6 glycine-HCl buffer. The eluate was neutralized as it came off the column by mixing with pH 9 borate buffer. The eluate and washings were tested for binding to labelled salbutamol by preparing serum dilution curves.

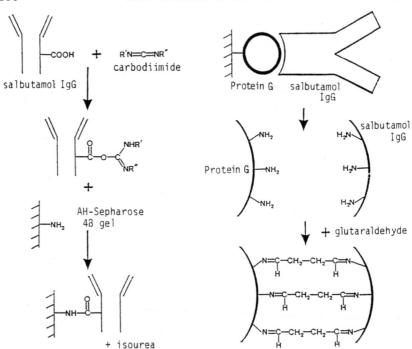

Fig. 3. The coupling of salbutamol IgG to AH-Sepharose 4B gel.

Fig. 4. Cross-linking salbutamol IgG to Protein G Sepharose 4 Fast Flow using glutaraldehyde.

The Protein G method gave much higher Ab titres in the eluate than did the immobilized salbutamol method. Ab's purified by Protein G were therefore used in all subsequent work. The glycine-HCl eluates containing the purified IgG were extensively dialyzed against water and then freeze-dried.

COUPLING OF PURIFIED ANTIBODIES TO AFFINITY GELS

Three coupling methods were investigated, using different affinity gels, to immobilize the purified Ab's.-
1. Coupling to AH-Sepharose 4B using water-soluble carbodiimides.
2. Cross-linking to Protein G Sepharose 4 using glutaraldehyde.
3. Coupling *via* the carbohydrate moiety to adipic acid gel.

The first method links Ab's to the gel by forming amide bonds between free amino and carboxylic groups, in a random manner, allowing no control over the Ab orientation (Fig. 3). The other two methods link the Ab's through the Fc region only, allowing the antigen-binding sites (FAB region) to be exposed. Glutaraldehyde, as used to minimize protein leakage from affinity supports [4], is employed as shown in Fig. 4. In the method involving the Fc-region carbohydrate moiety

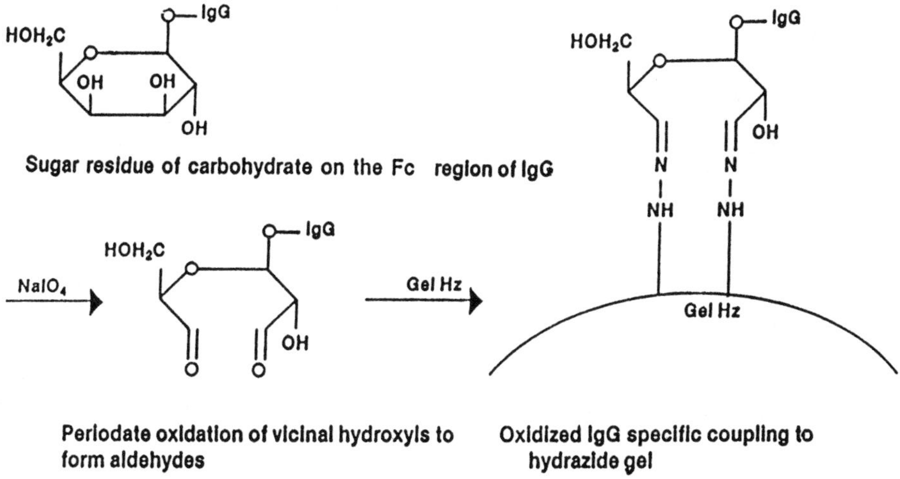

Fig. 5. Coupling Ab's to agarose adipic acid hydrazide (Gel Hz).

for linking IgG's (Fig. 5), vicinal hydroxyl groups are oxidized to aldehydes, using sodium periodate, and react with gel hydrazine groups to form stable hydrazones.

Ab coupling to AH–Sepharose 4B gel (Fig. 3) entailed adding a solution of the purified salbutamol Ab's (10 mg, 0.7 µmol) to the gel (100 µl) in a glass scintillation vial and adjusting the pH to 4.5 with 0.1 M HCl. A 5-fold molar excess of EDC was added. After mixing and standing at 4° overnight, the coupled gel was transferred to a polystyrene column and washed with pH 3.4 citrate buffer, pH 8.8 borate buffer and pH 7.4 phosphate buffer (5 ml of each).

Ab cross–linking to Protein G Sepharose 4 Fast Flow by glutaraldehyde (Fig. 4) involved allowing the salbutamol Ab's to bind to the gel before glutaraldehyde addition, to ensure that the cross-linking occurred between the Fc region of the Ab's and the binding sites of the Protein G. The gel (500 µl) was transferred to a disposable polystyrene column and washed with 5 ml of pH 7.4 phosphate buffer. A solution containing the purified Ab's from 1 ml of salbutamol antiserum was passed through the column. The gel was washed with 5 ml pH 7.4 phosphate buffer, followed by 5 ml pH 9 bicarbonate buffer, then removed. After glutaraldehyde addition (6 µl, 25% w/v; final concn. 0.03%) it was mixed at room temperature for 1 h on a rotary mixer, then transferred to a polystyrene column and washed with 5 ml pH 7.4 phosphate buffer.

Ab coupling to agarose adipic acid hydrazide (Fig. 5) entailed oxidation, by sodium periodate (1 ml, 10 mM; 0°, 20 min), of a solution containing the purified Ab's from 1 ml of salbutamol

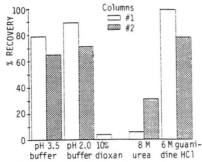

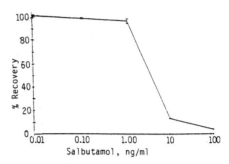

Fig. 6. Capacity of a column (of the glutaraldehyde cross-linking type), calculated from [3H]-salbutamol recovery.

Fig. 7. [3H]-salbutamol recovery from 2 columns (glutaraldehyde cross-linking type) with the different eluting agents shown.

antiserum. Aqueous glycerol (0.5 ml, 5 mM) was added to stop the reaction. The gel (Pierce; 500 µl) was added and, after mixing for 12 h at room temperature, the product was transferred to a polystyrene column and washed with pH 7.4 phosphate buffer to remove unreacted material.

DETERMINATION OF BINDING AND ELUTION CONDITIONS

Column capacities were determined with a range of concentrations of salbutamol in pH 7.4 phosphate buffer, spiked with a fixed amount of [3H]-salbutamol. After passage of 2 ml, the column was washed with phosphate buffer (2 ml) and the drug removed with eluting agent (2 ml); fractions were scintillation-counted to determine [3H]- recovery. These initial experiments indicated that the columns prepared by the glutaraldehyde method had [3H]-binding capacities (Fig. 6) greater than those prepared by the other methods; they were used exclusively in subsequent work.

To optimize the elution conditions, the following range of aqueous agents was tried:- citrate buffer pH 2; dioxan: 2, 4, 6, 8 and 10% (v/v); urea: 2, 4, 6 and 8 M; and guanidine hydrochloride: 1.5, 3.0, 4.5 and 6.0 M. As shown in Fig. 7, two of these agents gave good recoveries; but none were compatible with the GC-MS method for analyzing salbutamol. Methanol, which does not interfere, was investigated as an alternative eluting agent.

A recovery experiment was performed to determine the volume of methanol required to recover all the bound salbutamol. After passage of a solution of [3H]-salbutamol (0.5 ml) through the column, it was washed with phosphate buffer (0.5 ml). The drug was eluted with methanol, and the column immediately re-equilibrated with phosphate buffer. This procedure was repeated with varying amounts of methanol: with 0.5 ml, recovery was almost 100% (Fig. 8).

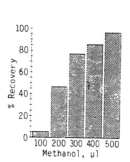

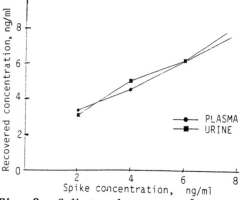

Fig. 8. ^{3}H-salbutamol
recovery from a column
of glutaraldehyde cross-
linked type, *vs.* volume
of methanol used to elute.

Fig. 9. Salbutamol recovery from
spiked plasma and urine following
immunoaffinity extraction and
determination by GC-MS.

Salbutamol in biological samples.- A final experiment
was performed, with assay by GC-MS, to extract salbutamol
from spiked plasma (0, 2, 4, 6 and 8 ng/ml) and urine. Each
plasma sample (500 µl) was diluted 1:10 with phosphate buffer
(pH 7.4, 0.07 M) and passed down an affinity column. After
washing with 2 ml phosphate buffer, the drug was eluted with
500 µl of methanol. The resulting extract was blown to
dryness under N_2; deuterated salbutamol (50 ng) was added
before drying down. Each dried extract was derivatized with
N,O-bistrimethylsilyl-trifluoroacetamide (BSTFA) and then analyzed by capillary column GC-MS (unpublished data). The results
were compared with those for freeze-dried standards. For
both plasma and urine there was good recovery with reasonable
linearity (Fig. 9) but with a basal level of ~1.5 ng/ml.

The cause of this background level of 'salbutamol' remains
to be determined, and the immunoaffinity/GC-MS method awaits
full validation against the existing GC-MS method with solid-
phase extraction, using deuterated salbutamol as an internal
standard. The data obtained do, however, already demonstrate
the potential usefulness of the technique for the quantitative
extraction of low concentrations of drugs from biofluids.

Acknowledgements

Thanks are expressed to Drs W.N. Jenner (Glaxo) and A. Tait
(Pitman Moore) for antiserum, and to Mr D. Higton (Glaxo) for
his coordination of GC-MS work.

References

1. Wofsy, L. & Burr, B. (1969) *J. Immunol. 103,* 380-382.

[Continued over

2. Farjam, A., Lingeman, H., Timmerman, P., Soldaat, A.,
   Brugman, A., van de Merbel, N., de Jong, G.J., Frei, R.W. &
   Brinkman, U.A.Th. (1990) in *Analysis for Drugs and
   Metabolites Including Anti-infective agents* [Vol. 20, this
   series] (Reid, E. & Wilson, I.D., eds.), Royal Society of
   Chemistry, Cambridge, pp. 365-370.
3. Tanner, R.J.N., Martin, L.E. & Oxford, J. (1983) *Anal. Proc.*
   20, 38-41.
4. Kowal, R. & Parson, R.G. (1980) *Anal. Biochem. 102*, 72-76.
5. O'Shannessy, D.J. & Quarles, R.H. (1985) *J. Appl. Biochem.*
   7, 347-355.

Note by Editor.- This article already had a terse outline in
Vol. 20 (p. 373), being an expansion of a contribution to the
1989 Forum. (Ref. 2. above gives Vol. 20 details.)

#ncD

NOTES and COMMENTS relating to

BIOANALYTICALLY EXPLOITABLE TECHNIQUES

Forum comments relating to the preceding main arts. and
to the 'Notes' that follow appear on pp. 333 & 334

Supplementary material contributed by Senior Editor,
complementing that on pp. 144 & 145, appears on p. 335;
some entries marked * in #ABC are also pertinent.

#ncD-1

A Note on

MASS SPECTROMETRY WITH ATMOSPHERIC PRESSURE IONIZATION

M.V. Doig

Department of Bioanalytical Sciences,
Wellcome Research Laboratories,
Beckenham, Kent BR3 3BS, U.K.

The range of bioanalytical problems that can be solved by MS has been limited by the development of suitable technology for sample introduction and ionization. When linking mass spectrometers to chromatographic assemblies there are four main factors that determine its success in bioanalysis: versatility, reliability, ease of interfacing to the column, and sensitivity. To achieve these qualities the designer of the interface must try to optimize the following: high analyte transmission to the MS source, and efficient transmission of the ionized molecules to the vacuum region for mass analysis. During the 1970's and 1980's the interface that satisfied most of these criteria was the capillary GC-MS interface using EI* or CI. The only limitation of this interface was its lack of versatility since, separation by GC dictates that the components must be both sufficiently volatile and stable to high temperatures. During the same period a variety of LC-MS interfaces were developed such as the moving belt, DLI, particle beam and thermospray ([1-3], & arts. in this and earlier vols.).

All these interfaces have been applied to bioanalysis. However, they are all interfaced to an ionization source that operates at reduced pressure and all suffer limited analyte versatility (polarity and mol. wt.) and sensitivity. The coupling of HPLC to MS *via* a source operating at atmospheric pressure was first reported by Horning & co-workers in 1974 but did not generate great interest until its commercialization during the 1980's by SCIEX in Canada (Thornhill, Ont.). The technique involved subjecting molecules to API, then introducing the ions into the high-vacuum region for mass analysis.

In general there has been exciting progress in producing ionization techniques suitable for directly coupling HPLC to MS. The innovation that appears to be the most successful

---

Abbreviations for mass spectrometry (MS) terms: AP(C)I, atmospheric pressure (chemical) ionization; EI, electron impact. *For the term* ion spray *an alternative version is* Ion Spray (CAPS.).

and versatile is API. The interfaces currently used for
API LC-MS are now reviewed, with some examples of bioanalytical
applications.

METHODS FOR API

Gas-phase ionization

The interface that is routinely most widely used for
linking HPLC to an API source that is dependent on gas-phase
ionization at atmospheric pressure is the heated pneumatic
nebulizer (Fig. 1). The HPLC effluent is promoted into the
gas/vapour phase by nebulizing gas in conjunction with heat.
This combination of heat and gas actually desolvates the
nebulized droplets to produce a dry vapour of solvent and
analyte molecules. Ionization of the solvent and air molecules,
e.g. N_2, O_2, H_2O and NO, is initiated by a corona discharge
at the discharge needle. Using a complex process of gas-phase
acid-base chemistry [4] these reactant ions then ionize the
analyte molecules. This form of ionization is more correctly
termed APCI. The main features that have led to the success
of the interface for bioanalysis is that it accepts flow
rates ranging from 0.1 to 1.5 ml/min. It can tolerate even
100% aqueous mobile phases, and the wide variety of volatile
salts, acids and bases that are commonly used to achieve
chromatographic separation by HPLC; moreover, although heat
is required, one temperature suits a range of flow rates
and mobile phase compositions. The only care that must
be taken is that additives to the LC eluent may entirely
change the CI process, leading to variations in the ionization
efficiency of the analyte molecules.

Liquid-phase ionization

An even greater increase in analyte versatility can be
obtained if the ions present or generated in the LC eluate
are directly transferred to the gas phase without using heat.
In a simple electrospray interface the eluate is introduced
into dry air or N_2 at atmospheric pressure through a metal
or glass capillary that is held at a potential of several
kV relative to the walls of the ion source [5]. This technique
operates best at flow rates of 5-10 µl/min, and the dispersion
of ions into the gas phase is difficult if the eluate contains
a high percentage of water. To improve the dispersion
into the gas phase the nebulizing action of the electric
field is assisted by a high-velocity gas flow. This concept
of pneumatically assisted electrospray is known as ion spray
[4] and can handle HPLC flow rates up to 200 µl/min. The
process by which the ions are emitted from the charged droplets
directly into the gas phase is thought to be a process of
ion evaporation. This process is described and reviewed
in [4].

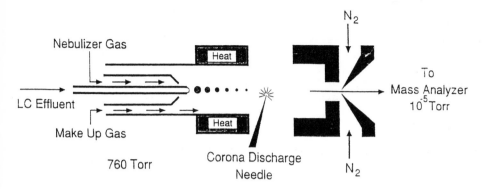

Fig. 1. Diagram of the heated pneumatic nebulizer (see text).

ILLUSTRATIVE APPLICATIONS

A.P. Bruins in Vol. 18 [6] indicated the potential of API-MS in bioanalysis. In his conclusions he indicated that API-MS would be as successful as thermospray only when it became available from traditional MS manufacturers. In 1989 SCIEX launched an LC-MS-MS computer combination called the API-III which, although dedicated to API-MS, represented long-overdue competition with the MS instrumentation and computing power available from more traditional manufacturers such as VG, Finnigan-MAT, Hewlett-Packard and Kratos. Several pharmaceutical companies including ours (Wellcome) and Pfizer have applied this technology to the measurement of drugs in serum, plasma and urine. Along with the inherent specificity of MS, reliability, ease of interfacing to the column and sensitivity have been achieved to an extent which we have not experienced since the advent of capillary GC-MS and which far exceeds the success of thermospray LC-MS. An example of such an application was recently published by Fouda and co-authors [7] for the analysis of CP-80,794, a modified peptide which inhibits renin secretion. They demonstrated intra-assay validation data over a concentration range of 0.05 to 10 ng/ml of 5.0% to 12.5% respectively for precision and 96% to 119% respectively for accuracy, with a chromatographic run time of 3.5 min.

In addition to its success for quantification, API-MS has been shown to be effective and sensitive for the structural elucidation and detection of drugs and metabolites in complex biological matrices, e.g. ionophores, quaternary ammonium salts, glucuronides and glutathione conjugates [8-10]. With other MS techniques these compounds have previously been insensitive or completely non-amenable.

Finally another work-area is mentioned to illustrate the potential and versatility of API-MS utilizing the ion-spray interface: mol. wt. information can be gained for large bio-molecules [11]. We have utilized ion-spray LC-MS-MS to look at IgG molecules (mol. wt. ~150 kDa) with a mass accuracy of ±0.01%. This has allowed us to examine alterations in the glycosylation of IgG molecules, to study drug interactions, and to optimize reagent synthesis for immunoassay development.

CONCLUSIONS

At the 1987 Forum [6], A.P. Bruins suggested that commer-cialization of API-MS would make it a successful technique for bioanalysis. During the last 3 years it has become apparent that this technique's versatility, ease of interfacing to LC, reliability and sensitivity has made a major impact on drug development within the pharmaceutical industry which far outweighs that of thermospray LC-MS. It can be used for the sensitive and specific quantitation of drugs and/or metabolites in biological fluids as well as for the structural elucidation of a wide range of biomolecules such as polypeptides, proteins and oligonucleotides.

References

1. Covey, T.R., Lee, E.D., Bruins, A.P. & Henion, J.D. (1986) *Anal. Chem. 58*, 1451A-1461A.
2. Arpino, P.J. (1989) *Mass Spectrom. Rev. 8*, 35-55.
3. Arpino, P.J. (1990) *Mass Spectrom. Rev. 9*, 631-669.
4. Bruins, A.P. (1991) *Mass Spectrom. Rev. 10*, 53-77.
5. Fenn, J.B., Mann, M., Meng, C.K., Wong, S.F. & Whitehouse, C.M. (1990) *Mass Spectrom. Rev. 9*, 37-70.
6. Bruins, A.P. (1988) in *Bioanalysis of Drugs and Metabolites, especially Anti-inflammatory and Cardiovascular* [Vol. 18, this series] (Reid, E., Robinson, J.D. & Wilson, I.D., eds.), Plenum, New York, pp. 339-351.
7. Fouda, H.G., Nocerini, M., Schneider, R. & Geduitis, C. (1990) *J. Am. Soc. Mass Spectrom, 2*, 164-167.
8. Conboy, J.J., Henion, J.D., Martin, M.W. & Zweigenbaum, J.A. (1990) *Anal. Chem. 62*, 800-807.
9. Schneider, R.P., Lynch, M.J., Ericson, J.F. & Fouda, H.G. (1991) *Anal. Chem. 63*, 1789-1794.
10. Bruins, A.P. (1991) *J. Chromatog. 554*, 39-46.
11. Smith, R.D., Loo, J.A., Edmonds, C.G., Baringa, C.J. & Udseth, H.R. (1990) *Anal. Chem. 62*, 882-899.

#ncD-2

A Note on

RESOLUTION OF COMPOUNDS ON UREA-BONDED CHIRAL HPLC COLUMNS: EFFECT OF DERIVATIZATION

[1]Robin Whelpton and [2]Dennis G. Buckley

Departments of [1]Pharmacology and [2]Chemistry,
Queen Mary and Westfield College,
Mile End Road, London E1 4NS,   U.K.

The preparation of urea-bonded chiral columns from amino-propyl columns by direct injection of isocyanates which (unlike those that feature in #ncD-3) are commercially available was described in Vol. 20 [1]. Two columns were prepared: 'phenyl', based on 1-phenylethylamine, and 'naphthyl', based on 1-naphthyl-amine. For chiral recognition there must be 3-point interaction between the phase and the analyte [2]. The most important interactions for phases of the present type are thought to be π-π bonding between aromatic rings, dipole stacking of –NH-CO-, and a steric effect dependent upon the configuration of the chiral centre. To ensure π-π bonding, DNB[*] derivatives of the analytes may be prepared, which is convenient for primary amines and amino acids. Alcohols are usually derivatized to the carbamate by 3,5-DNP isocyanate – a reagent which, unlike DNB Cl, is not commercially available. Trial of alterna-tives for alcohols was included in the derivatization studies now performed.

EXPERIMENTAL

Preparation of derivatives.- Three model compounds, 1-phenyl-ethylamine (**I**), 1-phenylethanol (**II**) and 2-phenylpropionic acid (**III**), were derivatized using standard techniques [3].   **I** was reacted with DNB Cl, 4-nitrobenzoyl chloride, 4-nitrobenzene-sulphonyl chloride or 4-nitrophenyl isocyanate.   **II** was reacted with DNBCl, 3-nitrophenyl isocyanate and 4-nitrophenyl isocyan-ate.   Amide derivatives of **III** were prepared *via* the acid chloride and subsequent reaction with 3,5-dinitroaniline or 4-nitroaniline. **III** was also reacted with 4-nitrophenylhydrazine, 2,4-DNP-hydrazine or 4-nitrophenacyl bromide. A number of racemic drugs, including amphetamine, tranylcypromine, phenylpropanola-mine, ephedrine and propranolol, were derivatized with DNB Cl.

[*]*Abbreviations.-* DNB(Cl), 3,5-dinitrobenzoyl (chloride); DNP, dinitrophenyl, with specification of isomer (2,4- or 3,5-) if not clear from the context; DNBPG, Pirkle 3,5-DNB-phenyl-glycine column; PPA, phenylpropanolamine.

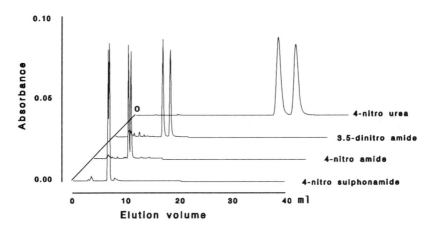

Fig. 1. Resolution of phenylethylamine derivatives on phenyl column. In Figs. 1-3 the columns (250 × 4.6 mm, prepared as in [1]), were eluted with hexane/dichloroethane/ethanol (50:10:1), 1 ml/min.

Chromatographic conditions [1] were as in the legend to Fig. 1. The derivatives were prepared in eluent (0.1 mg/ml) and 20 µl injected. Peak detection was at 254 and/or 330 nm.

RESULTS

All 4 derivatives of **I** showed at least some resolution on the phenyl column (Fig. 1). The least retained was the 4-nitro sulphonamide (k' = 1.6, α =1.06), whereas the 4-nitro urea (k' = 9.3, α = 1.14) was the most retained. Resolution was best for 3,5-dinitrobenzamide (α = 1.21) and adequate for 4-nitrobenzamide (α = 1.1). Of the 3 derivatives run on the naphthyl column (Table 1), both amides were well resolved (R_s: mononitro, 2.8; dinitro, 10.4); but the sulphonamide showed no resolution.

The DNB ester of **II** was not resolved on either column. Under the conditions used, the 4- and 3-carbamates were not completely resolved (R_s 0.9-1.3) on either column, but with naphthyl it was generally better. (Table 1: naphthyl values.)

The anilides of **III** were better resolved than those of **I** (phenyl: Fig. 2; naphthyl: Fig. 3 & Table 1). The 4-nitrophenyl and 2,4-DNP hydrazides were more strongly retained but less well resolved than the amides. The DNP hydrazide (R_s 1.33) was more resolved than the 4-nitro analogue (R_s 1.0) even though it had a lower k' on the phenyl column. The DNP hydrazide was almost completely resolved (R_s 2.0) on the naphthyl phase (Table 1 & Fig. 3).

Table 1. Chromatographic properties of derivatives of model compounds on naphthyl columns (k' = capacity factor of the enantiomer that elutes first). For **I**, DNB, etc., see title p.

Compound and derivative		k'	α	R
I,	4-nitro sulphonamide	3.3	1.00	0
amine	4-nitro amide	4.5	1.25	2.8
	DNB amide	12.0	1.92	10.4
II,	DNB ester	0.6	1.00	0
alcohol	4-nitro carbamate	2.8	1.10	1.3
III,	4-nitrophenacyl ester	0.8	1.00	0
acid	4-nitro anilide	6.8	1.46	7.9
	3,5-dinitro anilide	23.2	2.42	11.5
	2,4-DNP hydrazide	21.6	1.18	2.0

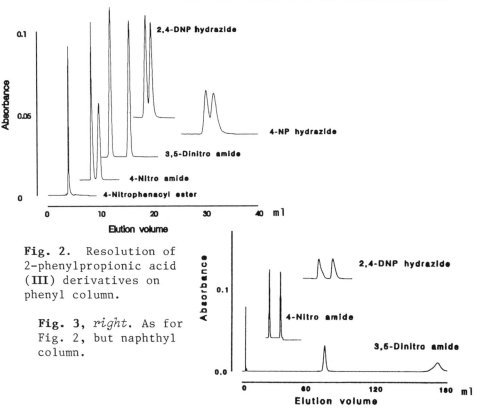

Fig. 2. Resolution of 2-phenylpropionic acid (**III**) derivatives on phenyl column.

Fig. 3, *right*. As for Fig. 2, but naphthyl column.

Drugs.- Of those tested, amphetamine, tranylcypromine, norfenfluramine and propranolol were partially resolved on the phenyl column (Table 2). PPA furnished DNB derivatives, *N*-DNB and *N,O*-diDNB. On shaking the latter in dichloromethane with aqueous NaOH, the ester was hydrolyzed to give the mono-derivative. This, but not diDNB-PPA, was resolved on

Table 2. Separation of drugs on urea-bonded columns (k' = capacity factor of the enantiomer that elutes first).

Compound	Phenyl: k'	α	Naphthyl: k'	α
amphetamine	2.1	1.07	12.3	1.23
norfenfluramine	2.3	1.08	15.5	1.00
tranylcypromine	4.5	1.05	24.8	1.11
propranolol	1.64	1.05	5.7	1.14
ephedrine	(not determined)		6.2	1.08
PPA, mono-DNB derivative	6.1	1.07	37.7	1.00
PPA, di-DNB derivative	3.9	1.00	48.4	1.03

the phenyl column. The converse was true of the naphthyl column, on which *N*-DNB had the lower k' but was not resolved (Table 2).

The enantiomers of amphetamine were well resolved on the naphthyl column (Table 2; α = 1.23). Surprisingly, however, the 3'-trifluoromethyl analogue, norfenfluramine, was not resolved on this phase; we had expected α to increase as it had for amphetamine and tranylcypromine.

DISCUSSION

Although underivatized analytes may show some resolution on the phenyl column [1, 4], it is generally agreed that chiral columns give more reliable results when the racemic analyte is derivatized. DNB Cl is used to prepare amides from amines, or 3,5–DNP isocyanate to give ureas or carbamates with amines or alcohols respectively. The latter reagent is not commercially available but has been prepared by treating 3,5-dinitroaniline with phosgene [5] or thermally decomposing 3,5-DNB azide [6]. We confirmed that 3,5-dinitro derivatives, particularly amides, are well resolved on these phase types. However, our results also show that other derivatives may be used, e.g. 4-nitroaryl compounds may be sufficiently resolved to allow use of more readily available reagents.

It would appear that these phases do not resolve ester derivatives, even the phenacyl ester that has two carbonyl groups capable of hydrogen bonding. That the anilides of **III** were better resolved than the corresponding amides of **I** shows the importance of the order of the -NH-CO- relative to the chiral centre, and may indicate that amide-linked columns would perform better. As far as we are aware it is the first time that hydrazide derivatization has been used to resolve chiral acids. Relatively high temperatures are required for this single-step reaction of acids with

2,4-DNP-hydrazine; hence the reaction is probably more suited to methyl esters, which should react more readily than the acids.   Phenylhydrazine would be appropriate for resolutions on columns with π-acid moieties, e.g. a DNBPG column.

The results with PPA are difficult to explain, particularly as on Spherisorb 5 Chiral 1 (from Phase Separations) both derivatives were resolved [7], the mono-DNB eluting last.   On the other hand, amphetamine was not resolved on this column. As the only difference between the commercial column and our phenyl column is that the former was prepared by bonding the appropriate triethoxysilane to native silica, success or failure may be due to subtle secondary interactions [cf. I.D. Wilson in Vol. 18, # NC(E)-9.- *Ed.*].

CONCLUSIONS

Our results show that derivatives, other than those based on DNB, may be used with these chiral columns, and that although the phenyl and naphthyl columns in some respects are similar, one or the other may prove to be the more suitable for a particular analyte.

Acknowledgement

We are indebted to Mr Brian King of Phase Separations Ltd. for provision of the aminopropyl columns and the Spherisorb 5 Chiral 1 column.

References

1. Whelpton, R. & Buckley, D.G. (1990) in *Analysis for Drugs and Metabolites including Anti-infective Agents* [Vol. 20, this series] (Reid, E. & I.D. Wilson, eds.), Royal Society of Chemistry, Cambridge, pp. 337-340.
2. Dalgliesh, C.E. (1952) *J. Chem. Soc. 137*, 3940-3943.
3. Smith, F.J. & Jones, E. (1948) *A Scheme of Qualitative Organic Analysis*, Blackie, London, 320 pp.
4. Whelpton, R., Jonas, G. & Buckley, D.G. (1988) *J. Chromatog. 426*, 222-228.
5. Oi, N. & Kitahara, H. (1983) *J. Chromatog. 265*, 117-120.
6. Pirkle, W.H., Mahler, G. & Hyun, M.H. (1986) *J. Liq. Chromatog. 9*, 443-453.
7. Whelpton, R. & Buckley, D.G. (1992) *Anal. Proc. 29*, in press.

#ncD-3

A Note on

DIRECT INJECTION OF CHIRAL ISOCYANATES, NOT MARKETED, ONTO AMINOPROPYL COLUMNS TO RENDER THEM CHIRAL

[1]Dennis G. Buckley and [2]Robin Whelpton

Departments of [1]Chemistry and [2]Pharmacology,
Queen Mary and Westfield College,
Mile End Road, London El 4NS, U.K.

With chiral columns prepared merely by injecting commercially available chiral isocyanates onto aminopropyl columns [1], we have demonstrated the resolution of racemic amines, alcohols and acids as amide, urea, carbamate or hydrazide derivatives containing mono- or di-nitro substituted aromatic rings ([1], & #ncD-2, this vol.). This accords with the hypothesis that dipole stacking of -HN-CO- and π-π bonding between aromatic rings are important interactions for chiral recognition. Paradoxically, compounds which do not conform to this model may also be resolved, e.g. TFAE[*]. As chiral phenothiazines have been partially resolved on the phenethyl phase [2], investigations are continuing with phases of this type. Two new phases have been prepared by the introduction of (i) a nitro group in the 4 position of the phenyl ring and (ii) a methylene group between the chiral carbon and the aromatic ring. The required isocyanates, **viz.** R-CH(CH$_3$)-NCO where R = 4-nitrophenyl (**NP**) or benzyl, were not commercially available; hence we prepared them from the appropriate amine.

EXPERIMENTAL

NP column.- To make the (R)-N-1-(4-**NP**)ethyl-N'-propyl-urea column, 1-(4-propyl)ethyl isocyanate was prepared as described below, and put direct onto the column.

NP + A column (**A** signifies inadvertent exposure to acetone; see below).- This was prepared from loose packing for comparison with **NP** and checking the completeness of the reaction with amino groups. (R)-1-(4-**NP**)ethylamine hydrochloride (1 g, 5 mmol)

[*]*Abbreviations, with Ed.'s variants of authors' designations* [ ], *for* columns: **Ph** = phenyl [HM4], **NP** = 4-nitrophenyl [HM7], **NP + A** = NP + acetone [HM10], **Bz** = benzyl [HM11]. DCE, 1,2-dichloroethane; DCM, dichloromethane; DNB, 3,5-dinitrobenzoyl; IPA, isopropyl alcohol; TFAE, 2',2',2'-trifluoro-1-(9-anthryl)-ethanol ('Pirkle test compound').

was suspended in dry toluene (10 ml) and triphosgene (0.9 g, 3 mmol) added.   The mixture was refluxed for 7 h, further triphosgene added (0.2 g) and the reflux continued for 15 h. The reaction mixture was filtered to remove an undesired 'urea' by-product, and the isocyanate solution in toluene added to loose column packing (Spherisorb 5 μm aminopropyl silica, 4 g) so that any 'urea' could be washed out.   The mixture was stirred for 15 min (room temp.) and refluxed for 4 h; after cooling, the silica was filtered off and washed with acetone (inadvertently; see later) and methanol to remove 'urea' by-product.   Elemental analysis showed that 79% of the available NH_2 groups had reacted with the isocyanate. Slurry-loading in methanol was then performed to furnish the column (250 x 4.6 mm).

Bz column.- For the (S)-N-1-methyl-2-phenylethyl-N'-propyl urea column, (+)-amphetamine sulphate (1.86 g, 10 mmol amphet-amine) was added to 1 M NaOH (20 ml) and extracted with DCM (2 x 50 ml) which on evaporation yielded 1.3 g amphetamine base. The base was dissolved in toluene (20 ml) and bubbled with HCl gas. Triphosgene (2 g, 6.7 mmol) was added and the solution refluxed for 20 h to give the isocyanate (IR peak 2263 cm^{-1}). The solvent was evaporated under vacuum and the residue injected (4 x 0.7 ml) onto a Spherisorb 5 NH_2 column (250 x 4.6 mm), to which DCM was applied at 0.5 ml/min.

Evaluation of columns.- The columns were tested using 1-phenylethylamine and 2-phenylpropionic acid as their DNB-amides with n-hexane/DCE/ethanol (50:10:1 by vol.) eluent, and TFAE with 2% (v/v) IPA in hexane as eluent.

RESULTS AND DISCUSSION

TFAE was well resolved on the phenyl column **(Ph)**, but the new columns gave disappointing results (Fig. 1).   On the nitrophenyl columns, **NP** and **NP + A**, the enantiomers chromato-graphed as a single, reasonably symmetrical peak. The retention on **NP** was slightly greater than on **NP + A**, but we do not know whether this reflects greater loading of the chiral phase or, as discussed below, treatment of **NP + A** with acetone. On the **Bz** column the retention was similar to the mean retention on **Ph** but there was no resolution, only a slightly distorted peak (Fig. 1).

1-Phenylethylamine, as the 3,5-dinitrobenzamide, was partially resolved on all three of the new columns (Fig. 2A; R_S 0.8-1.3). The retention times on **NP** and especially on **NP + A** exceeded those on **Bz**. The overall retention on **Bz** was similar to that on **Ph** but resolution was reduced.

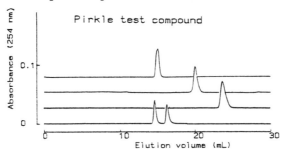

Fig. 1. TFAE run on 4 columns (see text; 1 ml/min). Columns, *from bottom:* **Ph** (phenyl); **NP** and **NP + A** (nitrophenyl); **Bz** (benzyl).

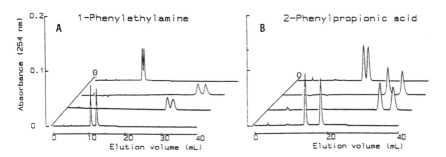

Fig. 2. Resolution of derivatives of (**A**) phenylethylamine, (**B**) 2-phenylpropionic acid (see text; 1 ml/min). Columns as in Fig. 1.

2-Phenylpropionic acid, as its 3,5-dinitroaniline derivative, showed acceptable resolution on all the columns (Fig. 2B). With this compound, **NP** and **NP + A** were very similar in respect of retention and resolution, but the resolution on **Bz** was inferior.

Use of isocyanates

Amines may be converted to isocyanates by the action of phosgene on the amine hydrochloride. Triphosgene, a relatively new reagent, is a solid and much easier to handle than phosgene gas. Although 1 mol of triphosgene is claimed to be equivalent to 3 mol of phosgene, we found that ~0.66 rather than 0.33 mol was required per mol of amine, presumably because of loss of phosgene as gas along with the HCl by-product.

A problem with isocyanates is that they are prone to hydrolysis, yielding the amine which will readily react with remaining isocyanate to form a urea. This was particularly true of the 4-nitrophenylethyl isocyanate, and the 'urea' had to be filtered from the reaction mixture and residual amounts washed from the loose aminopropyl silica before making up the column. The first wash was with acetone (in error as the wrong wash-bottle was used!) rather than methanol; hence the residual NH_2 groups reacted to give the imine (Schiff's base) as confirmed by elemental analysis. Initial results with **NP** were disappointing; but, suspecting that the problem might be 'urea' contamination, the column was washed with methanol until no more UV-absorbing material eluted. Because of the treatment with acetone it is difficult to draw conclusions about differences between **NP** and **NP + A** However, the similarity in retention and resolution with the propionic acid derivative may indicate similarity in the loading of the chiral stationary phase. The longer retention of TFAE on **NP** could be due to hydrogen bonding.

Chiral effectiveness of the stationary phase

Introducing the 4-nitro group increased the retention of all the test compounds but decreased resolution, that of TFAE being completely lost. We had hoped that increased π-bonding would increase resolution. Inserting a methylene group between the chiral centre and the phenyl ring hardly affected k' but reduced chiral recognition for these analytes.

Chiral phases of this type may be obtained *via* tri-ethoxy-silyl urea derivatives [3] and refluxing with native silica. However, we have chosen a route that the average analyst, equipped with an efficient fume cupboard, should be able to follow. Direct injection of isocyanate avoids the need for column-packing apparatus.

CONCLUDING COMMENTS

Use of triphosgene is a convenient way of producing isocyanates of sufficient purity to serve for making chiral columns with minimal 'work-up'. Whereas TFAE enantiomers did not resolve, derivatives of other model compounds did resolve. This finding shows how important it is to use a range of test compounds to evaluate chiral columns.

Acknowledgement

We thank Mr Brian King, Phase Separations Ltd., for provision of the aminopropyl columns.

References

1. Whelpton, R. & Buckley, D.G. (1990) in *Analysis for Drugs and Metabolites including Anti-infective Agents* [Vol. 20, this series] (Reid, E. & Wilson, I.D., eds.), Royal Society of Chemistry, Cambridge, pp. 337-340.
2. Whelpton, R., Jonas, G. & Buckley, D.G. (1988) *J. Chromatog. 426*, 222-228.
3. Pirkle, W.H. & Hyun, M.H. (1985) *J. Chromatog. 322*, 6295-6307.

#ncD-4

A Note on

THE ISOLATED ASPEC APPLIED TO ASSAY OF A NOOTROPIC AGENT, COMPARED WITH MANUAL OPERATION

Ann Strutton

Bioanalytical Department,
Hoechst Pharmaceutical Research Laboratories,
Walton Manor, Milton Keynes, MK7 7AJ, U.K.

Bioanalysis typically comprises three main activities: sample preparation, the analysis itself, and finally the reporting of results. Demands for increasing output coupled with top-quality have led to the application of automation and computerization to much of this field. In chromatography the analysis has been automated by means of autoinjectors, and reporting has been made faster and more accurate by the application of direct data transfer. However, sample preparation is still often performed manually although, in view of the repetitive and precise actions typically involved, automation would seem apt.

C88 3870

CAS 493 (parent drug) has CHO instead of COOH.

The automation available has evolved towards either total versatility (robotics) or dedicated systems such as the Gilson ASPEC[*] - a particularly economical means of automation. For the numerous **samples** of plasma and urine analyzed each year in our laboratory, predominantly we use liquid-liquid techniques. As the ASPEC is tailored for use with extraction cartridges, the following method involving SP[*] assay was chosen to be automated. The manual assay had already been developed and was in successful routine use. Here we compare the performances of the manual and ASPEC extractions.

EXPERIMENTAL

The compound, C88 3870 (formula above), is an important plasma metabolite of CAS 493, a novel nootropic agent under development by Cassella AG. As the compound contains both acidic and basic functional groups, SPE was the method of choice. A sensitive and specific assay entailing GC-NPD had been achieved.

[*]ASPEC denotes Automated Sample Preparation using Extraction Columns; the system is described and illustrated (J.B. Lecaillon *et al.*) in Vol. 20 of this series (eds. and publisher as for the present vol.). NPD, nitrogen-phosphorus detector; SP(E), solid-phase (extraction).

The method for extracting C88 3870 from plasma is as follows:

1. Internal standard is added to each plasma sample (1 ml) and mixed. Sorensen's citrate buffer (pH 3; 1 ml) is then added. The samples are centrifuged for 5 min to sediment any solid matter.

2. Bond Elut cartridges (100 mg, C-18) are pre-washed with chloroform and conditioned with methanol and citrate buffer (each 1 ml).

3. The supernatants are loaded onto the cartridges. These are washed with citrate buffer and toluene (1 ml of each).

4. The cartridges are transferred to tapered tubes. Analytes are eluted by adding chloroform (1 ml) and centrifuging for 1 min at 2000 rpm; this process is repeated once.

5. The eluted cartridges are discarded, and the eluates are dried down at 65° under a gentle N_2 stream.

6. At room temperature, the analytes are derivatized for 10 min with diazomethane in diethyl ether, which is then removed with a gentle N_2 stream.

7. The residue is redissolved in butyl acetate (50 µl) and transferred to autosampler vials. Aliquots (5 µl) are then analyzed by GC.

GC is performed on a Hewlett-Packard 5880 instrument, as follows:- NPD detector; capillary inlet system; 25 m × 0.31 mm fused-silica column coated with cross-linked 5% phenylmethyl-silicone (HP Ultra No. 2); He as carrier gas, with 12.5 psig inlet pressure giving a column flow of ~1.3 ml/min and linear velocity ~28 cm/min; injector port at 325°, oven at 240° and detector at 300°. Under these conditions the R_T's for C88 3870 and internal standard are typically 4.5 and 6.5 min; the limit of quantification is 20 ng/ml. Integration is performed on a Hewlett-Packard 3350 data-capture system using the internal standard mode.

The manual extraction is performed on two Spe-ed Mate 30 vacuum manifolds connected in tandem so that 60 samples can be extracted simultaneously.

ASPEC version of the extraction.- The manual method could not be completely transferred to the ASPEC because of the centrifugation in step 1. Moreover, at step 4 it was more convenient to elute the compounds into tapered tubes on the centrifuge so that the following step, a derivatization, could be performed in the same tubes. Accordingly, it was steps 2 and 3 that were automated; they represent ~40% of the time taken in performing the extraction manually.

The ASPEC is a cartesian robotic arm based around a Gilson 201 autosampler. The instrument is installed with a simple form of BASIC which allows operation through provided

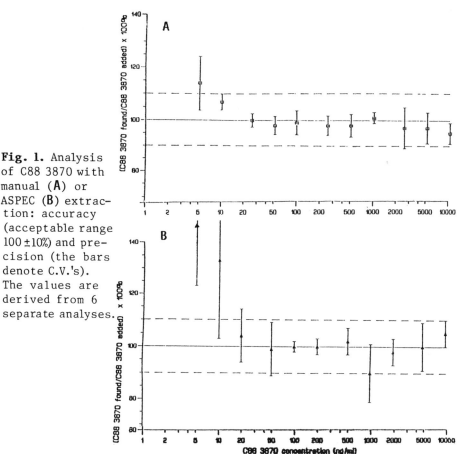

Fig. 1. Analysis of C88 3870 with manual (**A**) or ASPEC (**B**) extraction: accuracy (acceptable range 100 ±10%) and precision (the bars denote C.V.'s). The values are derived from 6 separate analyses.

programs which require answers to a set of questions; alternatively, a specific program can be written for each analysis. Through these programs parameters such as dispensing speeds and air volumes used to dry the cartridges may be defined. The instrument may be used either batchwise, each extraction stage being performed on multiple samples, or sequentially, each sample extraction being performed entirely separately.

RESULTS AND DISCUSSION

From the values shown in Fig. 1 it is evident that, across the entire concentration range of the assay, the ASPEC method was considerably worse than the manual method in accuracy (recovery) and precision. In an attempt to find the cause, firstly the volumetric accuracy of the ASPEC was examined. When pipetting known volumes the instrument was accurate to within ±2% and precise to within ±1%. Since this was the only real area in which the extractions differed, it became clear that the problem must lie with the extraction itself.

Analysis of blank samples showed that background non-analyte peaks were responsible for considerable variation at low C88 3870 concentrations. It emerged after investigation of all solvents and other materials that the interfering material arose directly from the extraction cartridges and was derivatized by the diazomethane. As it happened, this interference arose in a batch of cartridges which we were using for the ASPEC validation but which had not been used for the manual validation.

The second problem, due to analyte instability, appeared during the validation runs. At least some of these were of similar size to real routine runs (~60 samples). When running these large batches, it became clear that accuracy was falling during the sample preparation. Having established that the problem lay in instability, it was realised that this could be occurring in two stages: on the ASPEC (before or after loading onto the cartridge) or on the autosampler. Since in the sequential mode the ASPEC was taking ~11 min to process each sample, in a 60-sample run the last sample would have stood for 11 h before being loaded onto the cartridge and the first sample would have been dry on the cartridge for a similar time while the rest were being processed. C88 3870 was found to be stable when stored dry on the cartridge and in the autosampler, and was not light-sensitive; it was stable at pH 7, but turned out to be unstable in the pH 3 buffer.

This problem could have been avoided in a few ways. The extraction could have been split so that the ASPEC was conditioning all the cartridges while the initial, manual stage was being performed; this would have minimized the time spent in pH 3 buffer. Also, the use of cooled sample racks would have helped. Minimizing the extraction time by batchwise rather than sequential operation effectively halved the extraction time.

Concluding comments.- In respect of mechanical sample processing the ASPEC performed well. This was not an ideal assay to transfer to the ASPEC because the centrifugation step could not be automated; also, the final analysis being by GC, the autosampler capability of the ASPEC could not be used. Moreover, accuracy and precision were disappointing compared with the manual method, with two causes: variable backgrounds arising from the cartridges, and analyte instability. Instability was exacerbated by the ASPEC's slow processing time, ~10 samples/h, and had not arisen significantly in the manual assay. Evidently, when transferring an assay to an automatic system, it is important to check for analyte stability at every stage since automated assays are not necessarily faster.

ㅐncD-5

A Note on

SCINTILLATION PROXIMITY ASSAY - THE WAY TO
FULLY AUTOMATED RADIOIMMUNOASSAY

P. Linacre and S.E. Morris

Glaxo Group Research Ltd., Ware, Herts. SG12 0DP, U.K.

Traditional RIA* methodology is relatively time-consuming, labour-intensive and difficult to automate because it is essential to physically separate the Ab-bound and free radioligand at equilibrium. Therefore an assay system without the requirement for a phase-separation step would be a distinct advantage.

SPA, a new development, is a variant of RIA which enables detection of Ab-bound ^3H or ^{125}I (Auger electrons) in the presence of free radiolabel and without using conventional scintillant [1]. It therefore requires no separation step and has the potential for total automation. In SPA the separation stage is replaced by addition of an SPA reagent (Amersham Int[1].), comprising beads (fluomicrospheres) made from yttrium silicate which have been impregnated or coated with a material capable of fluorescence (fluor). The beads are covalently linked to second Ab's raised to the IgG's of the species used for primary Ab production.

In an SPA, at equilibrium, all first Ab would be immobilized by the second Ab (in excess) onto the fluomicrospheres (Fig. 1). Radiolabelled ligand that is bound to the primary Ab will then be in close enough proximity to the fluomicrospheres to allow the emitted β-particles to bombard the fluor. This results in the production of light which can be detected in a scintillation counter. Radioligand not bound to the Ab is located too far away from the fluomicrospheres to interact with the fluor, and any radiation emitted is lost to the aqueous environment.

Ranitidine assay is at present done by RIA using a ^3H radiolabel and a charcoal separation step. For our aim of converting the RIA method to the SPA format, the separation step was replaced by the addition of anti-sheep SPA reagent (the primary Ab having been raised in sheep). The assay was optimized using mini-vials which, after capping, were

*Abbreviations.- Ab, antibody; RIA, radioimmunoassay; SPA, scintillation proximity assay.

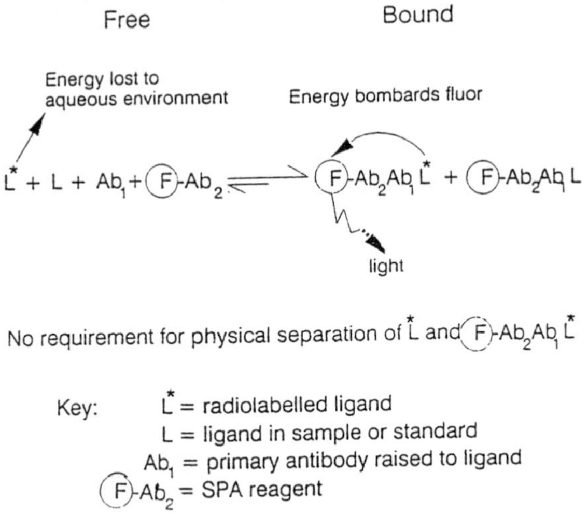
Free
Bound
Energy lost to aqueous environment
Energy bombards fluor
L* + L + Ab1 + (F)-Ab2 ⇌ (F)-Ab2Ab1L* + (F)-Ab2Ab1L
light
No requirement for physical separation of L* and (F)-Ab2Ab1L*
Key: L* = radiolabelled ligand
L = ligand in sample or standard
Ab1 = primary antibody raised to ligand
(F)-Ab2 = SPA reagent

Fig. 1. Principles of SPA.

incubated overnight (>15 h) on an orbital shaker at room temperature, then transferred to the counter. The optimal volume of SPA reagent was 200 µl per mini-vial (total reaction vol. 0.6 ml). The final assay protocol was much less labour-intensive than the RIA, needed only 30 min as compared with 120 min, and could be amenable to full automation using a Tecan Robotic Sample Processor.

Intra-assay comparisons between SPA and RIA (Table 1) showed that the two methods were equivalent. Both were precise (C.V. ≤6.5%) and accurate (bias ≤14%) over the working range. Work now in hand to improve the SPA involves scaling-down, replacing mini-vials by 200-300 µl microtitre plates. This will allow use of a Wallac Microbeta 1450 (6 detectors) to scintillation-count samples directly on the plates, improving throughput.

Table 1. Intra-assay accuracy and precision (n = 6). The analyte was spiked into diluent/bovine serum albumin, not plasma.

	RIA					SPA				
Nominal Conc.s (ng/ml)	2.0	5.0	10.0	20.0	50.0	2.0	5.0	10.0	20.0	50.0
Observed means (ng/ml)	1.7	5.2	10.1	20.2	51.0	2.3	5.1	9.9	19.7	50.8
% CV	4.2	6.0	3.4	4.1	1.5	5.9	6.5	3.0	1.2	4.0
% Bias	−14.0	5.0	1.0	1.0	2.0	12.0	2.0	−1.0	−2.0	2.0

Reference

1. Bosworth, N. & Towers, P. (1989) *Nature 341*, 167-168.

#ncD-6

A Note on

INTERPRETATION OF GC-MS STUDIES WITH STABLE ISOTOPES, IN RELATION TO PHARMACOKINETIC AND DRUG-METABOLIZING TESTS

[1]Jean-François Sabot, [2]Bernard Francois and [1]Henri Pinatel

[1]Laboratoire de Chimie Analytique II, Faculté de Pharmacie,
    Avenue Rockefeller, 69373 Lyon, France
[2]Service de Néphrologie, Hôpital Jules Courmont, 69310, Lyon

PHARMACOKINETICS

The need for stable isotope studies in pharmaceutical investigations is widely recognized, and is spurred by the need to accurately determine the conversion and breakdown of drugs. Determinations of metabolic clearance rates and drug quantification are important aspects of labelled drug use. Let us imagine a situation in which a drug (**D**) and in addition a tracer dose of the labelled homologue (**LH**) are given orally. Molecular-ion data from the mass spectrum of **D** (3 peaks) and **LH** (5 peaks) are shown in Fig. 1.

Specific and reliable assessment of circulating **D** levels is performed by collection of blood and extraction of **D**, **LH** and metabolites, prior to measurement by SIM[*] using GC-MS. Ionization of the molecules may be performed by electron impact, and and the ions of **D** and **LH** are selectively monitored. Diluted samples are extracted, derivatized if necessary, and subjected to GC-MS-computer analysis. The molecular ions relate to the mol. wts. of the parent compounds. The peak-area ratio **R**, for $(m/z_D)/(m/z_{LH})$, is calculated and the corresponding molar ratios are determined from standard curves prepared by mixing weighed amounts of natural **D** and of **LH**. A convenient procedure for calibration when using a stable-labelled form of the analyte is to use the equation for isotope dilution analysis. Pickup & McPherson [1] and Colby & McCaman [2] have considered the effects of isotopic contributions in both **D** and **LH**. Previously [3, 4] we discussed the choice of the best equation and the nature of the calibration curve. In pharmacological studies, two cases are encountered.-

#Where the amount of homologue (**LH**) used is *large* relative to **D**, an appropriate equation [1] is:

---

[*]*Abbreviations*.- SIM, selected ion monitoring; see text for AP, **D** (drug, *not* deuterium), **LH**, **R** and (coefficients) P, Q. In the example, **LH** has 2 ^{13}C atoms, i.e. it is $^{13}C_2$-**D**.

Fig. 1. MS molecular ions: peak intensities for **D** (*a, b & c* and **LH** (*d-h*). Coefficients are defined by equations [1]:
$P_D = a/(a+b+c)$
$P_{LH} = c/(a+b+c)$
$Q_D = d/(d+e+f+g+h)$
$Q_{LH} = f/(d+e+f+g+h)$.

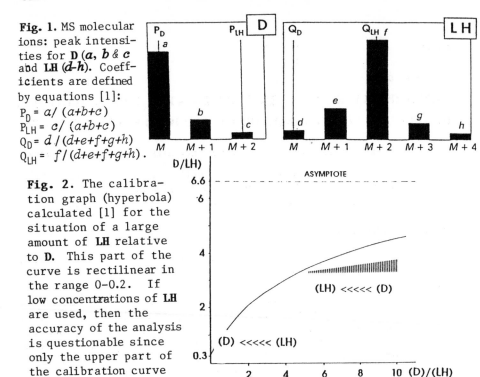

Fig. 2. The calibration graph (hyperbola) calculated [1] for the situation of a large amount of **LH** relative to **D.** This part of the curve is rectilinear in the range 0-0.2. If low concentrations of **LH** are used, then the accuracy of the analysis is questionable since only the upper part of the calibration curve is used.

$$R_{D/LH} = [(D/LH).P_D.F + Q_D.E]/[(D/LH).P_{LH}.F + Q_{LH}.E] \quad \text{(Eqn. 1)}$$

The calibration curve is shown in Fig. 2. (For E & F, see below.)

#Where **LH** is used at *low* concentrations, typically the response is non-linear. It is always necessary to draw experimental curves (hyperbola), but in this case equation #1 [1] may be modified to obtain the best reproducibility. We propose the plot of **R** (quantity of **LH**/quantity of **D**) *vs.* **LH/D,** the equation being:

$$R_{LH/D} = [Q_{LH}.E.(LH/D) + P_{LH}.F)/(Q_D.E.(LH/D) + P_D.F] \quad \text{(Eqn. 2)}$$

where $R_{LH/D}$ is the ratio of the measured ion intensity of **LH** to that of **D** and Q_{LH}, Q_D, P_{LH} and P_D are the 4 coefficients obtained by monitoring pure **D** and pure **LH**; LH/D is the mass ratio of the amount of **LH** to natural **D** (their molecular masses being F and E respectively). According to the amount of **LH** relative to that of **D,** equation #1 or #2 must be used to obtain a rectilinear calibration graph.

The first situation is encountered in 'reference' or 'absolute' methods to assay **D.** The second situation applies in determining the rate of disappearance of **D,** when **D** is given in addition to a tracer dose of **LH,** whose fraction may be <10-20% (w/w).

Fig. 3. Serum concentrations of AP in a volunteer after oral dosage of AP daily, with AP* on day 4 (see text).

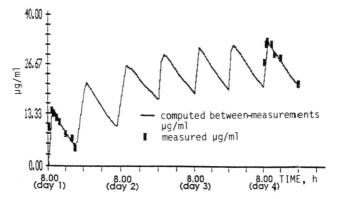

computed between-measurements μg/ml
measured μg/ml

40.00
26.67
13.33
0.00
μg/ml

8.00 (day 1)   8.00 (day 2)   8.00 (day 3)   8.00 (day 4)   TIME, h

ANTIPYRINE TEST

The antipyrine (AP; synonym: phenazone) test is used by clinical pharmacologists to determine the effects of environmental factors, especialy on hepatic drug-metabolizing enzymes *in vivo* [5, 6]. The present study [7] concerned the influence of renal insufficiency on the disposal and clearance of AP, tested with $^{13}C_1$-AP (AP*) after a steady-state plasma level of AP had been attained. To achieve this, AP was administered orally in 8 mg/kg doses at 8 a.m. and 8 p.m. for three consecutive days (Fig. 3). On the fourth day AP was given accompanied by a tracer dose of AP* (20% of AP, w/w). Whole blood samples were collected on day 1 and day 4 before and at several times after the morning dose. GC-MS with SIM allowed a number of different compounds to be quantified in a single analysis, as done here for AP, AP* and (added to the sample) internal standard, without derivatization. Since the concentration of AP* was low relative to AP, equation #2 above was applicable, the term **D/LH** being the AP*/AP concentration ratio to be determined (Fig. 4). Comparison with the peak area of internal standard as added to the plasma sample enabled the concentration of AP to be calculated and hence, from $R_{LH/D}$, the AP* concentration in the sample could be calculated. However, in practice the AP* concentration was calculated from the ratio $R_{LH/D}$ as observed, using the calibration curve obtained as above. Usually least-squares regression analysis is used to describe the relationship between signal and concentration. Here, the true relationship between $R_{LH/D}$ and **LH/D** is given by part of a half-hyperbola. However, for the low concentrations of AP* relative to AP (in the range 0-0.3) a straight line can be drawn, and the calibration curve is defined by its slope:

$$\Delta R / \Delta (\mathbf{LH/D}) = E \cdot (Q_{LH} \cdot P_D - Q_D \cdot P_{LH}) / F \cdot (P_D)^2$$

and by its y-intercept: $R_0 = (P_{LH}/P_D)$.

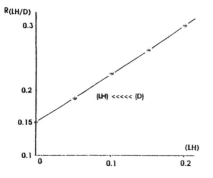

Fig. 4. Modified calibration curve, corresponding to $R_{LH/D}$ *vs.* **LH/D**. When low concentrations of **LH** are used, only the **LH/D** range 0-0.2 is considered; the calibration curve is then a straight line.

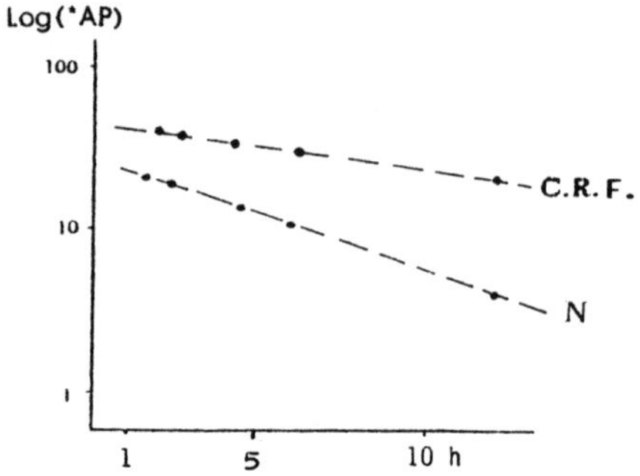

Fig. 5. Elimination phase of $^{13}C_1$-AP at day 4. AP half-life was in the range 5-12 h in normal subjects (**N**) and 14-40 h in patients with severe chronic renal failure (**C.R.F.**).

The rate of disappearance of AP* from plasma was determined by this approach. The half-lives were calculated from the linear part of the decrease of AP* (log.AP* = f(time); Fig. 5). It has been shown that the degree of chronic renal failure, as defined by serum creatinine levels and creatinine clearance values, influences the AP half-life.

In conclusion, the analytical method associated with the calculation procedure [7] shows adequate linearity, sensitivity and specificity.

References

1. Pickup, J.F. & McPherson, K. (1976) *Anal. Chem. 48*, 1885–1890.
2. Colby, B.N. & McCaman, M.W. (1979) *Biomed. Mass Spectrom. 6*, 225–230
3. Sabot, J.F., Derauz, D., Dechaud, H., Bernard, P. & Pinatel, H. (1985) *J. Chromatog. 339*, 233-242.
4. Sabot, J.F., Ribon, B., Kouadio, L.P.K., Pinatel, H. & Mallein, R. (1988) *Analyst 113*, 1843-1847.
5. Stevenson, I.H. (1977) *Br. J. Clin. Pharmacol. 4*, 261-265.
6. Vesell, E.S. (1979) *Clin. Pharmacol. Ther. 26*, 275-286.
7. Daumas, L., Sabot, J.F., Vermeulen, E., Clapot, J.M., Allegre, F., Pinatel, H., Mallein, R. & Francois, B. (1990) *Clin. Chem. Enzym. Comm. 3*, 191-199.

Editor's note.- A wide-ranging survey of stable-isotope scope, use and pitfalls, by G.E. von Unruh *et al.*, appeared in Vol. 14 [*Drug Determination......*, eds. as now; Plenum, 1984: pp. 27-37].

#ncD-7

A Note on

IDENTIFICATION OF DRUGS AND METABOLITES IN
URINARY CALCULI BY INFRARED SPECTROSCOPY

[1]J.F. Sabot, [2]Ph. Bernard, [2]M. Boucherat and [1]H. Pinatel

[1] Laboratoire de Chimie Analytique II, Faculté de Pharmacie, Avenue Rockefeller, Lyon, and [2]Laboratoire Central de Biochimie, Hôpital de l'Antiquaille, Lyon, France

The revolution in IR techniques over the last few years, notably Fourier Transform, has enormously increased their power and routine use for qualitative analysis of urinary calculi for organic and/or inorganic components [1-4]. KBr pressed-disk preparations are commonly used for such analyses. IR analysis of calculi also serves to provide information significant for diagnosis of drug stones. Thus, a calculus consisting of a drug and/or metabolites signifies crystallization of these compounds by oversaturation. Precise knowledge of the composition of such calculi can be important in deciding to adjust the dosage regimen or to exclude a drug from individual therapy. Urinary calculi are often multi-component, with a complicated microstructure resulting from their growth. Their main constituents are typically calcium oxalate – monohydrate (40% incidence)/dihydrate (25%), carbonatoapatite (18%), uric acid (10%), magnesium ammonium phosphate (2.5%), cystine (1%), and other compounds including drugs (0.7%).

Calculi from 1884 patients have now been analyzed by IR, and 14 found to contain drugs and/or metabolites. Absorption spectra of these calculi were compared with those of authentic compounds. The compounds discovered in the 14 were:- triamterene (**T**) + its metabolite hydroxytriamterene (as sulphate; **HTS**) in 5; glafenic acid and 4'-hydroxyglafenic acid along with protein in 2; *N*-acetylsulfadiazine [dosed as Adiazine] in 2; progesterone [Utrogestan] in 1; cyamemazine [Tercian] in 1, with crystals in skin; and 3 calculi containing unidentified compounds having spectra indicative of organic substances and able to fluoresce.

Worldwide, **T** has been identified in 450 calculi; the estimated incidence of **T** stone in **T** users is ~1:1500 [5]. Fig. 1 shows the IR pattern for one of our calculi, in which fluorimetry showed, as mmol/g, **T** 2.6 (67% w/w) and **HTS** 0.55. MS showed molecular ions characteristic of **T** and **HTS** (Fig. 2). TLC and, on the eluted spots, GC-MS revealed glucuronic acid, as also found in further calculi. **T** seems to be partly glucuronidated, but literature [6-8] indicates mainly unchanged **T** in calculi.

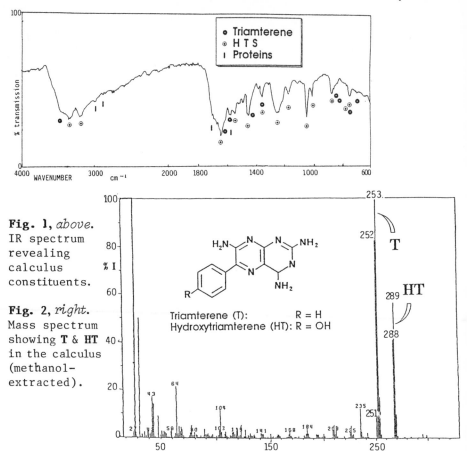

Fig. 1, *above.*
IR spectrum
revealing
calculus
constituents.

Fig. 2, *right.*
Mass spectrum
showing **T** & **HT**
in the calculus
(methanol-
extracted).

Triamterene (T):   R = H
Hydroxytriamterene (HT): R = OH

We reckon, then, that IR is an excellent way of identifying
drugs in calculi, even where too small (0.2 mg) to be studied
by other analytical methods.   Rarely, where a constituent
cannot be identified by IR, other physical means may be
used: NMR and MS have in fact been applied to calculi of
progesterone, cyamemazine and **T** (MS in the EI mode).

References

1. Beischer, D.E. (1955) *J. Urol. 73*, 653-659.
2. Oliver, L.K. & Sweet, R.V. (1976) *Clin. Chim. Acta 72*, 17-32.
3. Daudon, M., Protat, M.F., Reveillaud, R.J. &
   Jaechke-Boyer, H. (1983) *Kidney Int. 23*, 842-850.
4. Berthelot, M., Cornu, G., Daudon, M., Helbert, M. &
   Laurence, C. (1987) *Clin. Chem. 33*, 2070-2073.
5. Sorgel, F., Ettinger, B. & Benet, L.Z. (1986) *J. Pharm. Sci. 75*, 129-132.
6. Ettinger, B., Weil, E., Mandel, N.S. & Darling, S. (1979)
   Ann. Intern. Med. 91, 745-746.
7. Sorgel, F., Ettinger, B. & Benet, L.Z. (1985) *J. Urol. 134*, 871-873.
8. Carey, R.A., Beg, M.M.,   McNally, C.F. & Tannenbaum, P.
   (1984) *Clin. Ther. 6*, 302-309.

Comments on #D-2, G.J. de Jong - CHEMILUMINESCENCE DETECTION
                #D-3, D. Perrett - CAPILLARY ELECTROPHORESIS (CE)

de Jong, answering G.S. Land who asked about the robustness of the technique in routine use: all goes well once the method has been settled; groups in Sweden and Denmark have been applying the technique routinely for assays needing very high sensitivity.   **B. Law (answering D. Dell** who had asked why chemiluminescence isn't widely used): many, if not most, compounds of bioanalytical interest, e.g. β-blockers, do not give a chemiluminescent signal.   **Comment by de Jong:** the remedy lies in derivatization, suitably automated on-line. **de Jong, reply to K. Borner** who asked what instrument he recommended for chemiluminescence (CL) detection.- One can best use a fluorescence detector with a large optical aperture, e.g. Schoeffel FS-970 with a 2π steradian mirror in front of the flow cell (light source off).   Special CL detectors include the ATTO AC 2220 CL detector, having the flow cell close to the photomultiplier; more commercially available CL detectors are needed!   **de Jong, answering D.J. Perrett.-** Laser-induced fluorescence is quite similar to CL in sensitivity, but needs a costly laser and often entails derivatization.

Perrett remarked that he had observed a problem of oxidation in some samples due to high levels of H_2O_2.   **Response by de Jong.-**   We too have observed this phenomenon for a few analytes; it depends on the type of analyte and on the H_2O_2 concentration. Photo-initiated POO-CL [ref. 29 in #D-2] circumvents H_2O_2 addition, with some sacrifice of sensitivity. **W. Ritter commented** on a point which **de Jong** had mentioned, that analysis of a phenol dansylate entailed division for measurement of the dansylate moiety, because the phenol had electronegative substituents, and remarked that in this situation an excess of dansyl chloride (or hydrolysis products thereof), giving a background signal, should be avoided. **de Jong's response:** achieving a good chromatographic separation is paramount.   **Question by Perrett:** how do the capillaries behave in CL? - do you flush them out after each analysis? **de Jong's reply:** no! - flushing tends to increase the variation in migration time, and in any case is not necessary for biological samples.

D. Stevenson remarked that there are some concerns about how reliable CE is as a quantitative technique: are they well founded?   **Perret's reply.-** At present some equipment has proven unreliable, but with experience the situation is improving, as with any new technique.   CE should be as reliable as HPLC or GC in the near future.

Note by Ed.- To the examples in #D-3 of use of CE for drugs in biofluids, add a 1992 ref.:- p. 163, **Daunorubicin......** (2).

Comments on #D-5, I.D. Wilson - ^{19}F-NMR STUDIES
                    #**ncD-1**, M. Doig - MS WITH ATMOS. PRESS. IONIZATION
                    #**ncD-4**, A. Strutton - ASPEC *vs.* MANUAL PROCESSING

Wilson, replying to H. de Bree.- To convert a 'normal' NMR spectrometer into a ^{19}F-NMR machine, one needs merely a probe and amplifier, costing around £15,000.

Doig, answering C. Lindberg.- In routine quantitative analysis with the API system, both the heated nebulizer technique and the ion spray technique have been used. **B. Law enquired** what are the constraints on the LC eluent in respect of flow rate and aqueous/organic ratio. **Doig's reply:** the flow rate has to be low, 5-100 μl/min; neither methanol alone nor water alone is suitable, but use of non-volatile buffers, e.g. phosphate, may be feasible.

Law asked whether the centrifugation stage, which can't be automated in the ASPEC system, can be replaced by a filtration stage. **Strutton replied** that there is insufficient room in the instrument to insert a filter. **C. James commented** that the Zymark Benchmate does have the facility to use a filter. **W. Ritter commented** on a mentioned solvent, chloroform: it is banned in the U.S.A. and disfavoured in Germany (banned within Bayer AG). **Strutton replied** that there is no ban in the U.K., but harmonization within Europe might bring a change.

Automated sample preparation (*as in the* ASPEC)

S. Bass and some other discussants felt that automation of sample extraction saved little time, and wondered why companies devote so much effort to developing automated systems. **Strutton, answering C. James,** felt that the gain is significant but may be less than expected. This discussion took account of other pro-automation contributions that do not appear in this book.- (1) A. Churchill and J.C. Pearce (Glaxo, Ware) found the Gilson ASPEC system highly useful in method development for appraising different extraction columns. (2) C.A. James and R.J. Simmonds (Upjohn, Crawley) argued for robotics; *from their Forum abstract:*-
"..... sample preparation and extraction are still commonly performed manually...... [Reasons for slow adoption of robots] may be the relative complexity of the early instruments and the engineering back-up they required but also the modifications necessary to established methods and consequent re-validation. Newer laboratory robots such as the Waters Millilab and Zymark Benchmate are not expensive and easier to instal and program and their use in routine assay is very attractive."

Citations contributed by Senior Editor
- *some from* J. Chromatog., *abbreviated* J Chr
- *SEE ALSO some of the entries marked* * *in* #ABC *(assay 'compendium')*
- *drug-related citations generally pure analytes, not in biomatrix*
- *complementary refs., e.g. on column packings: SEE END OF Sect.* #B

Gel electrophoresis (displacement), exemplified by propranolol
- free or protein-bound (plasma dialyzed or solvent-extracted).-
Zelikman, I. & Hjertén, S.   (1989) Biomed. Chromatog. **3**,
161-165. Polyacrylamide gel with 2 layers, one small-pore;
flow-cell fluorimetry; a novel approach.

Reviews on CE.- (1) Karger et al. (1989) J Chr **492**, 585-614;
includes LIF & MS, and protein/oligonucleotide applications.
(2) Schwer, Ch. & Kenndler, E. (1990) Chromatographia **30**, 546-554.
(3) Campos, C.C. & Simpson, C.F. (1992) J. Chromatog. Sci.
30, 53-58.
Post-CE fluorescence **detection reactor**, e.g. adding OPA:
Rose, D.J. & Jorgenson, J.W. (1988) J Chr **417**, 117-131.

Robotics in **CE** and (including sample preparation) **HPLC.-**
Fouda, H.G. (1989) J Chr **492**, 85-108.
Automation of HPLC including solvent extraction and other
sample treatment: Turnell, D.C. & Cooper, J.D.H. (1989) J Chr **492**, 59-63.
Linkage to MS.- (1) **HPLC**, with frit-FAB probe for MS-MS:
Cappiello, A. et al. (1990) Chromatographia **30**, 477-483.
(2) **CZE, GC, HPLC, SFC** & **TLC** (overfiew): Schmid, E.R. (1990)
Chromatographia **30**, 573-576.   (3) **SFC**: Games, D.E. et al.
(1987) Anal. Proc. **24**, 371-372. *See also Sect.* #B (Garland).

SFC reviews: (1990) J Chr **505**, 3-525.

SFE applied to plasma: the sample is first adsorbed onto
a solid phase such as celite.- D.E. Games, *after his talk
at the Forum (no publication text), answering I.D. Wilson.*

On-line immunoaffinity sample pre-treatment in HPLC, with
oestrogen examples [cf. art. from same lab. in Vol. 20, p. 365].-
Farjam, A., Brugman, A.E., Lingeman, H.   & Brinkman, U.A.Th.
Analyst **116**, 891-896.

Cumulative Index of Analytes

with 'real-sample' entries, *prefixed* [, for Vols. 14, 16, 18 & 20
- *whose Indices remain useful for supplementary data, e.g.* R_T's
(Publication details on p. 170; titles at start of present book)

#**Vol. 12** (Plenum) has a **Cumulative Index** too, for Vols. 5, 7 & 10

=========

Key overleaf to the 10-category **chemical classification**
(collation based on some analytically relevant features).
Use of a compound as an internal standard is **not** indexed.
Hyphen '-' as in '17-' connotes *et seq.*, i.e. treatment in
depth. All pp. 148-169 entries refer to 'Compendium' (KEY: p. 148).

Prefixes to some page entries, *besides* ch = *chiral distinction:-*

Superscript, e.g. ¹, signifies that the study included
metabolite(s) of the listed compound (**see over**).
Subscript r signifies that pharmacologically relevant 'real'
samples were assayed, usually including plasma and/or urine.
Entries lacking this prefix mostly concern pure compounds.
Prefix p denotes a study comprehending a **precursor** or **prodrug**.
....

#**Inorganics** and **complexes**
as in Vol. 14 Index; **not** *one of
the 10 categories*

Aurothiolates: $_r$148 [**14:** 156,
  158 (incl. other Au complexes)
Cisplatin: $_r$163 [**14:** 142, 146-,
  161-, 199 (incl. other Pt cpds.)
[Cobalt complexes: **14:** 156, 158
(Des)ferrioxamine: **see Ib'**
Lithium: $_r$270
Nitrite & other inorganic
  anions: 267 [**14:** 350
Nitrate esters: **see**
  Ia, Glyc.... & Isosorbide
Phosphonates: **see Ia**,
  Belfosdil & Fotemustine,
  and **IIb'**, Bisphosphonates
[Thiocyanate: **14:** 350

CATEGORY I (no amino group, nor
cyclic N except maybe imide)

#**Ia:** acid *other than conjugate,*
  or ester *(criterion: SEE OVER)*

Acetretin: 80-, $_r$161
N-Acetylcysteine: 167 [**20:** 275
Arachidonic acid metabolites
  (diagram: 225; & see Leuko....,
  Prosta...., Thromb....): $_r$150
- antagonists: $_r$150, $_r$151
Arotenoids: $_r$63
Artemisinin: $_r$161
Arylpropionic acids ('Profens'):
  see individual entries in **Ia**
  & **IIIa** [**18:** 218; **20:** ch:374
Aspirin: $\frac{1}{r}$271, $\frac{1}{r}$274- [**14:** 94;
  18: 87
Belfosdil: $_r$152
[Benoxaprofen: **18:** 218
Benzoic acids: [**14:** ²49-
- with fluorine substituent:
  ¹²23- [**20:** ¹²375-
[Benzylpenicillin: **20:** 232
Bile acids: $_r$263 [**16:** ²318
Bumetanide: $_r$167
Busulphan: $_r$162

ASSIGNMENT 'CATECHISM' *(See previous p. for **other guidance**)*

Metabolites are not separately listed; the parent molecule's entry is preceded by a superscript: *Phase I* metabolite(s) denoted [1] or, if including *N*-desalkyl or desacyl, [1] (**bold**); *Phase II (conjugates)*, [2].

Parent molecules as indexed generally bear generic names as listed (with formulae) in the *Merck Index*.

Assignment as 'acidic' (to Ia, IIa or IIIa) applies where the pKa is <6; this excludes phenols, and **conjugates are excluded** since only the parent molecule is listed (prefixed [2]; see above). Also 'acidic' (notional) are **esters** yielding an acidic group in the *main* moiety if hydrolysed (as may happen *in vivo*).

Cyclic *N* (conducive to UV absorbance?) is never treated as 'amino'; it is 'imide' (possible category: Ic) if -CO-*N*-CO-, but otherwise *may* be basic.

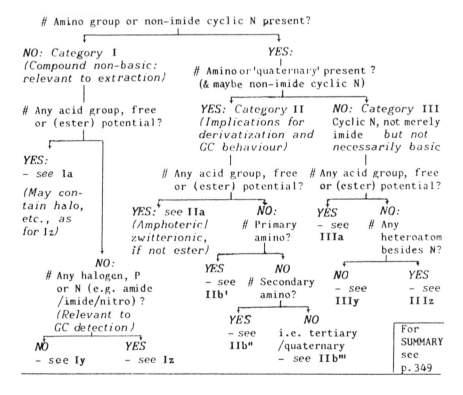

Amino group or non-imide cyclic N present?

NO: Category I
(Compound non-basic: relevant to extraction)

Any acid group, free or (ester) potential?

YES:
– *see* Ia
(May contain halo, etc., as for Iz)

NO:
Any halogen, P or N (e.g. amide /imide/nitro)?
(Relevant to GC detection)

NO   YES
– see Iy   – see Iz

YES:

Amino or 'quaternary' present?
(& maybe non-imide cyclic N)

YES: Category II
(Implications for derivatization and GC behaviour)

Any acid group, free or (ester) potential?

YES: see IIa
(Amphoteric/ zwitterionic, if not ester)

NO:
Primary amino?

YES   NO
– see   # Secondary
IIb'   amino?

YES   NO
– see   i.e. tertiary
IIb"   /quaternary
               – see IIb'''

NO: Category III
Cyclic N, not merely imide   but not necessarily basic

Any acid group, free or (ester) potential?

YES   NO
– see   # Any
IIIa   heteroatom
              besides N?

NO   YES
– see   – see
IIIy   IIIz

For SUMMARY see p. 349

NOTE: throughout the Index
'[' precedes entries for
previous vols., confined to
'real-sample' studies (*prefix* $_r$
omitted but implied).

For **key** see start, or p. 350

For key see start, or p. 350

#IIb': primary amino; no acid
  (unless conjugate) or ester

'[' precedes entries for
previous vols., confined to
'real-sample' studies (*prefix* r
omitted but implied).

CATEGORY III (cyclic N, not
merely imide; *no* amino)

#**IIIa**: acid *other than conjugate*,
or ester *(main moiety = acid)*

'[' precedes entries for
previous vols., confined to
'real-sample' studies (*prefix r*
omitted but implied).

─────────

'[' precedes entries for
previous vols., confined to
'real-sample' studies (*prefix* $_r$
omitted but implied).

For **key** see start or, for
short version, p. 350.

NOTE alphabetical sequence breaks
due to sub-listing of Phenothiazines *and (OVERLEAF)* Thioxanthenes

SUMMARY OF CATEGORIES *[Full KEY: see*
p. 238 & (prefixes) p. 237]

	I	II	III
Amino?	no	✓	no
Non-imide hetero-N?	no	maybe	✓
Acid or potential acid (not conjugate)?	✓ = Ia	✓ = IIa	✓ = IIIa
- no! (and not an ester)	Halo, P or N?	Primary amino?	Hetero atom besides N?
	- no: Iy	✓ = IIb'	- no = IIIy,
	- ✓ = Iz	If no: 2^y= IIb'' 3^y or 4^y =IIb'''	✓ = IIIz

Only parent compound listed; prefix
¹ if Phase 1 metabolites studied [or
¹ (bold) if dealkylated amino], & ² if
Phase II (conjugate); prefix $_r$ = 'real'
(e.g. plasma) sample. *Other prefixes:*
ch = chiral distinction;   p = prodrug.
'[' precedes entries for previous vols.,
confined to 'real-sample' investigations
(*prefix* $_r$ omitted but implied).

General Index

This Index deals mainly with features studied and with approaches and points of technique, indexed similarly to previous 'A' vols. (listed on p. 170) so as to facilitate back-searching. The preceding Analyte Index deals with compounds investigated; exceptionally, a few types are also listed below according to their nature, e.g. 'β-Blockers'.

In a page entry such as '17-', the '-' means 'et seq.', i.e. coverage in depth.

An Ix denotes Analyte Index. **Cm** denotes the 'Compendium' (pp. 148-169); °denotes allusion(s) on a particular page of **Cm**, wherein asterisks (*) draw attention to some entries of special interest.

An Ix = Analyte Index; Cm =
Compendium (pp. 148-169).
° denotes allusion(s) on a
particular page of Cm, wherein
asterisks (*) draw attention
to notable entries

An Ix, Analyte Index; Cm, Compendium (° denotes entries therein)